P9-AQH-593

Kathryn Fincher
Flower Show 1998

THE COMPLETE BOOK OF

# PLANT
# PROPAGATION

# THE COMPLETE BOOK OF
# PLANT
# PROPAGATION

*Editorial Consultant:* CHARLES W. HEUSER, Jr., Ph.D.

The Taunton Press

*The Complete Book of Plant Propagation*
First published in the United Kingdom in 1997 by Mitchell
Beazley, an imprint of Reed International Books Limited,
Michelin House, 81 Fulham Road, London SW3 6RB

Copyright © Reed International Books Limited 1997

All rights reserved. No part of this work may be reproduced
or utilized in any form or by an means, electronic or
mechanical, including photocopying, recording or by
any information storage and retrieval system, without
the prior written permission of the publisher.

Executive editor: Guy Croton
Executive art editor: Ruth Hope
Editor: Michèle Byam
Designers: Martin Bristow, Debbie Myatt
Production: Rachel Lynch
Artworks: Vanessa Luff and Charlotte Wess

ISBN 1-56158-234-4

Printed in China

# Taunton
**BOOKS & VIDEOS**

*for fellow enthusiasts*

The Taunton Press
63 South Main Street
Box 5506
Newtown, CT 06470-5506

# CONTENTS

# FOREWORD

The art of propagation is one that is often shrouded by mystique. While it is undoubtedly true that some gardeners have "green fingers" and appear to increase their garden stock effortlessly – it is more often the case that the apparently green-fingered gardener merely has a good grasp of the basic principles involved and has practised the necessary skills.

The main theme of this book is the elucidation of these basic principles and each chapter deals with the way they can be modified to suit the needs of various plant groups, whether they be the simplest of annuals grown from seed, or more horticulturally developed plants – like apples and pears – that we cultivate as grafted trees in our gardens.

One of the first principles is to identify the nature of the plant material that you wish to propagate, for this usually dictates the frequency of propagation and gives guidance regarding the sorts of techniques that may be appropriate. Plants may be short-lived and ephemeral, like annuals and biennials, thus needing annual propagation. Or they may be, to all intents and purposes, permanent garden features, as are many trees and shrubs, which need propagating only once or twice during the life of the garden. A plant may be a herbaceous clump-former that lends itself ideally to division, or it may be woody and amenable to increase by a variety of different sorts of cuttings.

A plant's hardiness also give clues to a suitable propagation environment: a plant that tolerates winter lows of 5°F (-15°C) and below, is likely to germinate, root and grow at lower temperatures than one that will not tolerate temperatures below freezing, or one that needs temperatures well above 32°F (0°C) to thrive. In the chapter on annuals and biennials and the chapter on herbs plants are listed as being either hardy or half hardy. A hardy plant is one that will tolerate temperatures around 32°F (0°C) without protection and half hardy plants are those that will do so but with some protection.

Some principles are immutable and apply across the whole range of propagating techniques and material – good hygiene and appropriate environmental conditions for rooting and growth, for example, are fundamental to success. Others, like the timing of taking cuttings, are variable, and success relies on following expert advice and applying one's own increasing experience.

It is nearly always true that there is more than one way to achieve a successful outcome and where one method fails, another will surely succeed. Half of the satisfaction of gardening is learning skills and applying them to the diverse plants that we grow in our gardens. It is, after all, the willingness to experiment and to learn from both successes and failures that is one of the marks of a good gardener.

# Annuals and Biennials

One of the joys of gardening with annuals and biennials is that, with such an enormous range available, their use in gardens is limited only by the imagination. Their ease of propagation and cultivation, diversity, and great adaptability has, deservedly, made them established favorites. Few other types of plants provide so much color, or give as much pleasure in the garden – one of the most rewarding activities is to grow plants that you have raised yourself from seed. Annuals and biennials need little outlay on time or materials and they can, within a matter of weeks, create glorious living tapestries.

◀ *Because annuals and biennials flower, set seed, and then die, plants such as* Tagetes, Cosmos, *and* Verbena *will always need to be propagated by seed.*

# WHY PROPAGATE
# ANNUALS AND BIENNIALS?

*Although, by their very nature, annuals and biennials are short-lived and must be propagated each year, they compensate by possessing an intensity of color and profusion of bloom that is unsurpassed by any other group of plants. Given the right conditions, they are also among the simplest of plant groups to propagate.*

**This brilliant border contains plants grown from seeds that have a variety of different needs for successful germination.**

Since true annuals and biennials are propagated from seed, the gardener does not have to make a decision on a preferred method of increase. There are, however, important considerations that dictate how, when, and where to sow, if success is to be guaranteed.

## RAISING HEALTHY ANNUALS

True annuals complete their life cycle in one season, flowering, setting seed, and dying in a few months. Hardy annuals, such as *Calendula* and *Nigella*, withstand some frost without damage and are sown directly in their flowering site in spring or in fall, for earlier flowers the following year.

Temperatures below 32°F (0°C) cause damage to half-hardy annuals, so they are usually raised under glass in spring, and then hardened off in a coldframe before being planted out once the danger of frost has passed. These

include *Lobelia erinus* and *Tagetes*, the African and French marigolds. Fast-growing half-hardy annuals may also be sown *in situ* in mid-spring.

Frost-tender annuals, such as *Cleome* (spider flower), need warm temperatures in order to thrive. They are sown – and grown on – under glass with additional heat, usually between late winter and early spring. Once they have reached a reasonable size, they are

hardened off gradually and are only planted out into their flowering positions when all danger of frost has passed and warm summer weather is well established.

## PLANNING FOR BIENNIALS

In their first year, biennials produce leaves and stems and, in the second, they flower, set seed, and die. Often as showy as annuals, their use demands a little more planning, since they need to be established in their flowering site in the season before blooming (usually late summer or fall). *Bellis perennis* (daisy) and *Myosotis* (forget-me-not) are well-known examples.

## PERENNIALS AS ANNUALS

Plants that are treated as annuals or biennials include some that are perennial under natural conditions. Pelargoniums, for example, are treated as annuals because they are not reliably winter-hardy in frost-prone climates. Garden pansies (*Viola* x *wittrockiana*), though cold-hardy, are naturally short-lived perennials; they bloom most abundantly in their first year, but deteriorate rapidly thereafter. A third group includes such plants as *Helichrysum petiolare*, valued for its foliage which is most attractive on young plants.

### Selecting for Double Flowers

*Matthiola or stocks have long been valued as cut flowers, especially in their double-flowered forms. Many modern cultivars produce a high proportion of double flowers; with some, the seedlings may be selected to produce almost 100% doubles. In some cases, only seedlings with yellow-green or pale-colored seed leaves give rise to double flowers. Sow the seed at 55–64°F (13–18°C) and, on emergence of the seed leaves, lower the temperature to below 50°F (10°C). The difference in color then becomes apparent and the dark-leaved seedlings should be discarded. In other cases, double flowers arise only from seedlings with distinctively notched seed leaves, so discard any unnotched seedlings.*

*The beautiful 'Cinderella Mauve' is a good example of a Matthiola cultivar with double flowers.*

Perennials used as annuals often need a long growing season to give of their best. Raise them from seed under glass in late winter with additional heat or, with plants such as *Felicia*, propagate them by stem-tip cuttings in late summer, and overwinter the young plants in frost-free conditions under glass.

( 1 )

**Pelargonium**
Pelargoniums are tender subshrubby perennials that have traditionally been raised by stem-tip cuttings overwintered under glass. Modern bedding pelargoniums offer the convenience of being raised annually from seed.

( 2 )

**Impatiens**
Frost-tender perennials like *Impatiens* are also raised annually from seed for bedding. Seed needs constant warmth for germination, and young plants should be hardened off and set out when danger of frost has passed.

( 3 )

**Begonia**
Begonias grown as annuals include the Semperflorens types that can be increased from seed or basal cuttings, and the tuberous Tuberhybrida types that may be grown from seed, basal, or stem cuttings.

( 4 )

**Calendula**
Hardy annuals, like pot marigolds, are among the easiest of plants that are grown as annuals. Seed can be sown in spring, or in autumn, to produce earlier flowers the following season.

# PROPAGATING TECHNIQUES
# ANNUALS AND BIENNIALS

*Raising annuals and biennials from seed is one of the simplest of propagating techniques, but success depends on an understanding of the various conditions required by different types of plants for germination and growth, and on the observation of a few basic ground rules — especially with regard to good hygiene.*

## SELECTING AND BUYING SEED

Before purchasing, check the seed packet for a "sow-by-date" to confirm that it is the current season's stock, thus ensuring freshness and viability. Most seed begins to lose viability or degenerate after a year, especially if not stored in cool, dry conditions.

The least expensive seed is produced by open, or natural, pollination. It yields offspring that exhibit natural variation — with slight differences in color, size, and habit. For most garden uses, this gives perfectly satisfactory results, and in informal plantings such subtle variations prove a positive advantage.

F1 hybrid seeds are produced by crossing two selected, pure-breeding parents to produce vigorous, uniform, and floriferous offspring that come reliably true to color form and habit. F2 hybrids result from crossing selected F1 parents, and are equally vigorous but slightly less uniform. Both are more expensive than ordinary seed, but are ideal where uniformity or increased vigor is required, as with bedding.

Some hybrids are offered as pelleted seed, where individual seeds are coated with a nutrient and fungicidal coat. They are also more costly, but may be space-sown and so do not need wasteful thinning later.

## Sowing in containers

Whether sowing in trays or pots, the technique is essentially the same. Always aim to sow thinly and evenly to permit adequate room for the seedlings to grow and develop freely. Overcrowding is a primary factor in damping off, and makes separating the seedlings at the thinning out stage very difficult without causing damage to the stems and roots.

► **1** Fill the container by mounding the germination medium loosely, slightly above the container rim. Tap the container to remove air pockets. Level off the medium with a straight edge so that its surface is flush with the container rim. Firm the medium gently so that the finished surface is about ½in (1cm) below the rim.

► **2** Sow the seed thinly and evenly, tapping it out gently from a folded sheet of clean paper held just above the medium surface to prevent the seed from bouncing.

► **3** Just cover the seed with a thin layer of sieved medium. To water, place the seed container in about ½in (1cm) of water in a shallow tray. Remove immediately when the water is seen at the medium surface.

► **4** Allow the tray to drain and cover with a sheet of glass or plastic to conserve moisture. Place a sheet of newspaper or shade netting on top of the glass or plastic to provide shade until germination occurs. Remove all coverings as soon as germination has taken place.

## WHEN AND WHERE TO SOW

The site and timing of sowing varies according to climatic conditions and the type of plant involved.

Hardy annuals are sown in spring, directly into their flowering site or in containers in an open frame. Delay sowing if the conditions are unfavorable – for example, if spring is very cold or wet. Most will germinate at temperatures of 45–54°F (7–12°C), so begin sowing once the soil temperature reaches the lower end of this range; slightly higher temperatures speed germination. Some hardy annuals may be sown in open ground in the fall and will overwinter safely in all but the most severe winters. Sown this way, they flower earlier the following year.

Half-hardy and frost-tender annuals are sown in controled conditions under glass. Sowings are made from late winter onward, in gentle heat. As a general guide, most half-hardy annuals germinate within the range 59–70°F (15–21°C); frost-tender plants between 64–75°F (18–24°C). Later sowings of half-hardy annuals can also be made in a coldframe, but will need additional protection if frost threatens. Insulate the frame with a thermal blanket and remove it during the day to admit light.

Biennials are raised in the greenhouse or coldframe from late spring to midsummer (depending on the individual species' speed of growth). They are grown on, in containers or a nursery bed, to be planted in their flowering site in the fall. Alternatively, sow outdoors in a prepared seedbed, and transplant later. Hardy biennials will overwinter in the open without damage, but in severe weather, protect with transparent covers or a thermal blanket.

## CHOOSING CONTAINERS

Unless large numbers of plants are needed, small containers, such as half pots or half trays will contain sufficient seedlings for all but the largest garden plot.

Biodegradable pots, whether the individual or modular, are ideal for sowing seedlings that will not transplant easily, since they can be planted out in their pots without root disturbance. Plastic plug trays are also used for planting out without causing disturbance to the seedlings' roots, as well as being convenient to use for sowing and thinning out, and less extravagant with germination medium.

## Sowing outdoors . . . . . . . . . . . . . . . . . . . . . . . . .

Prepare a seedbed by digging over to a spade's depth and removing any large stones and all weeds. Do this a few weeks before sowing to allow the soil to settle back again. Firm the ground lightly and evenly, but not when the soil is wet as this will cause compaction. Apply a base dressing of general fertilizer only if the soil is poor; too rich a soil will produce foliage growth at the expense of flowers. Rake over the prepared seedbed to produce a fine even tilth to a depth of 1–2in (2.5–5cm).

### Shallow furrows

► 1 Sowing seeds in shallow furrows has the advantage that rows of germinating seedlings can be distinguished from weed seedlings whose distribution is random. Mark out the furrows using a garden line stretched between stakes. Draw out a shallow furrow with the back of a hoe. A depth of ½in (1cm) suits most seeds; sow finer seeds more shallowly. If the soil is dry, water the furrow before sowing. Sow seed evenly and thinly along the furrow.

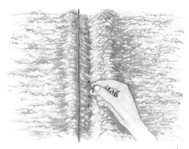
1

► 2 Gently cover the seeds by drawing soil over them with a rake. Firm the surface of the furrow with the back of the rake and water in.

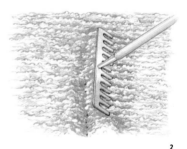
2

### Broadcasting

► 1 Rake the soil to an even tilth with parallel rake lines in one direction. Scatter the seed evenly and thinly over the surface.

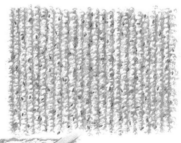
1

◄ 2 Rake in the seed in a direction at right angles to the first. Do this very gently to avoid dislodging the seed.

2

13

# GROUND RULES FOR SUCCESSFUL GERMINATION

The basic requirements for successful seed germination are appropriate light levels, sufficient air and moisture, and a suitable medium and temperature regime.

### Light

Light is a vital factor in germination process of all plants. The seeds of such annuals as *Impatiens*, *Begonia*, and *Petunia* need light to initiate the process, while other annuals require darkness, and many germinate in either light or dark. This varies from species to species, so check the seed packet before sowing. Germinated seedlings must have adequate light for development, if they are not to become weak and spindly. In the initial stages, however, they also need to be shaded from direct sunlight to avoid scorching the leaves.

### Sowing depth

Seeds range considerably in size, from begonias, with about 80,000 seeds per gram, to sweet peas with 10–12 seeds per gram. Seed size indicates the required sowing depth. As a rule, seeds are covered to their own depth with germination medium, sieved to ensure intimate contact – the smaller the seed, the finer the covering. Exceptions are seeds that need light to germinate; press these gently into the medium. To sow dust-fine seed evenly, mix with a little fine, dry sand and surface sow.

### Water

Seed needs adequate moisture to begin germination, and to maintain growth thereafter. Seed can be watered in with a watering can after sowing, but overhead watering can wash seed to the sides of the container, causing harmful overcrowding on germination.

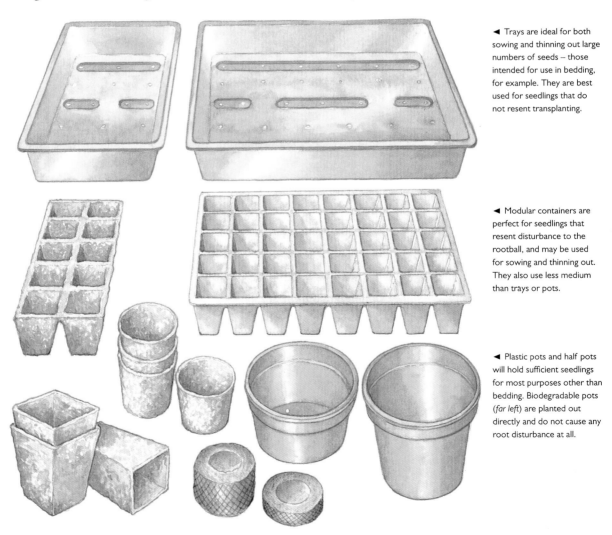

◄ Trays are ideal for both sowing and thinning out large numbers of seeds – those intended for use in bedding, for example. They are best used for seedlings that do not resent transplanting.

◄ Modular containers are perfect for seedlings that resent disturbance to the rootball, and may be used for sowing and thinning out. They also use less medium than trays or pots.

◄ Plastic pots and half pots will hold sufficient seedlings for most purposes other than bedding. Biodegradable pots (*far left*) are planted out directly and do not cause any root disturbance at all.

It is just as effective to water from below – before sowing for very fine seed, or after sowing for larger seed. Water is drawn up by capillary action and when it can just be seen at the surface, the medium is sufficiently moist. After sowing and germination, keep seeds and seedlings evenly moist, but never allow them to become waterlogged; if they dry out, they will not germinate or grow on successfully.

## Hygiene

Containers should be new or sterilized before use, and medium must be fresh. Do not sow seeds too thickly; seedlings need room to develop properly, and overcrowding can increase susceptibility to the soil-borne fungal infection known as damping off (botrytis), which is usually fatal. Risk of damping off is also increased by poor ventilation; maintain a flow of fresh air around the young seedlings at all times. Water with tap water rather than with rain water collected in a water tank, as this can be a source of fungal infection. Apply a proprietary fungicide before sowing, and when watering seedlings, as a preventative measure against damping off.

## Hardening off

Plants raised under controled and sheltered conditions under glass must be hardened off – or acclimated – before planting out. About six weeks before planting out, when the thinned out seedlings are growing well, move them to a cooler section of the greenhouse for about seven days, then transfer them to a covered coldframe. Gradually increase ventilation by opening the frame lights a little more, and for a little longer, each day. Keep an eye on the weather, and, if frost threatens, close the lights and insulate the coldframe. You should avoid producing any sudden changes of temperature, as this can lead to a check in growth, or in very severe cases, to the death of the young plants. After 2–3 weeks, the plants should be sufficiently hardened off and will be ready to plant out.

## Thinning

Thin the germinated seedlings to the recommended planting distances for the species concerned. Remove excess seedlings by hand, taking great care to avoid root disturbance to the remainder.

## Thinning out

Seeds sown in containers should be transferred, or thinned out, into larger containers before they become overcrowded.

► **1** When seedlings are large enough to handle, thin out into a more fertile growing-on medium to give them space to develop. First water the seed tray and allow to drain. Then tap it gently on the work bench to loosen the medium. Ease individual seedlings carefully from the seed tray using a spatula, retaining as much soil around the roots as possible. Handle them only by their seed leaves, to avoid damaging the roots, stems, and growing tips.

◄ **2** Make a hole in the tray of prepared medium, and place the seedling gently into it, so that it is at the same depth as it was in the seed tray. Firm in carefully with the spatula and water in using a watering can.

## Marking out an annual border

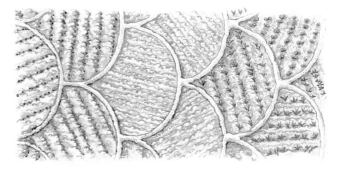

▲ To create an annual border, draw a plan on paper, arranging drifts of different varieties according to their color and height. Mark the outlines of informal drifts on the ground using fine sand. Then sow the seed in rows within the marked out drift areas. Align the rows in different directions in neighboring drifts. Seedlings are easier to weed when they are in rows, and the lines will not be apparent when the plants mature. Finally, carefully rake the soil over to cover the seeds. Firm and water in thoroughly.

# DIRECTORY OF
# ANNUALS AND BIENNIALS

*Below are key tips on a selection of annuals and biennials that are particularly suitable for propagation. Unless otherwise specified, follow the detailed instructions given under the propagation section on pages 12–15.*

## Key

| | |
|---|---|
| H | Height |
| S | Spread |
| 🌱 | By seed |
| ⚚ | Germination |
| ⚒ | By division |
| ☀ | Time to flowering |
| HA | Hardy annual |
| HHA | Half-hardy annual |
| FTA | Frost-tender annual |
| HB | Biennial |
| HHB | Half-hardy biennial |
| FTB | Frost-tender biennial |

## Amaranthus                    Amaranthaceae

About 60 species of annuals or short-lived perennials with pendent tassels of flowers. *A. caudatus* (Love-lies-bleeding, Tassel flower). Bushy annual with crimson tassels, to 2ft (60cm) long, from midsummer until frosts. 'Viridis' has green tassels. Easily grown in a sunny, sheltered moist site. Water freely in dry periods. HHA. H 2–4ft (60–120cm), S to 30in (75cm).

### PROPAGATION
🌱 under glass in early spring at 64–75°F (18–24°C).
⚚ 7–14 days.    ☀ 3–4 months.

## Antirrhinum

(Snapdragon)                    Scrophulariaceae

Up to 40 species of annuals, perennials, and semi-evergreen subshrubs with distinctive two-lipped flowers, often borne in showy spikes. *A. majus*, a short-lived perennial, is commonly grown as an annual.
*A. majus*. Bushy and branching with erect spikes of showy flowers from summer to fall. Among the many cultivars available there are dwarf forms, intermediates, and tall forms. Easily grown in fertile, well-drained soil in sun. HHA.

### PROPAGATION
🌱 surface-sow under glass in early spring, at 55–61°F (13–16°C).
⚚ 10–21 days.    ☀ 3–4 months.

## Arctotis                    Compositae

About 50 species of annuals and herbaceous perennials with daisy-like, brightly colored, often zoned flowers. The perennials are frequently grown as annuals.
*A. venusta*, syn. *A. stoechadifolia* (African blue daisy). Spreading perennial grown as an annual for its creamy-white flowers, with central blue discs, from midsummer to early fall. FTA. H 24in (60cm), S 16in (40cm).

## A. Harlequin Hybrids, (syn. A. × hybrida)
Similar to *A. venusta*, but with a wide color range. Perfect for borders and cut flowers. Grow in light, well-drained soil in sun. Deadhead regularly to prolong flowering. FTA. H 18–24in (45–60cm), S 12in (30cm).

### PROPAGATION
🌱 under glass in late winter to early spring, at 61–70°F (16–21°C). Alternatively, sow in fall and overwinter young plants under glass at min. 41–45°F (5–7°C).
⚚ 10–21 days.    ☀ 4 months.

## Bellis (English daisy)                    Compositae

Some 15 species of herbaceous perennials, with rosettes of spoon-shaped leaves and daisy-like flowers borne singly on long stems. *B. perennis* is grown as a biennial.
*B. perennis* (English daisy). Compact plant with single, semi-, or fully double flowers, from late winter to summer. The variety 'Pomponette' has pompon-like blooms. Deadhead to stop self-seeding. Grow in well-drained soil in sun or part-shade. HB. H and S to 8in (20cm).

### PROPAGATION
🌱 under glass from spring to midsummer at 50–55°F (10–13°C), or in furrows outdoors in early summer.
⚚ 7–21 days.    ☀ 6–8 months.
⚒ after flowering or in spring.

## Bidens                    Compositae

Some 200 annuals, herbaceous perennials, and deciduous shrubs with attractive flower heads on long stems. The short-lived perennials are usually grown as annuals in frost-prone areas.
*B. ferulifolia*. Grown as an annual for its ferny foliage and bright star-shaped, golden yellow flower heads borne from early summer to fall.

*Antirrhinum majus*

*Bellis perennis* 'Pomponette'

*Calendula officinalis*

*Callistephus chinensis*

*Centaurea cyanus*

**'Golden Goddess'** has finer leaves and larger flower heads and thrives in moist but well-drained soil and a sunny site.
HHA. H 12in (30cm), S indefinite.

**PROPAGATION**
🌱 under glass in early spring at 55–64°F (13–18°C).
🌿 10–21 days. ☀ 3 months.

### Bracteantha (Strawflower) Compositae
Some seven species of annuals and short-lived herbaceous perennials, usually grown as annuals. They have papery daisy-like flower heads that are often good for drying.
***B. bracteata,*** syn. *Helichrysum bracteatum.* Upright plant bearing everlasting, glistening flowerheads, from midsummer until frosts. Dwarf forms and taller cultivars are available. Easy in well-drained soil in sun. HHA. H 6–30in (15–75cm), S 12in (30cm).

**PROPAGATION**
🌱 *in situ,* in furrows, in mid-spring. Alternatively, sow in trays or peat pots, under glass, in early spring, at 61–64°F (16–18°C).
🌿 10–20 days. ☀ 4 months.

### Calendula (Marigold) Compositae
Up to 30 species of bushy annuals and evergreen perennials with aromatic leaves and daisy-like flower heads.
***C. officinalis*** (Pot marigold). Annual with flower heads in shades of orange, yellow, and cream. All are free-flowering from early summer to fall, and thrive in any well-drained soil in sun or part-shade. HA. H 12–24in (30–60cm), S 12–18in (30–45cm).

**PROPAGATION**
🌱 under glass in early spring at 55–64°F (13–18°C ), or in furrows *in situ* in spring or

fall. Provide protection for fall-sown seedlings in cold areas.
🌿 7–21 days. ☀ 3–6 months.

### Callistephus (China aster) Compositae
A single fast-growing, bushy annual species.
***C. chinensis.*** Rapidly growing variety that has single to double flower heads, in shades of white, pink, crimson, and indigo, borne from late summer until the first frosts arrive. A wide range of cultivars is offered, with compact or dwarf types and taller cultivars. They need a sheltered, sunny site in moist but well-drained soil. Provide support for tall cultivars and deadhead regularly to prolong the flowering period. China asters are susceptible to aster wilt and various other wilts; do not grow in the same site in consecutive years. HHA. H 10–24in (25–60cm ), S 10–12in (25–30cm).

**PROPAGATION**
🌱 under glass from early to mid-spring at 55–61°F (13–16°C).
🌿 10–14 days. ☀ 4 months.

### Centaurea Compositae
About 450 species of annuals, biennials, perennials, and subshrubs bearing rounded flower heads consisting of tubular, deeply cut florets. Leaves are sometimes silver-hairy.
***C. cyanus*** (Cornflower). Upright, slender-leaved annual with florets of blue, pink, white, rose, red, or purple around darker central discs. Flowers are produced from late spring to midsummer. Dwarf and taller cutivars are also available. Provide taller cultivars with light twiggy support. Grow in any well-drained soil in a sunny position. HA. H 8in–4ft (20cm–1.2m), S 6–10in (15–25cm).

**PROPAGATION**
🌱 *in situ,* in furrows or broadcast, from early to mid-spring or in fall. Alternatively, sow in peat pellets or peat pots. Avoid any root disturbance as far as possible.
🌿 10–14 days. ☀ 3–6 months.

### Clarkia Onagraceae
Some 36 species of thin-stemmed annuals with satiny, papery flowers in pale shades.
***C. amoena,*** syn. *Godetia amoena* (Satin flower). Upright annual with fluted, satin-textured flowers in shades of lilac to deep pink, throughout summer. HA. H 30in (75cm), S 12in (30cm).
***C. unguiculata,*** syn. *C. elegans* (Godetia). Slender annual bearing masses of miniature flowers throughout summer, in shades of salmon, pink, lavender, purple, red, or white. HA. H to 3ft (90cm), S 12in (30cm). Both are easily grown in moist but well-drained, preferably slightly acid soil in sun or part-shade. Provide light, twiggy support.

**PROPAGATION**
🌱 sow *in situ,* in furrows or broadcast, from early to mid-spring or in fall. Provide plastic covers as protection for fall-sown seedlings. They do not tolerate transplanting.
🌿 10–21 days. ☀ 3–6 months.

### Cleome (Spider flower) Capparidaceae
Some 150 species of bushy annuals and evergreen shrubs. The annuals are grown for their attractive, spider-like flowers with long prominent stamens.
***C. hassleriana,*** syn. *C. pungens, C. spinosa.* Upright, bushy annual with spider-like flowers in delicate shades of pink, rose, and white, throughout summer. Young plants are very

17

*Cleome hassleriana*

*Dianthus barbatus*

*Eschscholtzia californica*

susceptible to greenfly. Needs light, well-drained soil and a sheltered sunny site. FTA. H 3–4ft (90–120cm), S 18in (45cm).

**PROPAGATION**
🌱 under glass in early spring, at 64–68°F (18–20°C).
🌱 10–21 days. ☀ 4 months.

## Cosmos                                   Compositae

Some 25 species of annuals and herbaceous perennials with saucer- to cup- or bowl-shaped flower heads borne on long stems.
*C. bipinnatus* (Cosmos). Slender annual with delicately divided foliage and single or double flowers, some with fluted petals, in shades of white, rose, pink, and crimson, borne from midsummer until the first frosts. HHA. H 3–4ft (90–120cm), S 18in (45cm).

**PROPAGATION**
🌱 under glass from late winter to early spring, at 64–68°F (18–20°C).
🌱 7–14 days. ☀ 5 months.

## Dianthus (Pinks)              Caryophyllaceae

Over 300 species of annuals, biennials, perennials, and evergreen subshrubs with blue-gray leaves and rounded flowers.
*D. chinensis* (Indian pink). Compact, free-flowering biennials or perennials grown as annuals. They have single, fringed flowers borne throughout summer in shades of pink, crimson, scarlet, rose, and white, including bicolors with laced patterns or "eyed" centers. Plant in well-drained, neutral to alkaline soil in sun. HHA. H to 28in (70cm), S 10in (25cm).

**PROPAGATION**
🌱 under glass from late winter to early spring, at 64–77°F (18–25°C).

🌱 10–25 days. ☀ 4 months.
*D. barbatus* (Sweet William). Short-lived perennial which is grown as a biennial. It produces dense, flattened heads of small flowers, in shades of crimson, rose, carmine, salmon, pink, and white, including many bicolors, from late spring to early summer, or in mid- to late summer from early spring sowings. Excellent as border plants and cut flowers. Grow in well-drained, neutral to alkaline soil in a sunny site. HB. H 6–28in (15–70cm), S 12in (30cm).

**PROPAGATION**
🌱 in furrows in a prepared seedbed, between spring and midsummer or in fall. Alternatively, sow under glass in early spring at 59–64°F (15–18°C), to flower in the same year.
🌱 14–30 days. ☀ 5–9 months.

## Erysimum (Wallflower)              Cruciferae

Some 80 species of annuals, biennials, and mostly evergreen perennials, with racemes of flat flowers and lance-shaped leaves. Ideal for bedding and borders.
*E. cheiri*, syn. *Cheiranthus cheiri*. Short-lived, spring-flowering perennial grown as a biennial. The plants are compact and have usually sweetly scented flowers in shades of yellow, orange, scarlet, rose, pink, cream, and white. Grow in well-drained, neutral to alkaline soil in sun. HB. H 10–30in (25–75cm), S 12in (30cm).

**PROPAGATION**
🌱 in furrows in a seedbed, between late spring and early summer. Transplant to flowering site in fall. Alternatively, sow in fall and overwinter under plastic covers.
🌱 14 days. ☀ 9–10 months.

## Eschscholtzia

(California poppy)              Papaveraceae
Up to 10 species of annuals and perennials with divided, light green to blue-green leaves and bearing poppy-like, shallow, cup-shaped flowers, singly, often on slender stems.
*E. californica*. Spreading annual with finely divided foliage and satin-textured, single or semi-double flowers, in shades of gold, yellow, bronze, orange, scarlet, rose, or white, borne from midsummer until the first frosts. Easily grown in not too fertile, well-drained soil in sun. HA. H 12in (30cm), S 6in (15cm).

**PROPAGATION**
🌱 in furrows or broadcast, *in situ*, in mid-spring or early fall. Provide plastic covers over winter for fall-sown seedlings. They do not tolerate transplanting.
🌱 14–21 days. ☀ 3–7 months.

## Felicia                                   Compositae

Some 80 species of annuals, perennials, and evergreen subshrubs, or sometimes shrubs, with daisy-like, mainly blue flower heads.
*F. amelloides* (Blue daisy). Bushy subshrub, grown as an annual for its abundant pale to deep blue daisies, throughout summer. FTA. H to 2ft (60cm), S to 2ft (60cm).
*F. bergeriana* (Kingfisher daisy). Bushy, spreading annual with narrow gray-green leaves, bearing a profusion of yellow-centered, bright blue daisy flowerheads throughout summer. It is very wind tolerant. H and S to 8in (20cm). HHA. Both the above are easily grown in any well-drained soil in sun.

**PROPAGATION**
🌱 under glass in early spring, at 64–68°F (18–20°C).
🌱 14–30 days. ☀ 4 months.

*Felicia amelloides*

*Helianthus annuus*

*Lathyrus odoratus*

Stem-tip cuttings – for *F. amelloides* in late summer. Overwinter young plants in frost-free conditions under glass.
☀ 6–8 months.

## Gypsophila (Baby's breath)
Caryophyllaceae

More than 100 species of annuals and herbaceous, semi-evergreen, or evergreen perennials, often with a woody base, bearing delicate flowers in large, cloud-like masses.
**G. elegans.** Graceful open, branching annual, with gray-green leaves, bearing myriad tiny white flowers in summer. Variants with pink or carmine flowers are available. Grow in sun in light, very well-drained, preferably alkaline soil. Provide light twiggy support. HA. H 2ft (60cm), S 18in (45cm).

### PROPAGATION
❧ in furrows or broadcast, *in situ*, in mid-spring or early fall. Provide plastic covers over winter for fall-sown seedlings. Dislikes being transplanted.
❦ 7–21 days. ☀ 3–7 months.

## Helianthus (Sunflower)
Compósitae

Some 80 species of annuals and perennials. The genus is best known for its very tall, rough-leaved plants that produce the large showy flower heads so popular with children.
**H. annuus.** Rapidly growing annual bearing daisy-like flowers in mid- to late summer, each with a large central disc surrounded by golden- or lemon-yellow, orange,. or red-bronze ray florets. They range from dwarf types to giants. Grow in well-drained, fertile soil in a sunny site. Give tall cultivars support. HA. H 18in–10ft (45cm–3m), S 30in (60cm).

### PROPAGATION
❧ *in situ*, in furrows or broadcast, in spring. Alternatively, sow in plug trays or peat pellets under glass in late winter at 61–64°F (16–18°C).
❦ 7–21 days. ☀ 3–4 months.

## Helichrysum
Compositae

Some 500 species of annuals, herbaceous or evergreen perennials, and evergreen shrubs and subshrubs. Some are foliage plants, with insignificant flowers, others have pretty daisy-like or fluffy flower heads.
**H. petiolare,** syn. *H. petiolatum.* Freely branching shrub with trailing stems and heart-shaped, woolly-hairy leaves, grown primarily as a foliage annual in containers and hanging baskets, or as a houseplant.
**'Limelight'** has lime-green leaves.
**'Variegatum'** is cream-variegated. Grow in well-drained soil in sun. HHA. H to 20in (50cm), S 3ft (1m) or more.

### PROPAGATION
Semi-ripe cuttings – summer. Overwinter young plants in frost-free conditions under glass.
☀ 3–4 months.

## Impatiens
Balsaminaceae

Varied genus of 850 species of annuals and evergreen shrubs and subshrubs with succulent stems and leaves. Busy lizzie is very commonly grown for its flat bright flowers.
**I. walleriana** (Busy lizzie). Compact, spreading perennial grown as an annual. Modern hybrids, with green or bronze leaves, bear single or double flowers, in a wide color range, including red, pink, purple, salmon, orange, and white, from summer until the

first frosts. Easily grown in a sheltered site in light shade or sun, in moist but well-drained soil. FTA. H and S to 24in (60cm).

### PROPAGATION
❧ surface-sow under glass in late winter or early spring, at 64–70°F (18–21°C).
❦ 10–21 days. ☀ 4 months.
Stem-tip cuttings – summer. Overwinter young plants under glass at 50°F (10°C).
☀ 6 months.

## Lathyrus
Leguminosae

Some 150 species of often climbing, annuals and herbaceous or evergreen perennials valued for their showy, often scented flowers.
**L. odoratus** (Sweet pea). Climbing annual with often highly perfumed, pea-like flowers, some with frilled petals, available in reds, pinks, blue, lavender, salmon, cream, and white. There are also dwarf types, some of which are non-climbing, that need little or no support and can be used for hedges. Tall cultivars need training on canes or other supports. Easy in fertile, humus-rich, well-drained soil. HA. H 2–6ft (60cm–200cm) or more, S 12–18in (30–45cm).

### PROPAGATION
❧ *in situ*, in furrows, in spring or fall. Provide plastic covers for fall-sown seedlings in open ground, or place container-sown seedlings in a coldframe. Alternatively, sow in peat pellets, pots or sweet pea tubes under glass in late winter or early spring, at 59°F (15°C). Sweet peas have hard seed coats and benefit from chipping or abrading before sowing. They may also be soaked; do not cover with water and sow immediately the seed swells. If soaked too long they are prone to rotting.
❦ 10–20 days. ☀ 3–4 months.

*Limnanthes douglasii*

*Limonium sinuatum*

*Linum grandiflorum*

## Limnanthes    Limnanthaceae

About 17 species of low-growing annuals, only *L. douglasii* is usually grown.

*L. douglasii* (Meadow foam, Poached egg flower). Very free-flowering annual with finely divided foliage and cupped, white-edged, yellow flowers, throughout summer and fall. Attracts bees and other beneficial insects to the garden. Grow in any soil in sun. HA. H 6in (15cm), S 6in (15cm) or more.

**PROPAGATION**

*in situ*, in furrows or broadcast, in spring or fall. Self-seeds freely.

10–20 days.    3–4 months.

*[handwritten margin note: Wet soil / Sunny location]*

## Limonium    (Sea lavender)    Plumbaginaceae

Some 150 species of annuals, biennials, and deciduous and evergreen perennials and subshrubs with distinctive papery flowers often carried on leafless, winged stems. Grow in sun in well-drained soil.

*L. sinuatum.* Upright perennial grown as an annual with rosettes of basal leaves and winged stems with flat heads of tiny, funnel-shaped papery flowers from mid- to late summer or early fall. HHA. H 2ft (60cm), S 12in (30cm).

**PROPAGATION**

under glass in early spring, at 61–64°F (16–18°C). After planting out, water in young plants thoroughly and keep moist until well established. They are fairly drought tolerant once established.

10–20 days.    4–5 months.

## Linaria    (Toadflax)    Scrophulariaceae

About 100 species of annuals, biennials, and herbaceous perennials with two-lipped, spurred flowers and often gray-green leaves.

*L. maroccana.* Bushy, mound-forming annual bearing masses of tiny, snapdragon-like blooms in summer. Cultivars have flowers in bright shades, including many bicolors. Needs light, well-drained soil, and sun. Deadhead to prolong flowering. HA. H 10–18in (25–45cm), S 6in (15cm).

**PROPAGATION**

*in situ*, in furrows or broadcast, in spring.

14–21 days.    2–3 months.

## Linum    (Flax)    Linaceae

Some 200 species of annuals, biennials, perennials, shrubs, and subshrubs, some evergreen, with funnel- to saucer-shaped five-petalled flowers.

*L. grandiflorum* 'Rubrum' (Scarlet flax). Erect, slender annual with gray-green leaves and red-crimson flowers from midsummer to fall. Needs light, well-drained soil in sun. HA. H 16–18in (40–45cm), S 6in (15cm).

**PROPAGATION**

*in situ*, in furrows or broadcast, in spring or early fall. Dislikes root disturbance.

14–30 days.    3 or 8 months.

## Lobelia    Campanulaceae

Some 370 annuals, perennials, and shrubs with bright two-lipped tubular flowers. They range from trailing types to upright plants. Grow in moist soil in sun or partial shade.

*L. erinus.* Trailing to bushy perennial, grown as an annual, with masses of small, variously colored flowers, from summer to fall. HHA. H 4–10in (10–25cm), S 6in (15cm).

**PROPAGATION**

under glass, in late winter or early spring, at 61–64°F (16–18°C).

14–21 days.    3–4 months.

## Lobularia    (Sweet alyssum)    Cruciferae

Just five species of annuals or perennials with rounded clusters of tiny flowers.

*L. maritima*, syn. *Alyssum maritimum*. Annual forming compact mounds of gray-green foliage that are heavily studded with heads of dainty, sweetly scented, usually white flowers, in summer. Cultivars have flowers in shades of purple, lavender, and rose. Needs light, well-drained soil, and sun. Deadhead to encourage further flower production. HA. H and S 6in (15cm).

**PROPAGATION**

surface-sow under glass, in late winter or early spring, at 61–64°F (16–18°C), or *in situ* in spring.

14 days.    3–4 months.

## Matthiola    (Stock)    Cruciferae

Some 55 species of annuals, perennials and subshrubs with often gray-green leaves and usually scented, pastel-colored flowers.

*M. incana* (Brompton stock). Bushy, often short-lived perennial with spires of usually highly scented flowers in soft pure colors. Various cultivars are grown as annuals or biennials. Grow in moist but well-drained soil in a sheltered site in sun. HA/HB. H to 32in (80cm), S to 16in (40cm).

**Brompton stocks** have single or double flowers in spring and early summer. HB.

**East Lothian stocks** flower in spring or early summer, or in summer. HA/HB.

**10-week stocks** usually have double flowers in summer, about 10 weeks from sowing. There are dwarf and tall cultivars. HA.

**Column stocks** are tall, with double flowers specially bred for cutting. The best flowers are produced in a cool greenhouse. HB.

*Moluccella laevis*

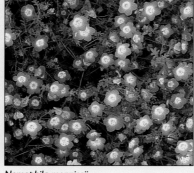
*Nemophila menziesii*

*Nicotiana alata*

Some strains can be selected at the seedling stage to give all double flowers. Refer to the individual seed packets for specific advice (*see also p.11*).

**PROPAGATION**

**Biennials**

❦ surface-sow under glass, in midsummer, at 55–64°F (13–18°C). Provide plastic covers or coldframe protection over winter.

❦ 10–14 days. ❀ 8 months.

**Annuals**

❦ under glass in early spring at 55–64°F (13–18°C).

❦ 10–14 days. ❀ 10–12 weeks.

## *Moluccella*      Labiatae

Only four species of annuals and short-lived perennials, with four-sided stems and unusual large conspicuous calyces that surround the tiny, tubular flowers.

*M. laevis* (Bells of Ireland, Shell flower). Erect annual, in late summer bearing spikes of tiny flowers with papery, pale green calyces. Grow in well-drained soil in sun. HA. H 24in (60cm), S 10in (25cm).

**PROPAGATION**

❦ under glass, in early spring. After sowing, pre-chill for two weeks in the bottom of a domestic refrigerator, then germinate at 61–64°F (16–18°C).

❦ 14–21 days. ❀ 4–5 months.

## *Myosotis* (Forget-me-not)      Boraginaceae

Over 50 annuals, biennials, and perennials bearing small flowers often with paler "eyes"

*M. sylvatica.* Tufted biennial with dense clusters of tiny, yellow-eyed, clear blue flowers in spring and early summer. Several cultivars of various heights are available,

mainly in shades of blue, but there are some white- or pink-flowered variants. Needs moist but well-drained soil in a sunny or lightly shaded site. HB. H to 12in (30cm), S to 6in (15cm).

**PROPAGATION**

❦ under glass, in early summer, at 59–68°F (15–20°C), or in furrows in a seedbed outdoors. Will self-seed, but cultivars may not come true.

❦ 14–30 days. ❀ 6–7 months.

## *Nemesia* ~~1½ + o2'~~      Scrophulariaceae

Over 50 species of annuals, perennials, and subshrubs with showy two-lipped flowers, the lower lip being most prominent.

*N. strumosa.* *Suttonii* Free-flowering, bushy annual with a profusion of flowers, from mid- to late summer, in shades of red, pink, purple, blue, yellow, orange, and white. Easily grown in moist but well-drained, preferably slightly acid soil, in a sunny site. Water freely in dry weather. HHA. H 8–12in (20–30cm). S 4–6in (10–15cm).

**PROPAGATION**

❦ under glass, in early to mid-spring, at 64–70°F (18–21°C).

❦ 7–21 days. ❀ 3–4 months.

## *Nemophila*  30"      Hydrophyllaceae

Some eleven annual, spreading, fleshy-stemmed species, sometimes with sticky hairs, bearing saucer- to bell-shaped flowers.

*N. menziesii,* syn. *N. insignis* (Baby blue-eyes). Low-growing annual with finely divided leaves and saucer-shaped, white-eyed, sky-blue flowers during summer. Needs moist but well-drained soil and sun or light shade. HA. H 6–10in (15–25cm), S 12in (30cm).

alba grandiflora
grandiflora
liniflora
marginata

**PROPAGATION**

❦ *in situ*, in furrows or broadcast, in spring or early fall. Dislikes any root disturbance. Will self-seed freely.

❦ 14–21 days. ❀ 3 or 8 months.

## *Nicotiana* (Tobacco plant)      Solanaceae

Nearly 70 species of annuals, biennials, perennial, and shrubs bearing long-tubed flowers with spreading mouths, usually opening only in the evening and at night.

*N. alata,* syn. *N affinis*. Rosette-forming perennial, grown as a free-flowering annual for its racemes of evening-opening, scented, funnel-shaped, pale yellow-green flowers, produced over long periods in summer. HHA. H 4–5ft (1.2–1.5m), S 12in (30cm).

*N. x sanderae* cultivars. Bear flowers in shades of purple, rose, crimson, lime-green, red, and white. They vary from tall, open-growing types to more compact variants. Flowers remain open during the day. HHA. H 16–36in (40–90cm), 12in (30cm). Both need fertile, moist but well-drained soil, in sun or part-shade.

**PROPAGATION**

❦ surface-sow under glass, late winter or early spring, at 64–70°F (18–21°C).

❦ 7–21 days. ❀ 4–5 months.

## *Nigella*      Ranunculaceae

Twenty species of annuals with feathery leaves and flowers often surrounded by a ruff of hair-like leaves, followed by decorative, sometimes inflated, seed pods.

*N. damascena* (Love-in-a-mist). Bushy annual with bright green foliage and blue, rose-pink, or white flowers in summer, followed by inflated seed pods. Good for cut flowers.

*Petunia* 'Fantasy Mixed'

*Rudbeckia hirta*

*Salvia* 'Red Arrow'

Easily grown in well-drained soil in a sunny position. HA. H 18in (45cm), S 10in (25cm).

**PROPAGATION**

🌱 *in situ*, in furrows or broadcast, in spring or early fall. Self-seeds freely.

🌱 10–25 days.   ☀ 3 or 8 months.

### Pelargonium (Geranium)   Geraniaceae

About 230 species of perennials, succulents, subshrubs, and shrubs, most of which are evergreen. They have often showy clusters of flowers and rounded to ivy-shaped, sometimes aromatic leaves. There are many cultivars of *Pelargonium*, including modern bedding types which are compact, with rounded, mid-green leaves, often with a bronze or green-bronze central zone. They bear heads of single or semi-double flowers in many shades, from spring to the first frosts. Many excellent $F_1$ and $F_2$ hybrids are available. These seed-raised cultivars bloom exceptionally freely in their first year from seed, and so do not have to be overwintered. Older cultivars, which cannot be guaranteed to come true from seed, are easily raised annually from cuttings. Protect double-flowered, old cultivar types from rain, or grow them indoors. Grow in fertile, well-drained soil with shelter and sun. FTA. H to 24in (60cm), S to 12in (30cm).

**PROPAGATION**

🌱 under glass, in late winter, at 70–75°F (21–24°C). Maintain a constant temperature after sowing as temperature fluctuations cause delayed or aborted germination.

🌱 3–21 days.   ☀ 3–4 months.

Stem-tip cuttings – spring, summer or early fall. Overwinter well-rooted cuttings at a minimum of 41°F (5°C).

### Petunia   Solanaceae

Up to 40 species of sticky-haired annuals or herbaceous perennials with delicate, scented, saucer- or trumpet-shaped, showy flowers.

*P.* x *hybrida*. Spreading, free-flowering annuals with single or double, trumpet-shaped flowers, in many shades, including veined, picotee, and star-marked types.

**Grandiflora** types have very large flowers that are easily spoiled by rain and wind;

**Multifloras** have smaller flowers, in greater profusion, and are more weather-resistant. Grow in light, well-drained soil, in a sheltered sunny site. Petunias bloom throughout summer. Deadhead to prolong flowering period. HHA. H 10–16in (25–40cm), S to 3ft (90cm).

**PROPAGATION**

🌱 surface-sow under glass, late winter, at 61–70°F (16–21°C). Do not discard the smallest seedlings when thinning out; they often have the most intense flower colors.

🌱 14 days.   ☀ 4–5 months.

### Phlox   Polemoniaceae

About 67 annuals, herbaceous and evergreen perennials, and a few shrubs with often profuse, salverform flowers.

*P. drummondii* (Annual phlox). Mound-forming, free-flowering annual bearing clusters of flowers in a range of bright colors, many with a distinct eye. They give a continuous display from late spring until the first frost. Grow in any well-drained soil in sun. HHA. H to 16in (40cm), S 10in (25cm).

**PROPAGATION**

🌱 under glass, in early spring, at 64–68°F (18–20°C).

🌱 14 days.   ☀ 3–4 months.

### Rudbeckia   Compositae

Some 20 species of annuals, beinnials, and herbaceous perennials bearing large daisy-like flower heads with cone-shaped centers.

*R. hirta*, syn. *R. gloriosa* (Black-eyed Susan, Gloriosa daisy). Short-lived perennial grown as an annual, bearing masses of large single or double flowers from summer to fall. They range from golden and lemon-yellow, to rich bronze and mahogany, including bicolors. Needs fertile, well-drained soil in a site with sun or part-shade. HA. H to 3ft (90cm), S 18in (45cm).

**PROPAGATION**

🌱 under glass, in late winter or early spring, at 61–64°F (16–18°C).

🌱 7–21 days.   ☀ 4–5 months.

### Salpiglossis   Solanaceae

Just two species of annuals or short-lived perennials with sticky-haired stems and leaves and trumpet-shaped flowers.

*S. sinuata* (Painted tongue). Upright annual bearing exotically colored, heavily veined, trumpet-shaped flowers from summer to fall, in many shades. Performs best in warm, dry summers as the flowers are easily spoiled by rain. They are also ideal as pot plants for growing indoors. Grow in a sunny, sheltered site in moist but well-drained soil. Deadhead to prolong flowering. HHA. H 18in (45cm), S 12in (30cm).

**PROPAGATION**

🌱 under glass, in late winter or early spring, at 64–75°F (18–24°C). May also be sown in fall or late winter for spring-flowering houseplants. Overwinter at a minimum of 61–64°F (16–18°C).

🌱 14–25 days.   ☀ 4–5 months.

*Tropaeolum majus*

*Verbena x hybrida*

*Viola x wittrockiana*

## Salvia                                   Labiatae

Some 900 species of annuals, biennials, herbaceous and evergreen perennials, and shrubs with sometimes very showy, brightly colored, two-lipped flowers and often hairy and aromatic leaves.

*S. splendens* (Scarlet salvia). Compact annuals with distinctive spikes of hooded, two-lipped, brilliant scarlet flowers borne well above the foliage from summer to fall. Cultivars with purple, white, salmon, and lavender flowers are also available. They are good for bedding, border edging, tubs and other containers. Grow in fertile and moist but well-drained soil in sun; provide the pastel-colored cultivars with some shade to preserve flower color. HHA. H 18in (45cm), S 10–15in (25–38cm).

### PROPAGATION
✿ surface-sow under glass, in late winter or early spring, at 59–64°F (15–18°C).
❧ 7–14 days.   ☀ 3–4 months.

## Tagetes (Marigold)            Compositae

About 50 species of annuals and herbaceous perennials which are strongly aromatic.

*T. erecta* (African marigold). Upright annuals, taller in habit than the other groups, bearing mainly double, pompon-like flowers, in yellow, gold, and lemon shades, throughout summer. HHA. H 12–18in (30–45cm), S to 18in (45cm).

*T. patula* (French marigold). Compact annuals with single or, more usually, double flowers in shades of yellow, gold, and mahogany, throughout summer. HHA. H 6–12in (15–30cm), S to 30cm (12in).

*T. tenuifolia*, syn. *T. signata.* (Signet marigold, Mexican marigold). Upright, branched

annuals with dainty, single flowers in golden or lemon-yellow and mahogany shades throughout summer. HHA. H 8–9in (20–23cm), S to 15in (38cm). All have finely divided foliage and are easily grown in well-drained soil in sun. Deadhead to prolong flowering and water freely in dry periods.

### PROPAGATION
✿ under glass, from late winter to early spring, at 64–68°F (18–20°C). Alternatively, sow *in situ*, in furrows, or broadcast, late in the spring.
❧ 7–21 days.   ☀ 4–6 weeks.

## Tropaeolum (Nasturtium)  Tropaeoleaceae

Up to 90 species of annuals and herbaceous perennials with hairless stems and leaves and showy, spurred flowers with spreading lobes.

*T. majus* (Garden nasturtium). Rapidly growing annual usually with a trailing or climbing habit. Modern cultivars, with a more bushy, non-trailing habit, are ideal for borders. The distinctively spurred flowers, produced through summer until the first frosts, are in many shades of yellow, red, or orange. Grow in sun or light shade in poor, well-drained soil; fertile soils encourage leaf growth at the expense of flowers. HA. H 3–10ft (1–3m), S 5–15ft (1.8–5m).

### PROPAGATION
✿ under glass, in early spring, at 55–64°F (13–18°C). Or *in situ*, in furrows, or broadcast, in mid-spring.
❧ 7–14 days.   ☀ 4–6 weeks.

## Verbena                           Verbenaceae

About 250 species of annuals, perennials, and subshrubs bearing dense terminal heads of tubular flowers with wide-spreading mouths.

*V. x hybrida.* Bushy perennial grown as an annual, with rounded clusters of small, salver-shaped, white-eyed flowers throughout summer in a brilliant color range, including scarlet, blue, purple, pink, peach, and white. Easily grown in moist but well-drained soil in sun. HHA. H 10–18in (25–45cm), S 12–20in (30–50cm).

### PROPAGATION
✿ under glass, in late winter or early spring, at 64–68°F (18–20°C).
❧ 14–25 days.   ☀ 4 months.

## Viola (Pansy, Violet)           Violaceae

About 500 species of annuals, biennials, and evergreen, semi-evergreen, and deciduous perennials and a few subshrubs. To a greater or lesser extent many flowers have the characteristic "face" of the pansy.

*V. x wittrockiana* (Pansy). Compact, short-lived perennials, grown as annuals or biennials, as flowering is best in young plants. Available in a wide range of colors, they flower either in winter and early spring or in late spring and summer. Grow in fertile, moist but well-drained soil in a sunny or partially shaded site. Deadhead to prolong flowering. HA/HB. H to 9in (23cm), S to 12in (30cm).

### PROPAGATION
✿ under glass, at 59–64°F (15–18°C) in mid- to late winter, for summer-flowering. Grow on in cool conditions at no more than at 50°F (10°C). Sow in mid- to late summer for winter flowering, either under glass or in a shaded coldframe. Sowing at higher temperatures than those recommended will prevent germination.
❧ 14–21 days.   ☀ 4–6 months.

# HERBACEOUS PERENNIALS

The herbaceous border in summer is one of the glories of the garden, but as well as providing color, shape, and texture at that time of year, perennials make valuable contributions during spring and fall, too. Even winter is brightened by plants such as hellebores and bergenias, and dried stems and seed heads offer continued interest in the bleaker months while also providing food for birds. In a mixed border, where shrubs or small trees feature throughout the year, interplantings of perennials do much to add color and excitement. To get optimum results from perennials, the gardener need learn only a few simple propagating techniques, for very little is required in the way of equipment.

◀ *Although many perennials are long-lived, most rely on regular propagation in order to keep them healthy and able to give of their free-flowering best.*

# WHY PROPAGATE
# HERBACEOUS PERENNIALS?

*Perennials lend themselves to a variety of propagation techniques, most of which are extremely easy to carry out. All that is required is the little forethought necessary to propagate in the year before planting and you will be able to plan the most extravagant of border schemes, secure in the knowledge that there are always plenty of plants.*

**Filling a busy perennial border can be expensive, but propagating from your own plants is practical, cheap, and fun.**

When you decide to increase your stock of perennials, there will be a range of techniques at your disposal. Some of these may be easier, or more appropriate to a certain plant group, than others, but all are worth mastering.

## OPTING FOR SEED

The simplest way of raising perennials is by sowing seed. Using this method, large numbers of plants can be raised economically, with little effort, and a huge range of seeds is available from catalogues. However, cultivars and hybrids seldom produce identical offspring to the parent, and some – particularly those with double flowers – never produce seed; these must be increased by other means.

## MAINTAINING PLANT HEALTH

Division is undertaken not only to increase stock, but also to ensure that plants stay healthy, with maximum flowering potential. Many perennials need routine division every third or fourth year for this reason.

The vast majority of perennials with fibrous roots or rhizomes, which spread from the original plant to form a clump, are perfect candidates for division. Division produces uniform offspring, and, since divisions have instantly functional roots and stems or growth buds, they grow rapidly. Perennials with a tap root or a single

stem are difficult to divide without irreparable damage; so, for any plant of this type, alternative methods – such as stem or root cuttings – are a safer, better option.

## PRODUCING IDENTICAL PLANTS

Cuttings produce identical plants to the parent and are ideal for plants that will not divide easily or that do not come true from seed. Stem-tip cuttings can be taken when the plant is growing strongly. Basal cuttings are less often used, perhaps because they are taken in spring when plenty is happening in the garden. With such plants as *Aster × frikartii* cultivars, they are the only practical method. The best candidates for root cuttings are plants with thick roots, whether tap-rooted, as with sea holly (*Eryngium*), or with searching-type roots, such as the plume poppies (*Macleaya*).

## LAYERING AND GRAFTING

Although layering is more commonly associated with woody plants, border carnations root readily from layers. Grafting also is more often practiced with woody material, the main exception being cultivars of *Gypsophila paniculata*, which are notoriously difficult to root from cuttings.

### Colorful Survivors

*A small group of perennials can be sown directly in their flowering site in spring and early summer, and will happily take their chances. Such attractive plants as* Alchemilla, Aquilegia, Euphorbia, Geranium, *and* Sidalcea, *as well as some penstemons and veronicas, all grow well in this way, if kept weed-free. The same plants can be self-perpetuating by self-seeding, lending a natural charm to a border if allowed to grow where they choose. Only species will come true from seed; seedlings of cultivars are likely to be inferior to their parents.*

*Aquilegia vulgaris is a good example of a perennial that maintains itself by self-sowing.*

 **Achillea filipendulina**
Four-year-old, well-established plants need dividing – otherwise they will become overgrown, and their centers start declining. When dividing, remove the center parts and re-plant the healthy, outside portions, thereby maintaining bright displays.

 **Phlox paniculata**
Phlox are alarmingly prone to eelworm. To guarantee a continuous healthy show, take root cuttings annually. Note: if you use this technique for variegated types, they will immediately revert to green; divide instead.

 **Alchemilla mollis**
A brilliant, prolific self-seeder, excellent as an edging plant, but the problem with it is that the seedlings get everywhere. To stop things getting out of hand, trim down all but one of the flowerheads in late summer, collect its seed, and sow in a coldframe. When mature, plant out exactly where required.

 **Stachys byzantina**
This is grown for its gorgeous velvety foliage, but it can quickly start looking quite ragged. Either divide annually, or take cuttings.

**Sedum spectabile**
This is one of the easiest plants to propagate. It divides readily: you do not even have to dig it up – just slice off a section of the plant with roots, re-plant, and it will take. The younger the plants you have, the better, because vigorous new growth guarantees strong color and scent, attracting masses of butterflies.

**Sisyrinchium striatum**
Each leafy shoot dies after producing a flowering shoot, but self-sown seedlings are abundant and can be transplanted into fresh soil when small to maintain the display.

# PROPAGATING TECHNIQUES
# HERBACEOUS PERENNIALS

*Techniques that increase the stock of perennial plants range from the simplest methods of division to the slightly more adventurous techniques of grafting. None of these methods is very difficult, most will give satisfyingly rapid results, and very little is required in the way of tools and equipment.*

## SEED

The best time to sow most perennial seed is in late winter or early spring. However, members of the families Primulaceae and Ranunculaceae, such as hellebores, always germinate freely if sown fresh, but much less so if they are spring-sown.

In the wild, germination is frequently delayed until conditions are perfect, and some seed has a built-in inhibitor that prevents germination until soil and weather conditions are suitable. Some seeds wait until spring rains provide sufficient moisture for growth,

### Sowing seed . . . . . . . . . . . . . . . . . . . . . . . . . . . . . . . . . . . . . .

Placing containers of seed in a plunge bed is a convenient way of storing until germination. It prevents them from tipping and spilling their contents, and reduces temperature fluctuations.

▶ **1** Sow the seed thinly and evenly on the surface of the seed germination medium and cover with a ¼in (5mm) layer of fiine sand. This prevents the seed drying out and makes it easier to water evenly and to remove weeds or moss.

▶ **2** Water the container and plunge in an open frame, shaded from direct sun. Do not cover unless there is excessive rain, but keep evenly moist until germination. If germination occurs before the last frosts, move the seedlings into a greenhouse.

Thin out the seedlings into individual containers as soon as the first true leaves appear, and place in a closed coldframe to re-establish. Then harden off and place in an open frame, until the seedlings are large enough to set out. If germination does not occur in the first year, retain the containers and keep them watered; seedlings may then appear the following year, or even the year after that.

and these often have tough seed coats that only break down when soaked. Lupins and many other members of the pea family (Leguminosae) are like this and, in cultivation, must be soaked or have their hard coats pierced by abrading or nicking. Others need chilling and then warmth to mimic winter followed by spring before they will germinate. Chilling can be achieved by placing the container in the bottom of the refrigerator (not the freezer) for up to 3 weeks. Most seed that needs chilling only requires exposure to temperatures of 34–41°F (1–5°C), and positively does not need freezing. The natural way is often the easiest and, for most perennials, sowing in late winter and exposing the containers to frost in an open frame is usually sufficient to break dormancy. If you are in any doubt as to the best time for sowing, sow half in the fall and half in spring.

Some robust perennials can be sown directly into the soil as for annuals and vegetables, and these also often provide self-sown seedlings. They can be thinned out into containers in the normal way, or left to develop and then moved to their flowering position. Unlike annuals, few hardy perennials need heat for germination, although many will germinate more quickly in warmth, at 59–68°F (15–20°C), in a propagator. Label the seed and seedlings at all stages of growth, ideally with the date of sowing and source of the seed, as well as with the name of the plant.

Thin out seedlings into individual containers as soon as the first true leaves appear, and place in a closed cold frame to re-establish. Then harden off and transfer to an open frame, until large enough to set out.

If germination does not occur in the first year, retain the container and keep it watered; seedlings may appear the following year, or even the year after that.

# DIVISION

Division is an essential part of perennial gardening and is used both to increase stock and to keep plants in good health with maximum flowering potential. Division produces offspring that are identical to the parent, which can seldom be guaranteed when new plants are raised from garden-collected seed. Nearly all perennials that spread from a central crown to form a mat or clump are perfect candidates for division. Those with a deep tap root or single central stem are generally unsuitable for division and, for these, stem or root cuttings are usually a better option.

The safest form of division is performed in the spring, just before new growth commences. This is especially the case in areas with hard winters, since spring division gives the plants a complete growing season to become well established, generate a new root system, and produce plenty of growth buds at the crown. Some spring-flowering plants may be divided in the fall, or immediately after flowering, as long as they are kept well watered if the weather becomes dry for prolonged periods.

Damp, warm, overcast weather provides optimum conditions for re-establishment. Always use healthy stock plants, as any pests or diseases will be carried over to the divisions. When dividing older plants, discard the old woody centers, re-planting only young, vigorous growth from around the edges of the clump.

## Techniques for division

There are several slightly different techniques of division, depending on the growth habit of the original plant and the number of divisions required.

The simplest way is to dig up the plant and either break it into pieces by hand or by inserting two forks back-to-back and levering them apart. This is ideal for fibrous-rooted plants whose roots are not entwined.

An alternative method of division is to disentangle the roots carefully and separate the clump into individual portions, each with a single shoot or crown. Such small divisions are best potted up and grown in a coldframe before planting out. Shake the soil from the roots and then gently manipulate individual crowns with the fingers until they fall apart. It is often easier, especially for plants grown in heavier soils, to wash the soil from the roots in a bucket of water; the individual

## Division . . . . . . . . . . . . . . . . . . . . . . . . . . . . . . . . .

There are several techniques for division, depending on the original plant and on the number of divisions that are required. The simplest is suitable for clump-forming and fibrous-rooted plants, such as asters.

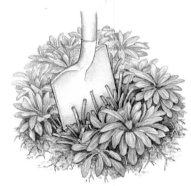

▶ **1** Lift the plant with a spade or fork, including the entire root zone if possible, when the new shoots reach about 2in (5cm) in length. Section large clumps with a spade, then tease out small pieces by hand, each with about half a dozen new shoots and a mass of fibrous roots. Discard the old material from the center of the clump.

▶ **2** Re-plant the divisions immediately, with the growing points at the same depth as they were originally. Set them in at spacings of about 12in (30cm), and then water in thoroughly.

▶ **3** If the rootball is very dense or tangled, lift the entire rootball and insert two garden forks back-to-back at the center of the clump. Use a levering action to prise the clump firmly but gently apart.

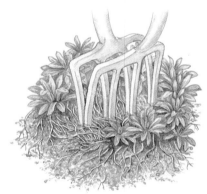

crowns should then become clearly visible and will separate very easily with minimal damage to the roots.

Tough-rooted plants, such as hostas, can be hosed clean of soil and the selected crowns separated with a sharp knife. A rather crude method is to slice them into two to four pieces with a sharp spade, but this wastes material and increases the risk of fungal infection entering the broken-off roots.

## Dividing rhizomatous perennials ...............

Division of bearded iris, as shown here, can be done in early fall but is best done just after flowering. They need dividing every three years or so to keep them flowering well.

► **1** After flowering, lift the clump of rhizomes, shake off excess soil, and separate into sections by hand. Select plump young rhizomes with healthy roots and discard the old rhizomes.

◄ **2** Section the plump young rhizomes into 2–4in (5–10cm) lengths with a sharp knife, ensuring that each has a healthy fan of leaves at one end. Trim the slender fibrous roots by one-third of their length.

► **3** Trim the leaves to about 6in (15cm) long; this will reduce wind rock when planted.

◄ **4** Set the rhizome into the soil so that the slender fibrous roots are buried, and the rhizome itself is buried to no more than half its depth. When replanting groups of iris rhizomes, make sure they are set so that all the leaf fans face in the same direction.

## Basal cuttings ...................................

Take basal cuttings from vigorous shoots as the first leaves unfurl. Cut off with a sharp knife, with a piece of harder, semi-woody material at the base. Trim off the lowest leaves and insert in pots or trays of moist rooting medium. Place in a coldframe out of direct sun. Keep the medium just moist. When rooted, after 3–5 weeks, pot up separately.

### Division of rhizomatous perennials

Plants with substantial rhizomes, such as irises or bergenias, can be cut into individual portions, provided that each section of rhizome has at least one growing point. Many rhizomatous plants, such as bergenias, can be cut up into relatively small sections, about 1½in (4cm) long, as long as each portion has a bud on it.

Larger divisions can be re-planted into the soil once it has been refreshed with well-rotted organic material. It is important that rhizomes are set back into the soil at the same depth as they were previously. For some genera, such as *Iris* (*see left*), the rhizomes grow naturally at or just above the soil surface and will rot if planted too deeply. If the weather is hot and dry, or the divisions are small, pot them up and keep them in a closed coldframe for 2–3 weeks. Then harden off before setting them out.

## CUTTINGS

Taking cuttings produces plants that are identical to the parent. It is also a useful technique for propagating a "sport" or mutation – a shoot that differs from all the others on the same plant, such as one that has produced variegated leaves. It is important only to use pest- and disease-free material when taking cuttings. Any disease present in the parent plant will be transferred to the new ones, and material infested with pests is likely to be weakened and will not establish well, if at all. Always cut cleanly and maintain hygiene levels. There are several different methods of taking cuttings.

### Basal cuttings

Basal cuttings are an excellent way of increasing many perennials. With some plants, such as the *Aster frikartii* cultivars, it is the only practical method and it is ideal for most other members of the family Compositae, such as *Artemisia* and *Tanacetum*, as well as the Labiatae, including *Nepeta* and *Stachys*. Basal cuttings are also particularly useful for plants with hollow stems, such as delphiniums and lupins, since the base of the new growth is solid and makes perfect cutting material, unlike the mature stems.

Basal cuttings are taken in spring as the new shoots emerge, but before the leaves are fully unfurled. The shoots used for cuttings are soon replaced and the plant is not harmed by the process.

Basal material is not so readily available once the plants have reached their full growth, and any that does appear is usually rather weak or etiolated because of the tall stems positioned around it. Plants such as violas can be induced to produce new basal growth later in the season, by shearing the plant over after the first flush of bloom. The resulting new shoots are ideal for basal cuttings.

A number of tender perennials, for example, chrysanthemums and argyranthemums, are traditionally propagated by basal cuttings. The old plants are lifted in the fall before being killed by the frosts, and overwintered in a frost-free shed or greenhouse, bedded in just-moist medium or peat substitute. They are usually given a little extra warmth in late winter or very early spring to force them into growth. When new basal shoots appear, they are removed and treated as ordinary basal cuttings.

### Pre-rooted cuttings

Taking basal cuttings is neither difficult nor a lengthy process, but there is a quicker variation – small, pre-rooted stems traditionally known as "Irishman's cuttings." It is often possible to find a few shoots that have already sent out one or two roots by searching around the margins of the clump. If necessary, trace the stem down below the surface of the soil. Sever the cutting below the rooted section and then pot up into proprietary rooting medium, being careful not to damage the existing roots. Proceed as for basal cuttings.

Plants that are amenable to this process include *Asters*, *Anaphalis*, and *Mimulus*, as well as many other mat- or carpet-forming, stem-rooting perennials.

### Stem-tip cuttings

Taking stem-tip cuttings is a productive means of increasing a wide range of perennials. Cuttings are taken from the tips of healthy, non-flowering, semi-mature or mature shoots, between early summer and fall. In genera such as *Penstemon*, which root very easily, shoots can be cut into a number of usable sections at almost any time during the growing period.

For most perennials, stem-tip cuttings strike easily without the need to stimulate rooting by using hormone rooting compound. Dipping the bottom ½in (1cm) of the cutting in rooting preparation is

### Stem-tip cuttings ...............................

Stem-tip cuttings are best taken early in the day, when plants are turgid (full of water). Place them in a plastic bag to reduce water loss and prevent wilting before insertion.

▶ **1** Using a sharp knife or pruners, cut the stem tip from the plant: this should be about 3–5in (7.5–12.5cm) long.

◀ **2** Trim the cutting to 2–3in (5–7.5cm) long, with a straight cut just below a node. Neatly trim the lower leaves from the bottom third of the cutting.

▶ **3** Insert the cuttings to the depth of the lowest leaves, around the edge of a prepared pot of rooting medium. Make sure that the leaves do not touch each other or the medium surface. Cover the pot with a sealed polyethylene tent, or place in a propagator, preferably with bottom heat at 59°F (15°C). Heating is not essential but speeds rooting. A mist unit is useful for large-scale propagation.

### Aftercare

Provide good light but shade from direct sunlight. Check regularly and remove any fallen leaves or rotted material as soon as seen. If left in place, they will spread infection to the healthy cuttings. Cuttings that root during the summer should be potted up immediately, but those that do not develop roots until late fall should be left in place until spring. Keep young plants in frost-free conditions over the winter.

nevertheless helpful because the compound contains a fungicide that helps to inhibit rots.

The gentle bottom heat provided by the propagator or soil-warming cables is not essential but speeds up the rooting process. A 59°F (15°C) temperature is suitable. For large scale propagation a mist unit is useful.

# ROOT CUTTINGS

As all gardeners know when digging out dandelions, ground elder, or couch grass, if just a little piece of root is left behind, it is likely to shoot. In essence, this is the same process as that of taking root cuttings, except that in this case, the gardener deliberately exploits the plant's capacity to regenerate itself from its root and produce a new individual that is identical to its parent.

The plants that respond best to this technique are those with thick roots, some of which are not amenable to increase by division or by cuttings. They can be

## Root cuttings .....................................

Taking root cuttings during the plant's dormant period is a suitable way of increasing those plants that do not lend themselves to propagation by division or cuttings.

◄ **I** Dig up the plant and remove a few pencil-thick roots. Wash the soil from the root cuttings and rub off small fibrous roots. Cut the roots into 2–3in (5–7.5cm) lengths, making a straight cut at the top (the end nearest to the crown) and a sloping one at the base, so that it is easy to see which way up to insert the cutting.

▶ **2** Plant the pieces of root vertically into a prepared pot of moist rooting medium, with the straight cut just below the level of the medium. Top-dress with a ¼in (5mm) layer of coarse sand. Label with the name and date.

◄ **3** For plants with very slender roots, the technique varies a little. Section the roots into 3–5in (7.5–12.5cm) lengths. Lay these lengths horizontally on the surface of a prepared tray of medium and just cover to their own depth with sieved medium. Mulch with a ¼in (5mm) layer of sand.

tap-rooted plants such as the sea hollies (*Eryngium*), or plants with thick, widely spreading roots, like the Californian tree poppies (*Romneya*). Some more slender-rooted plants, such as the Pasque flower (*Pulsatilla*), can also be propagated using this technique. Root cuttings are taken during the dormant period, in late fall or early winter.

There are several common reasons for failure of root cuttings. It is important to make sure that the root cutting is inserted with the correct orientation. Root tissue is programmed to produce shoot growth from its upper end and root growth from its lower end. This is why a root cutting is made with a straight cut at its top end and a sloping cut at the bottom end. It serves to ensure that the cutting is inserted the right way up. For plants with very tender roots the technique is modified by inserting them horizontally. Cuttings of slender roots must be longer than those of the usual thickness, as otherwise they do not contain sufficient food reserves for the development of new growth. If planted vertically, they are likely to rot before the new growth reaches the medium surface.

Overwatering commonly leads to failure. If kept too moist, the root cuttings are likely to rot before they can make any new growth. After insertion, place the tray or container of cuttings in a coldframe or greenhouse and keep it just moist; the pots will only need a little watering, if any, until new growth begins.

The last reason for failure is probably impatience. In spring, the roots will send up new topgrowth, but may do so before they are properly rooted. Before attempting to pot up, always check to make sure that the cuttings are well rooted. They are very unlikely to be successful if moved before a good network of fibrous roots has formed.

If only a few new plants are required, it is often possible to scrape away some soil from the margins of the plant and remove just a few roots, thereby not disturbing the parent plant.

A crude method of increase – effectively a means of taking root cuttings *in situ* – is sometimes used with oriental poppies (*Papaver orientalis*). Here, you need to sheer off the top-growth of the plant just below the surface of the soil with a spade. When new plants arise from the top of the severed roots, separate them and re-plant immediately.

# GRAFTING

Grafting is seldom used with herbaceous perennials, with the exception of the cultivars of *Gypsophila paniculata*, such as *G. paniculata* 'Bristol Fairy' and 'Flamingo', which are difficult to root from cuttings. The nature of the material does not lend itself to complicated grafts and a simple wedge graft, made in late spring, is sufficient. A two-year-old plant of the species is raised from seed to act as the rootstock.

Once the graft union has been made, transfer the pot to a propagator, with gentle bottom heat. Keep the medium just moist, watering from below if necessary. New roots and the graft union will form in 4–6 weeks. Wean gradually from the propagator and grow in a cool, frost-free place with good light but out of direct sun. Remove the tape when new growth resumes and pot on when the plants are well established.

# LAYERING

Layering is a technique that is mainly used for woody subjects, such as shrubs, but a few perennials can also be increased in this way, in particular *Dianthus*, such as pinks and border carnations. After flowering, prepare for layering by cultivating the soil around the plant to a depth of about 3in (7.5cm). Incorporate peat (or substitute) and sand to improve the drainage and aeration of the soil in the rooting area.

## Layering pinks and border carnations ..........

Layering is a traditional and simple means of increasing border carnations and pinks. After flowering, the plants produce vegetative shoots that will root within 5–6 weeks.

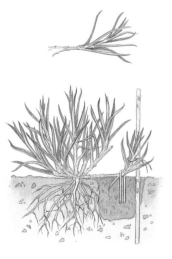

◄ Select a sturdy low-growing stem. Strip the stem of leaves, leaving about 4–5 pairs at the stem tip. Make a nick in the stem with a sharp knife to form a tongue that extends from the first bare, leafless node to just below the next node towards the parent plant. Bend the stem over onto the prepared soil surface and pin it down with a U-shaped piece of wire. Hold the stem tip upright by tying it to a vertical piece of split cane. Cover the stem with soil. When the stem has rooted, sever from the parent plant and either pot it up to grow on, or transfer to its permanent position.

## Grafting ..........................................

Grafting herbaceous material, like the *Gypsohila* cultivar shown here, is a fairly simple technique, although it does require forward planning to raise the rootstock material.

► **1** Lift the rootstock and select a single stout root, about ½in (1cm) thick, and cut about 5in (12.5cm) or so of root from the parent plant.

► **2** Trim off any small fibrous roots and reduce any large lateral roots to about ½in (1cm) long. With a straight top cut and a slanting bottom cut, trim the rootstock to 3–4in (7.5–10cm) in length. Make a vertical cut, ½in (1cm) deep, in the top of the rootstock. Set the rootstock in a prepared pot of rooting medium with the top 1in (2.5cm) above the medium surface.

► **3** To prepare the scion, take a sturdy non-flowering shoot, of ⅛in (3mm) in diameter, from the desired cultivar and trim to 1½–3in (4–7.5cm) long. Trim the base of the scion to a wedge shape so that it matches the cleft in the rootstock.

► **4** Insert the scion into the rootstock, matching up the cut surfaces. Bind with grafting tape and keep in a propagator with bottom heat until the graft has taken.

# DIRECTORY OF
# HERBACEOUS PERENNIALS

*Below are key tips on a selection of herbaceous perennials that are particularly suitable for propagation. Unless otherwise specified, follow the detailed instructions given under the propagation section on pages 28–33.*

## Key

| | |
|---|---|
| H | Height |
| S | Spread |
| 🌰 | By seed |
| ⚊ | Germination |
| ☀ | Time to maturity |
| ✎ | By division |

## Acanthus

(Bear's breeches)                    Acanthaceae

Some 30 species of mostly hardy herbaceous perennials, grown for their strongly architectural form and tall spikes of flowers, with papery hood-like bracts, in summer. The flowers are excellent for cutting. They tolerate most types of soil in sun or light shade, but grow best in deep, fertile soils. Those described are hardy.

*A. mollis.* Large, lobed, glossy, dark green leaves and white flowers with purple bracts. Late summer. H to 5ft (1.5m), S 3ft (1m).

*A. spinosus.* Narrow, deeply cut, spiny leaves, and white flowers with purple bracts from midsummer onward. The **Spinosissimus Group** has very deeply divided, spiny, grayish-green leaves with white spines and midribs. H to 4ft (1.2m), S 2–3ft (60–90cm).

### PROPAGATION

🌰 spring, in a coldframe. Sow the large seeds in individual pots. Germination is more rapid at 59–68°F (15–20°C) in a propagator. Do not transplant once the tap root has developed. Will self-sow if left on the plant and can be a nuisance. Dig up self-sown seedlings as soon as possible. Any part of the root left in the ground will re-shoot.
⚊ 14 days.    ☀ 3 years.

✎ spring or fall, with care. Divisions re-establish slowly.    ☀ 2 years.
Root cuttings – winter. Use deep pots to minimize root disturbance when planting out.    ☀ 3 years.

## Achillea (Yarrow)                    Compositae

About 85 species and many cultivars of hardy, evergreen, and herbaceous perennials, valued for their aromatic, often finely divided, gray-green or silver-gray foliage and corymbs of tiny, daisy-like flowers. Those described here tolerate a range of soils in sun or dappled shade. Most have flowers that are suitable for drying.

*A. millefolium.* Has tiny flowers throughout summer, in shades of lilac, lavender, cerise to deep red. H 24in (60cm), S 24in (60cm).

*Acanthus spinosus*

*Achillea 'Moonshine'*

*A.* 'Moonshine' has yellow flowers. H 24in (60cm), S 24in (60cm).

*A. ptarmica* (Sneezeweed). Bears white flowers throughout summer. **'Boule de Neige'** has double flowers; **'The Pearl'** has pompon-like blooms. H to 24in (60cm), S 12in (30cm).

*A.* 'Taygetea', syn. *A. aegyptica.* Evergreen with heads of yellow flowers in summer. H 24in (60cm), S 18in (45cm).

### PROPAGATION

✎ spring. Divide mature plants every 3–4 years to maintain vigor.    ☀ 14 months. Basal cuttings – spring, while shoots are still young. Valuable for short-lived, or woody-based species and cultivars, such as *A.* 'Moonshine' and *A.* 'Taygetea'.    ☀ 2 years.
🌰 fall or spring, in a coldframe. Plants may not come true but this is a useful method for mixed-colored forms.
⚊ 1–4 months.    ☀ 1–2 years.

## Anaphalis (Pearly everlasting)  Compositae

About 100 species of hardy, clump-forming perennials with silver foliage and white "everlasting" flowers during summer. They need moist but well-drained soil in full sun.

*A. margaritacea.* Produces yellow-centered flowers from midsummer to fall. H to 24in (60cm), S 24in (60cm).

*A. triplinervis.* Yellow-centered flowers in mid- to late summer above silver-gray leaves. H to 36in (90cm), S to 24in (60cm).

### PROPAGATION

✎ spring.    ☀ 15 months.
Basal cuttings – spring.    ☀ 15 months.
🌰 preferably sown fresh, fall or spring, in a coldframe. This is the quickest method of propagation. Germination is more rapid at 59–68°F (15–20°C) in a propagator.
⚊ 1–3 months.    ☀ 1 year.

Anemone × hybrida 'Max Vogel'

Artemisia absinthium

Aster novae-angliae

## Anemone — Ranunculaceae

Some 120 species of hardy to half-hardy perennials, with tuberous, rhizomatous, fibrous, or fleshy roots. Larger species, grown for their mounds of palmate foliage and cup-shaped flowers, are suited to an herbaceous border. Grow in moist but well-drained, fertile soil, in sun or dappled shade. (*See also Bulbs, Corms, and Tubers directory, pp. 64–65.*)

**A. × hybrida**, syn. A. × elegans, A. japonica (Japanese anemone). Hardy suckering perennial bearing semi-double, pale pink flowers in late summer and fall. **'Honorine Jobert'** has pure white flowers. H 4–5ft (1.2–1.5m), S 24in (60cm) or more.

### PROPAGATION

Root cuttings – early winter. Remove thicker roots and cut into 2in (5cm) lengths. ☀ 2 years.

✎ spring. Divide when the soil is not too dry and therefore not crumbly – success can be assured if divisions are re-planted without the soil falling off. Divide A. × hybrida with care. ☀ 6 months.

🌱 as soon as ripe, in a coldframe; germination can be erratic. Less common perennial species, such as A. rivularis, are best propagated by seed.

## Anthemis — Compositae

Some 100 species of annuals and mostly hardy perennials with feathery leaves and white, yellow, or orange, daisy-like flowers borne over long periods in summer. Grow in light, well-drained soil in sun. They are short-lived, so propagate regularly.

**A. punctata subsp. cupaniana.** Hardy and mat-forming with silver-gray leaves. White daisies in early summer. One of the easiest

plants to increase (see propagation). H 12in (30cm), S to 36in (90cm).

**A. tinctoria** (Golden marguerite, Ox-eye chamomile). A hardy species with golden-yellow daisies during summer. **'E.C. Buxton'** has pale lemon-yellow flowers. H 36in (90cm), S 36in (90cm).

### PROPAGATION

Basal cuttings – spring, 1½in (4cm) long. Some, such as A. punctata subsp. cupaniana, can simply have pieces snapped off and put in the ground where they will usually root. ☀ 16 months.

🌱 spring, in a coldframe. Cultivars may not come true. Some, such as A. tinctoria and A. sancti-johannis, self-sow. Transplant or pot up while still small.

🌿 2–3 weeks. ☀ 1 year.

## Artemisia — Compositae

Some 300 species of annuals, perennials, subshrubs, and shrubs. Many of the perennial species and cultivars are valued in the garden for their silver foliage; few have attractive flowers. Those described are hardy. Grow in well-drained soil in sun.

**A. absinthium 'Lambrook Silver'.** A woody-based perennial with deeply cut silver leaves. H to 36in (90cm), S 24in (60cm).

**A. lactiflora** (White mugwort). Clump-forming, with jagged, dark green leaves and panicles of creamy-white flowers from midsummer to fall. H 5ft (1.5m), S 24in (60cm).

**A. ludoviciana** (Western mugwort). Rhizomatous, with clumps of narrow, silver-white leaves. **'Silver Queen'** is lower-growing. **'Valerie Finnis'** is shorter still, with silver leaves. H 2–4ft (60–120cm), S 2ft (60cm).

### PROPAGATION

Cuttings – greenwood or heeled semi-ripe cuttings in late summer/fall. All cuttings are best taken from young growth and shoots should be free of flowers and buds.

A. absinthium takes best from late summer cuttings. Soft tip cuttings of A. a. 'Lambrook Silver' are taken in spring. Overwinter young plants at 41°F (5°C). ☀ 15 months.

✎ spring. Only suitable for clump-forming artemisias, such as A. ludoviciana. ☀ 15 months.

🌱 fall or spring, in a coldframe. Cultivars do not come true. 🌿 2–3 weeks. ☀ 1 year.

## Aster — Compositae

About 250 species of annuals, biennials, and perennials, bearing daisy-like flower heads with yellow discs. The hardy perennials, and their cultivars are valuable for their long flowering season, from late-summer to fall. Grow in deep, fertile, moist but well-drained soil in sun. A. amellus tolerates drier soil than A. novae-angliae and A. novi-belgii. Discard self-sown seedlings; they are rarely as good as their parents.

**A. amellus.** Bears corymbs of lilac-blue flowers in late summer/fall. Cultivars with pale pink to violet flower heads are available. H to 24in (60cm), S 18in (45cm).

**A. × frikartii.** Pale to dark blue-violet flower heads with a central orange-yellow disc. Late summer to fall. H 14–30in (35–75cm), S 12in (30cm).

**A. × novae-angliae** (New England aster). Rhizomatous perennial producing violet flower heads from late summer to fall. There are many cultivars, in white through shades

*Astrantia major*

*Bergenia cordifolia*

*Campanula glomerata*

of pink, deep red, violet, and lilac-blue.
H 4–5ft (1.2–1.5m), S 18–24in (45–60cm).
*A.* x *novi-belgii* (Michaelmas daisy, New York
aster). Produces violet flower heads, with
cultivars in colors from creamy white
through pink to deep crimson and violet
H 3–4ft (1–1.2m), S 18–24in (45–60cm).

**PROPAGATION**
✃ spring. Divide *A. novae-angliae* and
*A. novi-belgii* every 3 years to maintain vigor.
☀ 6 or 18 months.
Basal cuttings – spring. A few asters, in
particular *A. amellus* and *A.* x *frikartii* and
its cultivars, cannot be divided and must be
increased by basal cuttings.   ☀ 18 months.
🌰 fall/spring. Sow fresh. Pre-chill spring-sown
seed. Cultivars do not come true.
🌱 1–4 months.   ☀ 2 years.

## *Astrantia* (Masterwort)   Umbelliferae
Some 10 species of hardy perennials with
umbels of tiny flowers surrounded by
conspicuous papery, usually pale green
bracts. Flowers in early summer. Grow in
fertile, moist but well-drained soil, in sun.
*A. major* (Masterwort). Clump-forming with
deeply lobed leaves, and green or pink
flowers surrounded by white, often pink-
tinged bracts. Early to midsummer.
'Sunningdale Variegated' has yellow-
variegated leaves. H to 36in (90cm),
S 18in (45cm).

**PROPAGATION**
✃ fall or spring.   ☀ 14 months.
🌰 fall or spring. Sow fresh in a
coldframe. Cultivars are unlikely to come
true. Astrantias frequently self-sow, providing
a ready supply of seedlings.
🌱 3 weeks.   ☀ 1 year.

## *Bergenia* (Elephant's ears)   Saxifragaceae
Some eight species of mostly hardy,
rhizomatous, evergreen perennials, grown
for their leathery leaves and bell-shaped
flowers in spring. There are many cultivars
with flowers in white (*B.* 'Bressingham
White', *B.* '*Silberlicht*') through shades of pink
(*B.* '*Schneekönigin*') to crimson and red-
purple (*B.* 'Abendglut', *B.* 'Ballawley'). Useful
as ground cover in sun or part-shade, in
moist but well-drained soil.
*B. cordifolia*. Glossy dark green leaves, flushed
purple in winter, and pale to deep pink
flowers. Late winter or early spring.
H 24in (60cm), S 28in (70cm).
*B. purpurascens*. Green leaves tinted purple
beneath; purple flowers in spring. H 18in
(45cm), S 12in (30cm).

**PROPAGATION**
✃ fall or spring.   ☀ 1 year.
Rhizome cuttings – winter. Cut rhizomes into
1in (2.5cm) sections, each with a dormant
bud, which is often hidden under a scale.
Set horizontally in rooting medium and
keep moist. This method will give large
quantities of plants, but they will take longer
to come into flower.   ☀ 2 years.
🌰 fall/spring in coldframe. Pre-chill spring-sown
seed. Cultivars do not come true.
🌱 1–3 months.   ☀ 2–3 years.

## *Campanula* (Bellflower)   Campanulaceae
About 300 species of annuals, biennials, and
perennials, most with bell-shaped flowers
borne over long periods. Those described
are hardy. They need fertile, moist, but well-
drained soil, in sun or light shade.
*C. glomerata* (Clustered bellflower). White,
lavender, or blue-violet flowers, in summer.

'Superba' is taller with rich violet-purple
flowers. H 18in (45cm), S indefinite.
*C. lactiflora* (Milky bellflower). Has white
to pale blue, or blue-violet flowers, in
midsummer. 'Prichard's Variety' has violet-
blue flowers. H 4ft (1.2m), S 30in (75cm).
*C. persicifolia* (Peach-leaved bellflower).
Open spires of white or blue flowers. Early
to midsummer. H 18–30in (45–75cm),
S 12in (30cm).

**PROPAGATION**
✃ spring. Clump-formers, such as
*C. persicifolia*, and rhizomatous campanulas,
such as *C. glomerata*, are best increased by
division. Re-plant only young growth from
the edges of the clump.   ☀ 1 year.
🌰 fall or spring. Pre-chill spring-sown seed
for 3 weeks. Some self-sow. Cultivars do not
come true and should be increased
by basal cuttings.
🌱 2–6 weeks.   ☀ 1 year.
Basal cuttings – spring.

## *Centaurea*
(Hardhead, Knapweed)   Compositae
About 450 species of annuals and perennials.
A few are valued for their foliage and
flowers. Grow in well-drained soil in sun.
*C. hypoleuca*. Has divided, pale green leaves
and bears pink flower heads during summer.
'John Coutts' has deep pink flower heads.
H 24in (60cm), S 18in (45cm).
*C. montana*. Vigorous and rhizomatous,
and bears blue flower heads from early to
midsummer. H 18in (45cm), S 24in (60cm).

**PROPAGATION**
✃ spring or fall. Clump-formers
are easily divided. Divide regularly, every
2–3 years, to maintain vigor.   ☀ 1 year.

*Chrysanthemum x superbum*

*Delphinium 'Oliver'*

*Dianthus 'Dora'*

🌱 fall or spring. Pre-chill spring-sown seed for 3 weeks. Many self-sow.
🌿 2–6 weeks. ☀ 1 year.
Stem cuttings – fall (*C. cineraria*), overwinter under glass. ☀ 1 year.

## Chrysanthemum                    Compositae

Some 20 species of annuals and perennials, most of which are now placed by botanists in the genera *Dendranthema*, *Leucanthemum*, and *Tanacetum*, but are retained here for convenience. **C. x morifolium** (Florist's chrysanthemum) is now *Dendranthema × grandiflorum*. Most are half-hardy or frost-tender. The hardy perennials need moist but well-drained, fertile soil, in sun.
**C. rubellum**, syn. *Dendranthema zawadskii*. Has given rise to the **Rubellum Group** hybrids that bear daisy-like, single or double flowers, in a range of colors, from late summer to fall. Half-hardy. H 24–30in (60–75cm), S 18in (45cm).
**C. x superbum**, syn. *C. maximum*, (Shasta daisy). Now properly *Leucanthemum × superbum*. Masses of white daisies over long periods in summer. Hardy. H 3–4ft (90–120cm), S 18in (45cm) or more.

### PROPAGATION

Basal cuttings – winter or early spring, 2in (5cm) long. Most can be increased by this method, which is particularly good for *C. x morifolium*, cuttings of which should be taken under glass. ☀ 6 months.
⚒ spring or fall. Discard older material from center of the clumps. Increase hardier species by division in spring. ☀ 6 months.
🌱 spring, for Rubellum Group hybrids, at 59–68°F (15–20°C).
🌿 1 month. ☀ 6 months.

## Cortaderia (Pampas grass)        Gramineae

Genus of 24 species of large, tussock-forming, evergreen and semi-evergreen grasses grown for their impressive habit and large, long-stemmed, and silky-plumed inflorescences in late summer.
**C. selloana**. This hardy species forms tussocks of arching leaves and produces large, silvery, often pink-flushed, pyramidal plumes in late summer and early fall. Available in a wide range of forms of differing heights. It can be raised from seed, but inflorescences, which may take several years to be produced, are very variable.
H to 10ft (3m), S. 5ft (1.5m).

### PROPAGATION

⚒ mid-spring, as new growth begins. Separate sections from the extremity of the clump with a sharp spade, and either divide further by hand or re-plant as a clump. The leaf margins are razor-sharp, so wear gloves and remove foliage to make handling easier. ☀ 2–3 years.
🌱 spring. Cultivars will not come true. ☀ 3 years.

## Delphinium                     Ranunculaceae

About 250 species of hardy annuals, biennials, and perennials, and many garden hybrids, grown for their mainly blue flowers, although cream and pink-flowered variants are also available. The **Elatum Group**, the most common in cultivation, produce one main spike of flowers, with a central eye or "bee," in early and midsummer.
The **Belladonna Group** bears branched spikes of flowers in early and late summer. Grow in fertile, moist, but well-drained soil in a sunny position.

### PROPAGATION

Basal cuttings – spring, 2–3in (5–7.5cm) long, with gentle bottom heat. For the Elatum and Belladonna Group hybrids, this is the usual means of increase. ☀ 14 months.
⚒ spring. Cut through the crown with a sharp knife, leaving each section with roots and at least one bud. Pot up and grow on. ☀ 14 months.
🌱 late summer to fall, or spring, at 59–68°F (15–20°C). Erratic germination.
🌿 1–4 months. ☀ 6–12 months.

## Dianthus

(Carnation, Pink)            Caryophyllaceae

About 300 species of annuals, biennials, and often short-lived perennials. The many cultivars of hardy, evergreen border carnations and pinks are the most suitable for herbaceous borders. Cuttings are the most common method of increase; seed is only used for annual and biennial species, or if raising new cultivars.

### PROPAGATION

Stem-tip cuttings – summer, of non-flowering shoots. Pinks can be increased by "pipings." Pull the stem so that it parts at a leaf joint, and treat as ordinary cuttings. ☀ 1 year.
Layering – summer, after flowering. For border carnations and pinks that do not root easily from cuttings. ☀ 1 year.

## Dicentra                       Fumariaceae

Some 19 species of hardy annuals and perennials, which may be tap rooted, tuberous, or rhizomatous. Dicentras are grown for their fern-like foliage, and for their pendent, heart-shaped flowers. A number of named cultivars is available with flowers of

*Dicentra spectabilis*

*Echinops bannaticus* 'Taplow Blue'

*Epimedium* x *rubrum*

creamy white, and shades of pink to deep carmine or crimson. Grow in moist, fertile, leafy soil in partial or dappled shade.

**D. eximia.** Rhizomatous, with narrowly heart-shaped, white, pink or pink-purple flowers, in late spring, often repeating through to fall. H 24in (60cm), S. 18in (45cm).

**D. spectabilis.** Fleshy-rooted, with pink and white flowers in late spring and early summer. H 3–4ft (1–1.2m), S 18–24in (45–60cm).

**PROPAGATION**

✎ early spring. Clump-formers, such as *D. eximia*, are easily divided. *D. spectabilis* is best increased by cuttings. ☀ 1 year.

🌱 fall or spring, in a coldframe or at 59–68°F (15–20°C). Pre-chill spring-sown seed for 3 weeks. *D. spectabilis* and its white form, 'Alba', can both be grown from seed; the latter usually comes true.

❦ 1–3 months. ☀ 1 year.

Cuttings – take root or basal cuttings of *D. spectabilis* in spring. ☀ 2 years.

## Echinacea (Cone flower)      Compositae

Some nine species of hardy, rhizomatous, spreading perennials, grown for their daisy-like flowers. Grow in fertile soil, in sun or part-day shade. An easy genus to increase, all species can be propagated in the same way.

**E. purpurea**, syn. *Rudbeckia purpurea*. Has red-purple ray florets around a central golden-brown disc, from midsummer to fall. Cultivars and seed-raised hybrids are available with white, purple, or crimson flowerheads, some with orange or orange-brown discs. H to 5ft (1.5m), S 18in (45cm).

**PROPAGATION**

Basal cuttings – spring. ☀ 1 year.

🌱 fall or spring, sown fresh.

*E. purpurea* cultivars are unlikely to come true. All tend to self-seed.

❦ 1–4 months. ☀ 1 year.

✎ spring or fall. They resent root disturbance and may be slow to re-establish.

☀ 6–18 months.

## Echinops (Globe thistle)      Compositae

About 120 species of mostly hardy annuals, biennials, and perennials, grown for their foliage and globular flower heads. Grow in well-drained, not too fertile soil, in sun. Division of the strong thick roots is difficult.

**E. bannaticus.** Woolly stems, spiny gray-green leaves, and blue flower heads in late summer. **'Taplow Blue'** has brighter blue flower heads. H to 4ft (1.2m), S 2ft (60cm).

**E. ritro.** Spiny leaves with white hairs and blue flower heads in late summer. **'Veitch's Blue'** has darker blue blooms in several flushes. H 2–3ft (60–90cm), S 18in (45cm).

**PROPAGATION**

Root cuttings – winter, 1½–2in (4–5cm) long. ☀ 18 months.

🌱 fall/spring, in a coldframe or seedbed, or at 59–68°F (15–20°C). Self-sows freely. Pot up seedlings in deep pots to minimize root disturbance. Offspring of cultivars are variable. ❦ 1–4 months. ☀ 18 months.

## Epimedium      Berberidaceae

Some 25 species of hardy, rhizomatous, evergreen or deciduous perennials, grown for their foliage and flowers. Grow in fertile, moist, but well-drained soil in partial shade.

**E. x versicolor** tolerates sun and drier conditions. If grown in leafy, woodland-type soil, epimediums spread to form large stands that are easily divided. In spring, it is best to

wait until they have flowered. Alternatively, divide them in fall so that they can settle down before the next flowering time.

**E. grandiflorum**, syn. *E. macranthum*. Has green leaves, bronze-flushed when young, and white, pink, yellow, or purple flowers in spring. H to 12in (30cm), S 12in (30cm).

**E. x rubrum.** Has dark red and yellow flowers in spring, and red leaves when young, and in fall. H 12in (30cm), S 12in (30cm).

**E. x versicolor.** Produces coppery young leaves, with pink and yellow flowers in spring. H 12in (30cm), S 12in (30cm).

**PROPAGATION**

✎ fall or after flowering. ☀ 1 year.

🌱 fall. Spring sowing may delay germination; pre-chilling for 3 weeks helps.

❦ 1 year. ☀ 1 year.

## Erigeron (Fleabane)      Compositae

About 200 species of hardy annuals, biennials, and perennials grown for their daisy-like flower heads in summer. The named cultivars have a color range from pink (**E. 'Dimity'**), lilac-pink (**E. 'Charity'**), to violet (**E. 'Dunkelste Alle'**, syn. 'Darkest of All'). Many will layer if trailing stems are covered with soil. Grow in well-drained soil, in sun.

**E. karvinskianus**, syn. *E. mucronatus*. Produces white flowers that age to red-purple. H to 12in (30cm), S 36in (90cm) or more.

**PROPAGATION**

Basal cuttings – spring, 1–2in (2.5–5cm) long, with bottom heat, at about 59°F (15°C). ☀ 14 months.

✎ spring. Divide clump-forming species every 2–3 years to keep vigor. ☀ 6 months.

🌱 spring, in a coldframe. Seed is most satisfactory for the species, especially

*Eryngium giganteum*

*Gentiana asclepiadea*

*Geranium 'Ann Folkard'*

*E. karvinskianus*, which self-sows prolifically, producing many usable seedlings.
🌱 2–3 weeks.   ☀ 14 months.

## Eryngium (Sea holly)   Umbelliferae

About 230 species of mostly hardy, usually prickly annuals, biennials, and perennials which, despite belonging to the cow parsley family, look more like thistles, with attractive foliage and umbels of often metallic-hued flowers. The perennials are tap rooted, making division difficult. Grow in well-drained, fertile soil, in sun.

*E. giganteum.* Large, metallic-blue flowers with spiny, silvery bracts in summer. H 36in (90cm), S 18in (45cm).

*E. x oliverianum.* White-veined leaves, and bright blue flowers from midsummer to fall. H 36in (90cm), S 18in (45cm).

*E. x tripartitum.* Many small blue-violet flowers with silvery gray-blue bracts, in midsummer and fall. H to 36in (90cm), S 20in (50cm).

**PROPAGATION**
Root cuttings – winter, 2in (5cm) long. If the plant is expendable, lift and take off the roots, otherwise dig down and carefully remove a few roots from the side of the clump. *E. giganteum* should be increased from seed.   ☀ 18 months.
🌰 fall (preferably) or spring, in a coldframe, or at 59–68°F (15–20°C). Prick out into deep pots and plant out as soon as possible to minimize root disturbance.
🌱 1–4 months.   ☀ 18 months.
🔧 spring. Do this extremely carefully to avoid damaging the deep, thick tap root. Root cuttings are preferable.
☀ 2 years.

## Euphorbia (Spurge)   Euphorbiaceae

Some 2,000 species, ranging from annuals to trees. Care should be taken when cutting or handling these plants as they produce a latex-type sap that can cause skin irritation and is dangerous and painful if it gets into the eyes. The hardy species include clump-forming perennials, like *E. amygdaloides* and *E. polychroma*, shrub-like perennials, such as *E. characias*, and prostrate perennial succulents, like *E. myrsinites*. Several, such as *E. griffithii* and *E. sikkimensis*, have glassy, red or pink young stems.

**PROPAGATION**
🔧 spring. Rhizomatous species and those with a spreading or clump-forming habit are easily divided.   ☀ 12–18 months. Cuttings – spring or fall. Basal (*E. polychroma*), tip (*E. palustris*), and lateral (*E. griffithii*) cuttings can all be tried as some are more successful than others with certain species. Dip cut surfaces in charcoal to reduce bleeding.   ☀ 18 months.
🌰 in fall or spring, in a coldframe. Offspring of cultivars are often very close, if not identical, to the parent. Pre-chilling and then some warmth, 59–68°F (15–20°C), may speed germination. Move self-sown seedlings when still small.
🌱 1–4 months.   ☀ 12–18 months.

## Gentiana (Gentian)   Gentianaceae

About 400 species of annuals, biennials, and perennials, the majority of which are considered alpines *(see p. 83)*. The most commonly seen of the larger herbaceous perennials, *G. asclepiadea*, is best grown from seed, although it can be divided carefully in spring, and basal cuttings are sometimes

successful. It needs humus-rich, fertile, moist, but well-drained soil, in part shade.

*G. asclepiadea* (Willow gentian). Bears trumpet-shaped, blue, occasionally white flowers, from mid- to late summer to early fall. H to 36in (90cm), S 18in (45cm).

*G. lutea* (Yellow gentian) has a thick tap root and can only be increased from seed.

**PROPAGATION**
🌰 fall, in a coldframe. Pre-chill spring-sown seed, then germinate at 59–68°F (15–20°C). Needs light for germination.
*G. lutea* takes several years to reach maturity.
🌱 1–4 months.   ☀ 18 months.

## Geranium (Hardy cranesbill)   Geraniaceae

Some 300 species of annuals, biennials, and perennials of a range of heights and colors from blue through purples and pinks to white. Forming the backbone of any flower garden, many have a long flowering period. They grow in most soils in sun or shade; smaller species require good drainage.

**PROPAGATION**
Between them, they are propagated by the complete range of techniques including layering in summer (e.g. *G. procurrens*). The clump- or mat-forming types can be increased by division in spring. Many can be increased by cutting off part of the overground rhizome and potting it up. *G. 'Ann Folkard'* and similar, scrambling hybrids can be increased by stem cuttings taken in summer. *G. sanguineum* can be grown from root cuttings. Most species can be increased from seed in fall or spring, and many cultivars, although not all, produce plants that are close to their parents, i.e. *G. wallichianum* 'Buxton's Variety'.

39

*Geum chiloense*

*Gunnera manicata*

*Helleborus orientalis*

## Geum           Rosaceae

About 50 species of hardy perennials, grown for their single or semi-double flowers, produced throughout the summer. The best way to increase existing stock is by division but many cultivars, such as *G*. **'Lady Stratheden'** and **'Mrs. J. Bradshaw'**, come reasonably true from seed. Most are easily grown in fertile, well-drained soil, in sun.

**PROPAGATION**

✂ spring. Re-plant directly. In dry weather, pot up and plant out once the plant has re-established. ☀ 1 year.

❦ fall or spring. For best results, sow in a propagator at 59–68°F (15–20°C) or germination is likely to be slow or erratic. ❧ 1–4 months. ☀ 14 months.

## Gunnera (Chilean rhubarb)    Gunneraceae

Some 40–50 species of giant, rhizomatous and diminutive, creeping perennials.
*G. manicata*. Almost hardy, but needs a deep mulch of leaf litter in cold areas. Grow in moist, fertile, humus-rich soil, in sun or dappled shade. Provide shelter from cold winds. Division is the best method of increase. H 6ft (1.8m), S 7ft (2.1m).

**PROPAGATION**

✂ spring. Many gunneras are huge and so are impossible to dig up and divide in the conventional sense. For propagation, cut pieces from the edge of the clump. Pot up in a large pot and in a fibrous medium and keep moist.
☀ 2–3 years.

❦ fall, when ripe, at 59–68°F (15–20°C). ❧ 1 month. ☀ 4–5 years.

## Gypsophila        Caryophyllaceae

Approximately 100 species of mostly hardy annuals and perennials, few of which are in cultivation. Most species can be easily grown from seed, but cultivars of *G. paniculata* need to be increased vegetatively. Unfortunately, they are difficult to grow from cuttings and need to be saddle-grafted. Grow in deep, well-drained, slightly alkaline soil in sun.
*G. paniculata* (Baby's breath). Hardy perennial, with gray-green leaves, bearing clouds of tiny white flowers in summer.
**'Bristol Fairy'** has double white flowers.
**'Flamingo'** has pale pink blooms.
H 3–4ft (90–120cm), S 3–4ft (90–120cm).

**PROPAGATION**

❦ spring at 59–68°F (15–20°C), or in a seedbed (germination takes a little longer).
❧ 2–4 weeks. ☀ 18 months.
Cuttings – spring or early summer, of young side shoots, with a heel. ☀ 14 months.
Saddle-grafting – summer, onto seed-raised rootstock of *G. paniculata*. Appropriate for 'Bristol Fairy' and 'Flamingo'. ☀ 1 year.

## Helenium (Sneezeweed)    Compositae

Some 40 species of annuals, biennials, and mostly hardy, fibrous-rooted, clump-forming perennials with daisy-like flower heads over long periods in summer. Cultivars are available in shades of yellow (**'Butterpat'**), orange, (**'Wyndley'**), bronze-mahogany (**'Moerheim Beauty'**), and crimson-red (**'Bruno'**). All perennials and their cultivars are most easily increased by division.

**PROPAGATION**

✂ fall or spring. Discard the woody center. Divide routinely every 2–3 years.
☀ 3 or 15 months.

## Helleborus (Hellebore)    Ranunculaceae

Some 15 species of rhizomatous perennials with cup- or saucer-shaped flowers in winter or early spring. Most are hardy or nearly so, and thrive in leafy, humus-rich, fertile soil that does not dry out, in partial shade. Many named cultivars are available, and hybrids often occur where several species are grown together. *H. argutifolius* and *H. foetidus* are difficult to divide and should be raised from seed. Most cultivars of *H. orientalis* will not come totally true from seed, although many will come close to the orginal. For true reproduction, division is the safest method. This is also true for *H. niger*.
*H. niger* (Christmas rose). From early winter to spring, produces white flowers, that are sometimes flushed pink. **'Potter's Wheel'** has large, pure white flowers with a green eye.
H to 12in (30cm), S 18in (45cm).
*H. orientalis* (Lenten rose). Bears green-white flowers, flushed purple to pink, midwinter to spring. H 18in (45cm), S 18in (45cm).

**PROPAGATION**

❦ fall, as soon as ripe, in a coldframe. Prick out in spring. Pot up self-sown seedlings and grow on before re-planting.
❧ 1 month. ☀ 2 years.

✂ fall; or spring, after flowering. Separate into individual crowns with a sharp knife. Pot up and grown on in light shade. Slow to re-establish. ☀ 2 years.

## Hemerocallis (Daylily)     Liliaceae

Some 15 species of mostly hardy perennials, some of which are evergreen or semi-evergreen. Grown for their flowers, which last only a day but are produced throughout the summer. There are numerous named

*Hemerocallis* 'Pink Prelude'

*Hosta fortunei*

*Kniphofia* 'Little Maid'

cultivars, which need to be increased by division. They prefer fertile, humus-rich, moist but well-drained soil, in sun.

**PROPAGATION**

✎ spring. Tease apart by hand. Re-plant larger pieces and pot up the smaller ones.

☀ 15 months.

🌰 fall or spring. Species only. Pre-chill spring-sown seed for up to 3 weeks.

🌱 2 months.  ☀ 2–3 years.

## Heuchera                Saxifragaceae

About 55 species of evergreen or semi-evergreen perennials, grown for their flowers and foliage. Most in cultivation are hardy. Species can be grown from seed or divided. Most forms divide very easily and can be planted straight back into the soil as long as it is not too dry. Named cultivars and hybrids are increased by division; *H. micrantha* var. *diversifolia* 'Palace Purple' can be grown from seed. Other cultivars do not come true.
*H. micrantha*. Lobed leaves marbled green and gray. It bears tiny white flowers in early summer. **var. *diversifolia* 'Palace Purple'** has bronze-purple leaves. H 24in (60cm), or more, S 18in (45cm).
*H. sanguinea* (Coral bells). Lobed, rounded leaves, marbled dark and pale green, and bears red flowers, in summer. H 12in (30cm), S 12in (30cm).

**PROPAGATION**

✎ fall or spring. Divide by hand, or cut with a knife; ensure that each piece has buds and roots. Pot up and grow on until established.

☀ 14 months.

🌰 spring, in a coldframe. Germination is quicker at 59–68°F (15–20°C).

🌱 1 month.  ☀ 14 months.

## Hosta (Plantain lily)        Funkiaceae

About 40 species of hardy perennials with numerous cultivars and hybrids, grown mainly for their foliage; many also have racemes of fragrant flowers in summer. Leaves range from small to large and come in all shades of green, through to blue-green, blue-gray, and gray-green, some variegated with gold, cream, or silver.

**PROPAGATION**

✎ spring. Lift plants and wash all the soil from the roots. Cut into individual crowns with a sharp knife and pot up the divisions.

☀ 2 years.

🌰 fall or spring. Species only.

🌱 1–3months.  ☀ 2 years.

## Iris                        Iridaceae

About 300 species of rhizomatous and bulbous perennials (*see also Bulbs, Corms, and Tubers directory, pp. 68–69*). The hardy, rhizomatous, summer-flowering bearded irises have sturdy rhizomes on or near the soil surface and fans of sword-shaped leaves. There are many cultivars, with large flowers having distinct standards and falls, with a "beard" of hairs at the center of the each fall petal. The other major group is the beardless irises, which include ***I. spuria, I. sibirica*** and their cultivars, with delicate flowers and narrow leaves.

**PROPAGATION**

✎ for rhizomatous irises, divide in summer, immediately after flowering. Divide clump-forming irises, such as *I. sibirica*, in spring, just as or just before new growth begins.

🌰 fall, preferably as soon as ripe, or spring, in a coldframe. Species only.

🌱 1–3months.  ☀ 3–4 years.

## Kniphofia (Red-hot poker, Torch lily)
Liliaceae/Aloaceae

Some 68 species of rhizomatous, clump-forming perennials with upright, cylindrical racemes of small tubular flowers. It is mostly cultivars that are grown, and they must be increased vegetatively.

**PROPAGATION**

✎ late spring. Tease crowns apart, cutting if necessary, then pot up and grow on until they are large enough to be planted out.

☀ 16 months

🌰 spring, in a coldframe. Species only.

🌱 2–4 weeks.  ☀ 3–4 years.

## Lychnis (Catchfly)        Caryophyllaceae

Some 20 hardy species, most of which are perennials, grown for their flowers, and, in some species, for their silver foliage. They thrive in well-drained, fertile soil, in sun. Gray-leaved species tolerate dry soils. Most species produce copious seed and will self-sow, providing plenty of useful seedlings, many of the color forms coming true.
*L. chalcedonica* (Maltese cross). Bears brilliant scarlet flowers in early or midsummer. H to 4ft (1.2m), S 12in (30cm).
*L. coronaria* (Rose campion). Gray leaves and a succession of magenta flowers in late summer. It is short-lived and best propagated each year. **'Alba'** has white flowers. **'Oculata'** has pink-eyed, white flowers; both come reasonably true from seed, unless pollinated by the magenta-flowered species. H 32in (80cm), S 12in (30cm).

**PROPAGATION**

🌰 spring, in a coldframe, seedbed, or *in situ*. Many self-sow freely.

🌱 2–4 weeks.  ☀ 3–16 months.

41

*Lychnis coronaria*

*Miscanthus sinensis*

*Paeonia 'Lovely Rose'*

🌱 spring. Division is only suitable for clump-forming species, such as *L. chalcedonica*.
☀ 1–2 years.

## Meconopsis (Blue poppies)    Papaveraceae

Some 43 species of hardy annuals, biennials, and perennials, noted for their blue poppy-like flowers; there is also yellow (*M. chelidonium*) and dusky red (*M. napaulensis*). Many perennials are short-lived or die after flowering. Most produce abundant seed and readily self-sow. Move seedlings as soon as possible, before the tap roots begin to form. A few, such as *M. grandis*, can be divided after flowering. Likes a cool, moist atmosphere and is difficult to grow in dry areas. Grow in leafy, open, moist but well-drained, neutral to slightly acid soil, in dappled shade.
*M. betonicifolia* (Tibetan blue poppy). Bears pendent, bright blue flowers in early summers. H 4ft (1.2m), S 16in (40cm).
*M. grandis* (Himalayan blue poppy). Bears nodding, blue to red-purple flowers in early summer. H to 4ft (1.2m), S 2ft (60cm).
*M.* x *sheldonii*. Bears pale to deep blue flowers in late spring to early summer. H to 5ft (1.5m), S 2ft (60cm).

### PROPAGATION

🌰 as soon as ripe, in late summer. Overwinter seedlings in a well-ventilated cold greenhouse or frame; they are very susceptible to damping off. *Meconopsis* seed needs light for germination; do not cover.
🌱 1–4 months.    ☀ 18 months.

## Miscanthus    Gramineae

Some 20 species of mostly hardy, clump-forming perennial grasses, grown for their foliage and silky flower panicles in late summer. Grow in moist but well-drained, not-too-fertile soil, in full sun. The cultivars are most often grown; division is the best method of increasing them.
*M sinensis*. Bluish-green leaves and silvery, hairy flower panicles in late summer or fall. Smaller cultivars include **'Variegatus'**, with leaves striped white and pale green, and **'Silberfeder'** with silvery, pink-flushed flower panicles that persist through winter. H to 12ft (4m), S 3ft (1m) or more.

### PROPAGATION

🌱 mid-spring, just as new growth begins. Large clumps can be divided by spading off a section and then dividing this by hand.
☀ 2 years.
🌰 spring, in a coldframe. Not suitable for cultivars.
🌿 2–4 weeks.    ☀ 2–3 years.

## Nepeta (Catmint)    Labiatae

About 250 species of hardy and half-hardy perennials with aromatic leaves and spikes of flowers. The most common tolerate a range of well-drained soils, in sun or light shade.
*N.* x *faassenii*, syn. *N. mussinii*. Aromatic, gray leaves and spikes of lavender-blue flowers throughout summer. **'Six Hills Giant'** is similar, but much larger, reaching, 3ft (90cm) tall. H 18in (45cm), S 18in (45cm).

### PROPAGATION

Basal cuttings – spring. Or, shear over after flowering, and take cuttings from the resultant new growth in late summer.
☀ 15 or 10 months.
🌱 spring. Water until established.
☀ 15 months.
🌰 spring, in a coldframe or seedbed. Cultivars may not come true.    ☀ 15 months.

## Paeonia (Peony)    Paeoniaceae

About 33 species of mostly hardy, long-lived, tuberous-rooted perennials and shrubs, most of which are cultivated, as well as thousands of named cultivars, derived mainly from *P. lactiflora*. They have single, semi-double, or double flowers in early summer, in white and pink to rich crimson. Needs deep, well-cultivated, fertile soil that is humus-rich and reliably moisture-retentive, in sun or light dappled shade.

### PROPAGATION

🌱 fall or early spring. Divisions re-establish slowly. Prepare the planting site thoroughly; incorporate organic matter. Keep well watered until re-established.
☀ 2–3 years.
Root cuttings – winter.    ☀ 3 years.
🌰 as soon as ripe, in fall or early winter. Suitable for species only. Retain pots undisturbed if germination does not occur in the first spring.
🌿 up to 2–3 years.    ☀ 3–4 years.

## Papaver (Poppy)    Papaveraceae

About 70 species of annuals, biennials, and perennials grown for their silky, bowl- or cup-shaped flowers. Grow in fertile, well-drained soil in a sunny site.
*P. orientale* (Oriental poppy). Has divided leaves and bowl-shaped, orange-red flowers, in early summer. Many cultivars are available, with a color range from white (**'Perry's White'**), scarlet (**'Allegro'**), to crimson (**'Beauty of Livermere'**).

### PROPAGATION

Root cuttings – in fall or early winter, 2in (5cm) long. Either dig up the whole plant and discard it after taking cuttings, or dig

*Papaver bracteatum*

*Pulmonaria augustifolia*

*Rudbeckia fulgida*

down beside it carefully and remove only a few roots. Leaves appear before new roots; do not pot on until well-rooted.
☀ 2 years.
❦ spring, in a coldframe. Transplant seedlings when small to minimize root disturbance. Cultivars of *P. orientale* are unlikely to come true from seed but can be increased from root cuttings.
❧ 2 weeks. ☀ 2 years.

## Penstemon Scrophulariaceae

Some 250 species of perennials and subshrubs, grown for their foxglove-like flowers which, in most border perennials, are borne from midsummer to fall. Cultivars with a wide color range, and of varying hardiness, are available. The hardiest include *P.* 'Andenken an Friedrich Hahn' (wine-red), 'Appleblossom' (pale pink), and 'Evelyn' (rose-pink). The larger cultivars are grown mainly as perennials, while many of the species are grown as alpines. They are some of the easiest plants to root. Grow in well-drained, fertile soil in sun *(see also Alpines and Rock Garden Plants directory, p.80).*
*P. campanulatus,* syn. *P. pulchellus.* Bears violet flowers in early summer. Hardy to about 23°F (-5°C), it is a parent of many modern hybrids. H to 24in (60cm), S 18in (45cm).

### PROPAGATION
Cuttings – early or midsummer. Any part of the stem will root. Take stem-tip cuttings at any time the plant is in growth.
☀ 1 year.
❦ fall or spring. Species only. Cultivars will not come true and should be increased by cuttings. Many species self-seed freely.
❧ 1–4 months. ☀ 3–15 months.

## Phlox Polemoniaceae

Some 67 species of annuals, evergreen or herbaceous perennials, and a few shrubs. The hardy border phlox, cultivars of *P. maculata* and *P. paniculata,* are grown for their large fragrant flowers, borne in midsummer. Grow in moist but well-drained, fertile soil, in sun or partial shade. Phlox are very susceptible to stem eelworm; do not take propagating material from infested plants.

### PROPAGATION
⚒ fall or spring. Divide healthy clumps only, and discard old material from the centers of the clumps. ☀ 1 year.
Basal cuttings – spring. From healthy clumps only. Stem eelworm will be passed on if present on parent plants. ☀ 1 year.
Root cuttings – in early fall or winter. Root cuttings are usually free from eelworm infestation. ☀ 1 year.

## Pulmonaria (Lungwort) Boraginaceae

Some 14 species of hardy, rhizomatous perennials, grown for their funnel-shaped flowers, in shades of blue, pink, violet, red, or purple; some also have attractive spotted foliage. Grow in moist, leafy soil, in dappled or partial shade. All pulmonarias are best divided, although they grow readily from seed. In clumps of cultivars which have self-sown, different flower colors are likely to be present. Take cuttings from rogue plants if you like the color; otherwise remove them.
*P. angustifolia* (Blue cowslip). Intense, deep blue flowers in spring. H to 12in (30cm), S 18in (45cm).
*P. rubra.* Bears red to salmon-pink flowers in late winter and spring, above green leaves. H to 16in (40cm), S to 36in (90cm).

### PROPAGATION
⚒ spring, after flowering. Wash soil from the roots and the plant will fall into natural divisions. Remove old rootstock and shorten the leaves. Pot up and plant out once they are growing strongly. ☀ 1 year.
❦ spring, in a coldframe. Species only. Cultivars self-sow, but are rarely true to type.
❧ 1 month. ☀ 1 year.

## Rudbeckia (Coneflower) Compositae

About 20 species of annuals, biennials, and hardy, rhizomatous perennials grown for their flower heads with cone-shaped central discs, produced over long periods in summer. Grow in moist but well-drained soil, in sun or dappled shade. Division is the most satisfactory way of increasing these plants. Cuttings work well but may take longer. Seed is becoming increasingly available commericaly but often produces slight variations in flower color and pattern.
*R. fulgida* (Black-eyed Susan). Bears orange-yellow flower heads with a dark brown central disc. Late summer to fall. H 36in (90cm), S 18in (45cm).
*R. laciniata.* Bears pale yellow flower heads with a greenish disc, from midsummer to fall. 'Goldquelle' is shorter, with double, lemon-yellow flower heads.
H 5ft (1.5m) or more, S to 3ft (1in).

### PROPAGATION
⚒ spring or fall. ☀ 18 months.
Basal cuttings – spring, 2in (5cm) long. Older shoots often already have a few roots attached to them. ☀ 18 months.
❦ spring, in a coldframe. Species only. Pre-chilling will help germination.
❧ 2–3 weeks. ☀ 8 months.

*Salvia × superba*

*Sedum spectabile*

*Sidalcea malviflora*

## Salvia (Sage)                                    Labiatae

About 900 species of annuals, biennials, subshrubs, and perennials, some rhizomatous or tuberous-rooted. They are grown for their two-lipped, tubular flowers and their aromatic foliage (*see also Herb directory, p.169*). The hardy border perennials prefer light, not-too-fertile, moist but well-drained soil, in sun. Commercially available seed can be used for propagating the species and some selected clones. Some salvias, such as *S. sclarea*, happily self-sow, providing enough seedlings for most purposes.

*S. nemorosa.* Dense spikes of white, pink, or violet-purple flowers in midsummer. Leaves are notched and wrinkled. **'Lubecca'** and **'Ostfriesland'**, are more compact with violet and blue-violet flowers respectively. H to 3ft (1m), S 2ft (60cm).

*S. × superba.* Spikes of intense blue-violet flowers from midsummer to fall. H 3ft (90cm), S 18in (45cm).

*S. × sylvestris.* Has wrinkled, scalloped leaves, and spikes of pink-violet flowers in early to midsummer. **'Mainacht'**, syn. 'May Night', has indigo-violet flowers with purple bracts. H 30in (75cm), S 12in (30cm).

### PROPAGATION

✎ spring. Tuberous-rooted *S. patens*, which is not reliably hardy, can be lifted in fall; divide before re-planting in spring. ☀ 1 year.
Cuttings – basal in spring, stem tip in early summer. Woody-based perennials, semi-ripe cuttings in late summer. Also take semi-ripe cuttings of more tender species to protect against winter loss. Overwinter semi-ripe cuttings in a frost-free frame or greenhouse. ☀ 1 year.

🌰 surface-sow in spring, in a coldframe. Suitable for species only. For more tender species, such as *S. patens*. Cultivars rarely come true.
🍃 2–3 weeks.  ☀ 1–2 years.

## Sedum (Stonecrops)                    Crassulaceae

Genus of some 400 species of fully hardy to tender annuals, biennials, perennials, and shrubs, many of which are small alpine plants (*see also Alpines and Rock Garden Plants directory, p.85*). The larger, hardy species and their cultivars are valued for their handsome leaves, in spring and summer, and for their small, star-shaped flowers, in late summer and fall. They prefer not-too-fertile, well-drained, neutral to slightly alkaline soil, in sun.

*S. spectabile*, syn. *Hylotelephium spectabile* (Ice plant). This has fleshy, gray-green leaves and wide, slightly domed heads of pink flowers, in late summer and fall.
**'Brilliant'** has deep pink flowers.
**'Septemberglut'**, syn. 'September Glow', has rich pink flowers. H 18in (45cm), S 18in (45cm).

### PROPAGATION

Cuttings – spring, summer, or fall. Stem or basal cuttings root easily. Leaf cuttings also grow readily; insert one or more leaves vertically in rooting medium in a propagating case; gentle bottom heat speeds rooting.
☀ 16 months.
✎ spring.  ☀ 16 months.

## Sidalcea (Prairie mallow)              Malvaceae

Genus of around 20 species of hardy annuals and perennials valued for their silky-petaled, cup-shaped flowers. Those most often grown are cultivars or hybrids and are best

increased by vegetative means, such as division, to ensure that plants come true. They need light, moist but well-drained, fertile soil, in a sunny position.

*S. malviflora* (Checkerbloom). Produces clumps of glossy leaves and spires of pink flowers in midsummer. **'Elsie Heugh'** has pink flowers. **'Loveliness'** is lower-growing, with soft pink flowers. H to 4ft (1.2m), S 18in (45cm).

### PROPAGATION

✎ spring or fall. Lift and wash away the soil, then divide by hand, using a sharp knife if necessary. Split larger clumps with back-to-back forks.  ☀ 14 months.
🌰 spring or fall, in coldframe. Species only; cultivars may self-sow, but do not come true.
🍃 2–3 weeks.  ☀ 1 year.

## Thalictrum (Meadow rue)   Ranunculaceae

Some 130 species of hardy and frost-tender, tuberous or rhizomatous perennials. Several are hardy border perennials that are valued for their sometimes glaucous foliage and their tiny flowers with large fluffy clusters of stamens and pistils. Grow in moist but well-drained soil in dappled shade, or in sun where soil remains moist.

*T. aquilegifolium.* Divided, pale green leaves and bears fluffy heads of rose-lilac flowers, in early summer. H 3ft (1m), S 18in (45cm).
*T. delavayi*, syn. *T. dipterocarpum*. Dainty, pale green foliage on purple-flushed stems, and panicles of white or lilac flowers in mid- to late summer. H 4ft (1.2m), S 2ft (60cm).

### PROPAGATION

✎ spring. Re-plant immediately or pot up small divisions and grow on. Spreading plants, such as *T. minus*, are easily divided.

*Tradescantia* Andersoniana Group

*Veronica spicata*

*Yucca gloriosa* 'Variegata'

Small clump-formers are less easy.
☀ 1 year.
🌱 as soon as ripe or in spring, in a coldframe. Many, such as *T. aquilegifolium*, self-sow. Move seedlings while still young.
⚘ 2–3 weeks. ☀ 1–2 years.

## Tradescantia     Commelinaceae

Genus of some 65 species of evergreen perennials valued for their three-petaled flowers with conspicuous stamens, borne from early summer to fall and often also for their attractive leaves. Those featured are hardy. They prefer moist but well-drained, fertile soil, in sun or dappled shade.
**T. Andersoniana Group.** This hardy selection forms clumps of fleshy, jointed stems and lance-shaped leaves, usually green, often purple-tinted. Cultivars come in a range of colors; **'Isis'** and **'Zwanenburg'** have deep blue flowers; **'Osprey'** has white flowers with blue filaments. **'Karminglut'**, syn. 'Carmine Glow' has rich carmine-pink flowers.
H 18–24in (45–60cm), S 18–24in (45–60cm).

### PROPAGATION
⚒ fall or spring. ☀ 1 year.

## Veratrum     Scrophulariaceae

Genus of some 45 species of hardy, rhizomatous perennials mainly grown for their foliage; they also produce spires of small, star-shaped flowers. The leaves are large and ovate and bright emerald-green as they emerge in spring. Grow in deep, moist but well-drained, fertile soil in partial shade, with shelter from winds. Growing from seed is not difficult – just lengthy. For the first year of germination all growth is below ground,

so do not throw away the pot. Take care in handling *Veratrum* as all parts are toxic.
**V. album** (False or White hellebore). White, sometimes green-tinted, flowers in summer. H 6ft (1.8m), S 2ft (60cm).
**V. nigrum.** Produces maroon flowers that appear to glow in the sun, in mid- to late-summer. H 4ft (1.2m), S 2ft (60cm).

### PROPAGATION
🌱 summer or fall. Sow as soon as ripe in a coldframe. Grow on in a nursery bed.
⚘ spring. ☀ 3–4 years.
⚒ fall or early spring. ☀ 2–3 years.

## Veronica (Speedwell)     Scrophulariaceae

Genus of 250 species of annuals and woody-based, sometimes rhizomatous, perennials. Hardy species are grown for spikes of white, pink, or blue flowers in summer. Grow in moist, drained, fertile soil in sun or dappled shade. Most are easy to increase by division.
**V. gentianoides.** Spikes of blue flowers in early summer, and rosettes of glossy leaves. H to 18in (45cm), S 18in (45cm).
**V. longifolia.** Produces tall, dense spikes of blue to lilac-blue flowers, in late summer. H 4ft (1.2m), S 12in (30cm).
**V. spicata.** Spikes of blue flowers throughout summer, above narrow, hairy leaves. **subsp. incana** has silver-hairy leaves and deep violet-blue flowers. H 12in (30cm) or more, S 18in (45cm).

### PROPAGATION
⚒ fall or spring. Discard the woodier central portions. ☀ 1 year.
🌱 fall or spring, in a coldframe. Many of the larger species, such as *V. longifolia*, self-sow and provide a ready supply of seedlings.
⚘ 1–3 months. ☀ 1–2 years.

## Yucca     Agavaceae

Some 40 species of evergreen, woody-based perennials, shrubs, and trees, many of which are frost-tender. They are grown for their creamy-white flower panicles, up to 8ft (2.5m) tall, and for their architectural habit. Hardy species need sharp drainage and full sun. Outside of North America, yuccas must be hand-pollinated for seed production.
**Y. filamentosa** (Adam's needle). Hardy species with stiff, narrow leaves that have curly threads at the margins. Flowers in mid- to late summer. H 32in (80cm), S to 5ft (1.5m).
**Y. flaccida.** Hardy with short, spreading rosettes of narrow, blue-green leaves. Flowers in mid- to late summer. H 20in (50cm), S to 5ft (1.5m).
**Y. gloriosa** (Spanish dagger). Less hardy than *Y. flaccida*, but with sharp drainage tolerates temperatures to about 23°F (-5°C). It has blue-green to dark green leaves. **'Variegata'** has leaves with yellow margins. Flowers in late summer. H 6ft (1.8m), S to 6ft (1.8m).

### PROPAGATION
🌱 spring, at 55–64°F (13–18°C). Subtropical and tropical species will need higher temperatures to germinate.
⚘ 1–2 months. ☀ 4–5 years.
Cuttings – fall. Take stem cuttings from side shoots, and root in a propagator. Keep frost-free in winter. Pot up in spring.
☀ 6 years.
Root cuttings – winter, 2in (5cm) long, planted vertically. Rather than lifting the whole plant, dig down beside it and remove a few thick roots. ☀ 6 years.
⚒ spring. Suitable for yuccas with a suckering habit, such as *Y. filmentosa*.
☀ 2 years.

45

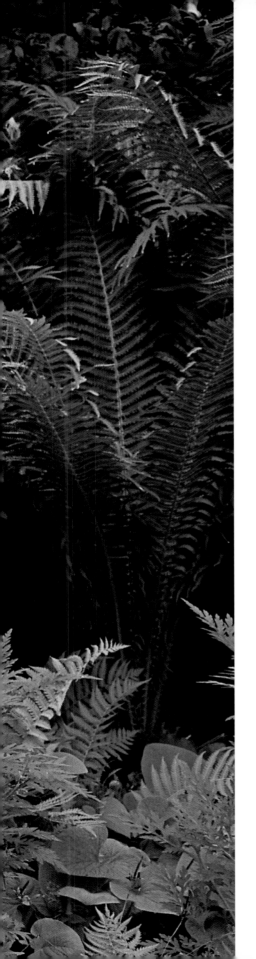

# ERNS

Ferns have long been popular as house and conservatory plants, but in recent years hardy ferns have once again become more widely grown in gardens. Ferns complement almost any garden style – whether planted in a shady mixed border for contrast, or grouped in a dark, damp corner where little else will grow, their wonderful form and texture is almost indispensable. In addition, evergreen ferns have the advantage of giving some interest all year round, many evergreen species remaining handsome until flattened by snow. The natural reproduction of ferns is quite different to that of flowering plants, and propagating them yourself provides an opportunity both to increase stocks and to observe a fascinating process at first hand.

◄ *Massed plantings of* Matteucia struthiopteris *provide a wonderful effect in a shady border; this particular fern can be increased by division or spores.*

# WHY PROPAGATE
# FERNS?

*Despite their bold, architectural shapes and subtle range of shades, ferns remain one of the most under-utilized of all the plant groups available to modern gardeners. While a single specimen can look quite magnificent, they are best grown in clumps. Having mastered the basic propagation techniques, you will soon be able to rear any number of plants.*

**The ferns shown here and opposite can be increased by the full range of propagation techniques.**

All ferns likely to be grown in the home or garden are perennials and, given the correct growing conditions, are exceptionally long-lived. So quite unlike other perennials, which need propagating to maintain health, vigor, and flowering, the main reason for propagating ferns is simply to increase their numbers.

Ferns are certainly not difficult to propagate. Sexual reproduction involves a recombination of the genes that determine physical appearance, so the offspring are seldom identical to either their parent or each other. In many cases this is of no importance, but selected garden forms often do not come true from spores and must be increased by vegetative methods.

## INCREASING BY DIVISION

Where only a few additional specimens are required, this is the best method of increasing stocks. Division of mature plants inevitably causes some disturbance and is best carried out during spring. Species with creeping rhizomes, like *Adiantum venustum*, can be cut into sections and most will grow as long as they have some growing points. Species that form an upright crown, or shuttlecock, of fronds must be separated so that each individual crown remains intact if division is to be successful.

## OPTING FOR SPORES

For mass multiplication of ferns it is most efficient to propagate from spores – a mature plant can produce one thousand million spores in one season! Increasing from spores is not difficult and the process itself is fascinating.

The life cycle for sexual reproduction is clearly divided into two generations. The generation we see as the fern plant is known as the sporophyte, which produces spores usually on the underside of the fronds. Spores germinate and grow into a free-living structure, the prothallus or gametophyte, which is the second generation. It is on this structure that the male and female sexual organs are found and gametes produced. As the male gametes must swim in a film of water beneath the prothallus to fertilize the female egg cells, an adequate moisture level is essential during this stage of reproduction.

After fertilization, the new fern is produced. The new plant parasitizes the now redundant prothallus, using its nutritious tissues to get a good start in life until the new plant can produce its own root and the first leaf begins to function to provide it with a food supply.

## ADDITIONAL RESOURCE

Mass production of ferns is sometimes possible from bulbils, and these offspring will almost invariably be identical to the parent. Bulbils tend to be produced most prolifically on *Polystichum setiferum* cultivars in wet summers or on adequately watered plants.

Interestingly, there is never any outward sign of a bulbil on the leaf bases of the Hart's-tongue fern, *Asplenium scolopendrium*, yet when treated correctly, nearly every leaf base will produce at least one fern.

### "Living Antiques"

*Some ferns have mutated to produce attractive crested or finely divided plumose fronds. They seldom come true from spores and must be propagated vegetatively, which is a slow process. With some of the best cultivars, division is only possible when one or more side crowns is produced. Some cultivars have been propagated by this method since the Victorian times. Identical to the original plants, they are often referred to as "living antiques."* Polystichum setiferum *'Plumosum Bevis'* and Athyrium filix-femina *'Clarissima' are good examples.*

Polystichum setiferum
'Plumosum Bevis'
with unfurling fronds.

**① Polystichum munitum**
This species produces spores that ripen and may be sown in late summer. It forms shuttlecock crowns and can also be increased by the careful division of side crowns in spring.

**② Athyrium niponicum var. pictum f. metallicum**
A form that occasionally produces bulbils, and may be divided. Spores produce green- and colored-leaved offspring in almost equal numbers, so select the most highly colored variants.

**③ Asplenium scolopendrium Crispum Group**
These cultivars are generally unsuitable for division and produce variable offspring from spores. They can, however, often be induced to form bulbils on the base of the fronds.

**④ Polystichum setiferum Divisilobum Group**
Unlike the parent species, these cultivars often form bulbils along the frond midrib which can be sown to produce offspring identical to the parent plant.

# FERNS

*The techniques of increasing ferns by division or from bulbils differ only slightly from those used when dividing perennials or increasing bulbs from bulbils. Spore propagation, is, however, unique to the fern family. It may appear complex at first but, in reality, the steps of the procedure are actually very simple.*

## DIVISION

Most ferns can be propagated by division at the beginning of the growing season. Species with creeping rhizomes, such as *Adiantum venustum* and *Polypodium* species, can be cut into a number of sections. Most pieces will grow if you make sure that there are two or three healthy fronds on each division, with a 2in (5cm) ball of soil around each plantlet.

Many ferns have erect rhizomes with an upright crown, or shuttlecock, of fronds. Separate crowns with a knife, or prise apart using two forks. It often takes at least a year for the parent plant to recover.

Re-plant large divisions immediately. Pot up small ones in well-drained loamless potting medium, with added slow-release fertilizer. Protect for three months in a sheltered spot, until established, and do not over-water.

## BULBILS

Some ferns produce bulbils along the midribs of the fronds. With hardy ferns, these fronds may be layered onto the soil surface in fall – still attached to the parent plant – and left until spring, when the rows of little plantlets can be pricked out into seed trays.

### Bulbils on leaf bases

Some ferns, notably *Asplenium scolopendrium* and its cultivars, can be propagated from leaf bases. This is very useful for sterile cultivars, such as *A. scolopendrium* 'Crispum'. In spring, wash the soil from an established plant to expose the old leaf bases. They may appear dead, but snap them off at their point of attachment to the main rhizome and you will see live tissue. Surface sterilize a tray of seed medium as if sowing spores (*see opposite*). Wash the leaf bases, then stick them in the medium bottom up (with the end that was attached to

the rhizome pointing upward). Re-plant the parent. Place the tray in a clean plastic bag and keep in a light place at 59–68°F (15–20°C). After three months, one or several bulbils should develop on each leaf base. They will be ready to plant out the following spring.

### Bulbils . . . . . . . . . . . . . . . . . . . . . . . . . . . . . . . . . . .

Tender ferns, such as *Asplenium bulbiferum*, and hardy ferns whose fronds cannot be layered to the soil, may be propagated by pegging down fronds individually into seed trays.

▲ **1** Cut the frond from the plant and peg it into a seed tray of low-nitrogen potting medium or peat.

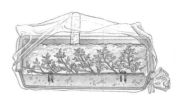

▲ **2** Place the tray in an inflated, sealed plastic bag, then keep at 59–68°F (15–20°C) in good light.

▲ **3** When the young plantlets are 1½–2in (4–5cm) high, with two to three tiny fronds, prick out into potting medium. Lift the plants with a spatula, ensuring that each has a good clump of soil attached to the rootball.

▲ **4** Pot up singly into 3in (8cm) pots, setting the rootball into a dibbled hole in the center of the pot, at the same level as it was in the seed tray. Young plants are usually ready to plant out in 6–9 months, in early summer.

# PROPAGATING FERNS FROM SPORES

## Collecting spores

Inspect the underside of the fronds for sori (the spore-bearing structures) from midsummer onward. These are usually covered by a flap of tissue (the indusium). If, using a hand lens, you can see that the shiny globules beneath it, the sporangia (the spore-containing structures) have not shed their spores. If the sporangia are pale, they are probably not ripe, so wait a few days until they darken. If they are shaggy, they have almost certainly shed their spores, so select another frond.

Debris from old sporangia can be confused with fresh spore. If in doubt, scrape the "spore" onto a piece of white paper and examine it with a hand lens. Spores can be recognized by the thousands of minute grains of uniform size.

When the sporangia are ripe, break off a section of frond and put it in a clean, dry envelope; after 2–3 days, dust-like spores will have dropped to the bottom. Clean any debris from the spores, as it may lead to contamination. In a draft-free room, tip the spores onto a piece of paper, hold at 45° and tap gently. Pieces of debris move down rapidly, spores very slowly: practice will soon tell you when the spores have separated. Return them to a fresh envelope or sow immediately.

## Spore storage

Many spores can be stored in a cool place for several months, but *Osmunda* and *Lygodium* species must be sown within a few days of collection. Storage in a refrigerator can prolong the viability of all spores, sometimes for several years.

## Sowing spores

Fill a seed germination pan with moistened medium, and place the pan in an oven at low heat to pasteurize. The temperature of the medium should be raised and held at 140°F (60°C), or a little above, for 30 minutes. Cover immediately with foil to prevent contamination by airborne fungi, moss, or fern spores. Spores can be sown when the medium has cooled. Choose a draft-free room and a clean surface that can be swabbed down with alcohol. Spores are best sown as soon as possible after collection, when viability is highest. If possible, sow spores of the same species from different sources, mixed

## Spores · · · · · · · · · · · · · · · · · · · · · · · · · · · · · · · · ·

Most ferns ripen spores in late summer. Success in propagation depends on following the steps below carefully, and maintaining high standards of hygiene throughout.

◀ **1** Prepare a label with the fern name and date of sowing, and set this on one side of the container. Tap the spores from a clean sheet of paper to give a very thin covering to the medium surface. Place the container immediately into a new plastic bag and seal it. Keep at a temperature of 59–68°F (15–20°C), in indirect light.

▶ **2** About 6 weeks later, a green film will appear on the surface of the medium. By the following spring, the prothalli should be easily distinguishable. To grow them on, you will need to lift small patches with the point of a knife.

◀ **3** Set out the patches of prothalli ½–¾in (1–2cm) apart, in depressions in a prepared pot of seed germination medium. Return to a clean, sealed plastic bag. Keep the culture closed until tiny, erect fronds appear. Harden them off gradually.

▶ **4** Thin out the tiny ferns into a seed tray or plug tray of seedling medium. Water gently and cover with a plastic dome until the small ferns are established. After a few months, recognizable ferns about 2in (5cm) high can be potted on into individual pots.

together. This helps to ensure cross-fertilization, thus increasing the genetic diversity so that the resulting ferns should be more vigorous.

From sowing to planting out usually takes 2–3 years. If the culture is contaminated with moss, tease it out with tweezers and water from below with a solution of potassium permanganate. If algae invade, some ferns may yet survive. Although contamination may spread rapidly, it can be checked with a fungicide.

# DIRECTORY OF
# FERNS

*Below are key tips on selected ferns that are particularly suitable for propagation. Unless otherwise specified, follow the detailed instructions given under the propagation section on pages 50–51.*

## Key

| | |
|---|---|
| H | Height |
| S | Spread |
| B | By bulbils |
| ✂ | Division |
| Sp | By spores |
| PO | Time to planting out |

## Asplenium (Spleenworts)    Aspleniaceae

About 700 species of evergreen and rhizomatous, terrestrial, epiphytic, or rock-dwelling ferns. Relatively few are cultivated, but these include **A. bulbiferum** (Hen-and-chicken fern) and **A. nidus** (Bird's nest fern). Both are commonly grown as houseplants. **A. scolopendrium** (Hart's-tongue fern) and **A. trichomanes** (Maidenhair spleenwort) are both hardy species and are ideal for a well-drained, shady border.

### PROPAGATION

✂ not normally recommended for this group. If attempted, pot up divisions into a sandy, sharply drained propagating medium. PO 3–4 months.

Sp ripen in late summer. Spores from the cultivars of *A. scolopendrium* will give rise to variable progeny. PO 2–3 years.

B leaf bulbils for *A. bulbiferum* can be grown in the usual way. Bulbils are the best method of multiplying cultivars of *A. scolopendrium*. They are very rarely found on the fronds, but can almost always be stimulated to grow on the leaf bases (*see p. 50*). PO 6–12 months.

## Athyrium    Woodsiaceae

About 100 species of deciduous, terrestrial, or rock-dwelling ferns with erect or creeping, sometimes branched, rhizomes and variable attractive fronds. **A. filix-femina** (Lady fern) and its cultivars, and **A. niponicum var. pictum f. metallicum** syn. 'Pictum' (Japanese painted fern), both of which are hardy, are the most commonly grown. They need damp or wet, shady conditions and are ideal for pool sides.

### PROPAGATION

✂ this is the only way to propagate the cultivars called "living antiques" that ensures the continuation of a true clone. PO 6 months.

Sp ripen in late summer. Cultivars will produce variable progeny. *A. niponicum var. pictum f. metallicum* produces approximately 50% green-leaved and 50% purplish-leaved offspring. PO 1–2 years.

B very rarely produced, but, with luck, can be spotted with a hand lens on the pinnules or the pinnule stalks. Harvest bulbiferous leaves and keep warm in a closed case or plastic bag. PO 1–2 years.

## Blechnum    Blechnaceae

Some 200 species of rhizomatous and evergreen, terrestrial or epiphytic ferns. Some are grown as houseplants: **B. gibbum** (Miniature tree fern) for example. **B. spicant** (Hard fern), **B. chilense**, syn. B. tabulare, and **B. penna-marina** are all hardy plants suitable for growing in the garden. They prefer a site in leafy, acid soil in a damp, shady border.

### PROPAGATION

✂ spring, for creeping species like B. penna-marina and, to a lesser extent, B. chilense. Species with upright rhizomes, like B. spicant, can be separated with care. PO 3 months.

Sp ripen in late summer. Use an acidic propagating medium for spore sowing. PO 2–3 years.

## Dryopteris (Male fern, Buckler fern, Wood fern)    Dryopteridaceae

About 200 species of rhizomatous terrestrial ferns. They are mainly deciduous but may stay green in sheltered sites. There are also many cultivars, most of which are hardy and suited to growing in the garden in well-drained or damp positions. They can tolerate modest amounts of direct sunshine.

### PROPAGATION

✂ as growth begins in spring. Most have upright rhizomes and, with age, produce side-crowns around the side of the main crown, which can be separated in the usual way. PO 3–6 months.

Sp late summer. Spore-raised progeny from cultivars of *D. affinis* (Golden male fern)

*Asplenium scolopendrium*

*Blechnum gibbum*

Matteuccia struthiopteris

Osmunda regalis

Polypodium vulgare

come true to type. Cultivars of other species come fairly true, except for the crested cultivars which are less reliable. PO 2–3 years.

## Gymnocarpium                 Woodsiaceae

Genus of five species of hardy, terrestrial, deciduous ferns with long, creeping rhizomes which may colonize areas rapidly where conditions suit them.
*G. dryopteris* (Oak fern). With distinctive triangular fronds, to 7in (18cm) long, this species establishes rapidly in sites of dappled shade, in leafy, well-drained soil. H 8in (20cm), S indefinite.

### PROPAGATION

✎ all species have creeping rhizomes and are easily split up. PO 3–6 months.
Sp easy from spores, but division is so successful that spore propagation is rarely necessary. PO 2 years.

## Matteuccia                 Woodsiaceae

Between two and four species of hardy, deciduous ferns with erect or creeping rhizomes and short, trunk-like bases.
*M. struthiopteris* (Ostrich fern). The most widely grown of the genus. It is ideal for wet, shady borders or shady poolsides. H 5½ft (1.7m), S 3ft (1m).

### PROPAGATION

✎ Although *M. struthiopteris* has an erect crown, or "shuttlecock," it also produces a creeping rhizome. New crowns are produced at random, 6–12in (15–30cm) away from the parent plant. Separate and re-plant in a new site immediately. PO immediately.
Sp early spring. Stumpy, unhealthy-looking,

black spore-bearing fronds are produced at the center of the shuttlecock in late summer. In fall, the outer green fronds wither rapidly but the sporing fronds stand erect all winter. The spores ripen in late winter. Sow immediately or store in a refrigerator. PO 2–3 years.

## Onoclea                 Woodsiaceae

Genus of one or two species of hardy, rhizomatous, deciduous ferns.
*O. sensibilis* (Sensitive fern). This is the widely grown species with triangular or more lance-shaped, divided, pale green fronds. It is suitable for growing in wet and shady positions in the garden. H 24in (60cm), S indefinite.

### PROPAGATION

✎ spring. *O. sensibilis* spreads vigorously, and is easily divided. PO immediately.
Sp early spring. Sporangia are borne on specialized fronds that resemble pipe cleaners – dark sticks covered with small beads near the tip – and are produced in late summer. Spores are green and ripen in late winter. Sow immediately or store in a refrigerator. PO 2 years.

## Osmunda                 Osmundaceae

About 12 species of hardy and frost-tender, rhizomatous, and sometimes trunk-forming, deciduous ferns.
*O. regalis* (Royal fern). The most commonly grown, it is hardy, forming impressive large "shuttlecocks" of pale blue-green fronds. It makes a magnificent large specimen for wet areas and pool sides, and will tolerate direct sun if given adequate moisture at the roots. H 6ft (1.8m), S 12ft (4m).

### PROPAGATION

✎ spring. Many species, including *O. regalis*, produce a huge rootstock that can be cut into pieces for growing on. Ensure that individual crowns are clearly separate as failures are not uncommon. PO immediately. Sp produced in early summer in specialized clustered structures, either at the tip of otherwise normal fronds, as in *O. regalis*, or on specialized fronds, as with *O. cinnamomea*. Spores are green and should be sown immediately or stored in the refrigerator. PO 2–3 years.

## Polypodium                 Polypodiaceae

Some 75 species of terrestrial, epiphytic, or rock-dwelling ferns, with creeping, often surface rhizomes. The genus includes a few hardy species and their cultivars. Most are wintergreen but some, e.g. *P. cambricum*, syn. *P. australe*, become leafless during the summer months. They thrive in well-drained borders, with or without direct sun.

### PROPAGATION

✎ The commonest method of increase. All species have creeping rhizomes that can be readily cut. Use a free-draining medium. PO immediately.
Sp late summer or fall for most species; spores of *P. cambricum* ripen through midwinter. Sporangia of *P. cambricum* are yellow when ripe, making quite a feature in the winter garden. The cultivars do not come reliably true from spores.
B very rare but occasionally produced all over the upper surface of the fronds of *P. vulgare* 'Elegantissimum'. They are difficult to strike but worth trying in low-nutrient medium or peat substitute.

# ULBS, CORMS, AND TUBERS

In gardens, the term "bulb" is generally used
to include corms, tubers, and rhizomes. Plants
with these subterranean storage organs have a
relatively brief period of growth and flowering –
the majority in spring and early summer –
followed by a long dormancy below ground.
This means that bulbs can be used to prolong
the season of interest by enhancing the dull
periods before other plants peak or when they
are past their best. They can also be used in
formal spring displays, as well as in containers,
indoors and out. However, bulbs create the most
memorable effects when naturalized *en masse* in
short turf or beneath trees and shrubs – which
is reason enough to propagate your own.

◄ *Important bulbous plants, such as
tulips and hyacinths, are propagated
by a range of techniques that are
specific to bulbs.*

55

# WHY PROPAGATE
# BULBS, CORMS, AND TUBERS?

*Whether naturalized in grass, incorporated in a border scheme, or planted in containers, for maximum impact bulbs should be grown in generous clumps. Since they can be expensive, and the choicer varieties are often only available in limited quantity, it is simple good husbandry to master the technique of multiplication for yourself.*

Few sights can match the joy that the first bulbs of spring bring, heralding the start of another gardening year with their fresh colors and often rich perfumes. Of course, bulb displays are not confined to this early flourish, and throughout the season there are choice specimens to covet. They are all very different, but have one thing in common – the bigger the clump, the better they look. By propagating your own bulbs, such massed plantings can be achieved very cheaply. From seed, bulbs usually take 4–7 years to reach maturity, or even longer in large species. Fortunately for those gardeners who lack the necessary patience, there are a number of alternative techniques that are usually much quicker in producing results.

## GROWING FROM SEED

Bulbs are expensive. Seed collected from your own plants is free and, having been produced under local conditions, is more likely to succeed in your garden. Given suitable growing conditions, most species will produce seed, and the offspring will usually be the same as the parent plant. Hybrids and named cultivars, even if fertile, rarely reproduce themselves, and the offspring will display mixed combinations of characteristics. Nevertheless, such seed can be sown for interest and may produce offspring that are unique to your garden.

The mixture of bulbs and cormous plants shown here will need a variety of techniques to propagate.

## DIVIDING CLUMPS

Many bulbs produce offsets that can be separated off to increase stocks. Division is essential for some species, since if clumps become overcrowded they eventually cease flowering. Lift such clumps when dormant, separate loose bulbs, remove old debris, and re-plant parent and offspring into fresh soil or a new site.

## CUTTINGS AND CUTTAGE

Several more specialized techniques may be used to increase the stock of a wide range of bulbs. Scaling, chipping, scoring, scooping, sectioning, and cuttings are all options that involve causing intentional damage to a specific part of the bulb, which in turn responds by producing offsets or bulblets from dormant buds. It includes scaling, chipping, scooping, scoring and sectioning. All of these techniques demand perfect hygiene and although they are commonly used by commerical growers, they can, with practice, be mastered to great advantage by the amateur gardener.

*Tulip Breaking Virus*
*This virus is spread by sap-sucking aphids, and causes feathering, streaking, or flaming on the ground color of the petals of tulips. Its effects are most famously seen on the Rembrandt tulips depicted by the Dutch painters of the 17th century. The virus is not transmitted by seed, and these old varieties can be maintained only by propagating from offsets. The Rembrandts will, however, form a source of virus infection and, for this reason, must be grown in*

*isolation away from other tulips and other susceptible bulbs, such as lilies.*

The feathered markings of *Tulipa* 'Striped Bellona' are typical of those induced by the tulip breaking virus.

They are a very effective way of bulking up the numbers of your stocks, and ideal where massed plantings are needed.

Taking cuttings from true bulbs in not an option, since their leaves and stems do not have the ability to form root initials. However, cuttings are a very economical way of increasing tuberous-rooted plants, such as dahlias. If induced in to growth early in the year, basal cuttings taken from the resulting shoots will produce flowers later the same season.

**① *Tulipa* species**
Unlike the highly bred cultivars, which either do not produce seed or do not come true, many small *Tulipa* species can produce flowers from seed in their third or fourth year.

**② *Anemone blanda***
The knobbly tubers divide naturally to form large clumps. For more rapid increase, section the tubers into ½in (1cm) pieces when dormant, but do not allow them to dry out.

**③ *Muscari***
Many of the grape hyacinths in cultivation are true species, and are easily increased by seed and often self-sow. They will flower in their second or third year.

**④ *Tulipa* 'Red Riding Hood'**
As with most other *Tulipa* cultivars, this produces offset bulblets at the base of the mother bulb. Increase your stock by separating bulblets and grow them on in a nursery bed.

**⑤ *Narcissus***
Only species and new cultivars of *Narcissus* are raised from seed. Cultivars are propagated by separation of offset bulbs in early or late summer, and larger bulbs can be sectioned or twin-scaled.

57

# PROPAGATING TECHNIQUES
# BULBS, CORMS, AND TUBERS

*Division is one of the simplest means of bulb propagation and raising plants from seed is another option, producing large numbers of offspring, although these can take some time to reach flowering size. However, some bulbs do not readily produce offsets or set seed, and for these there is a range of alternative propagation techniques.*

## WHAT IS A BULB?

The term "bulb" is often rather loosely applied to a range of plants that have a specialized, subterranean food storage organ and a periodic growth habit which makes it possible for them to survive, while dormant, an inclement period of either severe cold or high temperatures, without moisture. On the return of suitable growing conditions, the stored nutrients allow the plant to make immediate and rapid growth. This rapid growth ensures that the plant can flower and set seed in the short time available before the next season's dormancy is enforced by the return of unsuitable growing conditions, whether they are cold and wet or hot and arid.

Included in this category are true bulbs, corms, tubers and rhizomes. Since all these food-storage organs also contain growth buds, at the base of the scales or on the surface of corms, rhizomes and tubers, the gardener is able to use them as material for a wide range of vegetative propagation techniques.

### Bulb types .................................................................................

Bulbs, corms, tubers, and rhizomes are all subterranean food storage organs, but their appearance and structure – and hence the methods of propagation appropriate to each – vary considerably. You will need to know which type you are dealing with in order to select the correct propagation technique.

**1** A bulb is a compressed stem, and its fleshy scales are modified leaves. The scales have growth buds at their bases and enclose flower bud initials that will produce the coming season's flowers. The papery outer tunics are the dried remains of the previous season's scales. Some bulbs, like lilies, are composed of many scales, while others have only two or three, as in *Fritillaria*. Other examples of scaly bulbs include *Hippeastrum*, *Hyacinthus*, *Narcissus*, and *Lilium*.

**2** A rhizome is a horizontally growing stem, usually narrow and elongated, which sends up leaves or flower stems at intervals from buds produced along the rhizome or at its tip. Many grow below the soil surface, while others, such as *Iris* and some rhizomatous anemones, are partly exposed.

**3** A corm is the swollen base of a stem in which reserve food materials are stored. It is usually rather flattened and solid, with buds on its upper surface at the base of the previous year's growth. A new corm (or corms) is produced annually on top of the old one, which withers away. Examples of cormous plants are *Crocus*, *Gladiolus*, and *Freesia*.

**4** A tuber is the swollen part of an underground stem. Usually rounded, it may have various protrusions or depressions containing growing buds in shallow axils, often spirally arranged over the surface, as in *Tropaeolum*. The tubers of *Anemone* are hard, knobbly, and somewhat flattened. Some tubers have buds only at the top, where old growth has died back, as in *Dahlia* and *Cyclamen*.

# BULBS FROM SEED

Relatively few bulbous plants raised from seed will flower within a year, but *Tigridia pavonia*, *Dahlia*, and *Freesia* are among those that do. The seeds of these plants must be sown in warmth, in late winter or very early spring, and good growing conditions must be maintained, with appropriate feeding in the later stages. For most hardy bulbous plants, however, seeds should be sown in late summer or early fall, as the process of germination is initiated by the falling temperatures at this time of year.

## Sowing seed . . . . . . . . . . . . . . . . . . . . . . . . . . . . .

Sowing seed collected from the bulbs in your garden is an inexpensive way to produce large numbers of plants, although the young bulbs will take several years to reach flowering size.

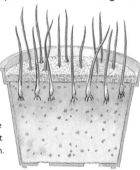

▶ **1** Fill a clay or plastic pot with medium to within about 1in (2.5cm) of the rim. Sow the seeds evenly and thinly on the surface of the medium, and cover with sieved medium to a depth roughly five times the diameter of the seed. Top with a layer of coarse sand to just below the pot rim, to deter algae and liverwort growth, and to prevent disturbance of the seed by heavy rain. Label with the plant name and date. Plunge the pot to its rim in the open ground, or in an open frame with a sand base, in a position out of direct sun.

### Aftercare

Once the seedlings have germinated, grow them on in good light, with shade from direct sun, and keep them evenly moist until the leaves show signs of dying back. Stop watering and keep dry during the dormant period. Resume watering only when signs of new growth are visible. Natural rainfall usually provides sufficient moisture, but in dry periods additional water may be required. Keep a portable glass or plastic frame at hand to cover during severe weather and dormancy.

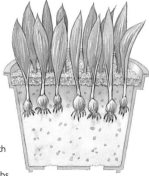

▶ **2** If growth has been vigorous, pot the small bulbs into trays or individual pots of potting medium in the late summer of their first year, then replace the pots in the frame. Most bulbs will take two growing seasons before they are ready to be potted on. Leave them in their pots and apply a half-strength balanced liquid feed during their second growing season. Smaller bulbs will be ready to be planted out at the end of their second year and may bloom in their third. Most of the larger hardy bulbs may be planted out in the garden at the end of their third year.

## Division . . . . . . . . . . . . . . . . . . . . . . . . . . . . . . . . .

Detaching the smaller daughter bulbs from large bulbs such as *Crinum* is essential to maintain vigor and flowering, as well as being a useful way to increase your stocks.

▶ In spring, just before growth begins, lift the clump, retaining as much as possible of the massive root system. Shake off loose soil, and remove any soil that is compacted using a jet of water from a garden hose. Separate large individual bulbs from the clump by hand, or if necessary use back-to-back forks to prise tangled clumps apart. Gently pull away offsets from the base of the mother bulb, retaining as many of the roots as possible. Pot up the offsets singly into sandy potting medium, with the growing point positioned just above the surface of the medium, and grow on for one or two seasons before planting out in their permanent flowering site. Re-plant the parent bulbs directly into the flowering site, setting them out at appropriate spacings.

The majority of bulb seed does not require any pre-sowing treatment. A few plants, such as some *Iris* species, have hard, impermeable seed coats that benefit from nicking or chipping so that water can penetrate to initiate germination. To do this, use a sharp knife to remove a small piece of the seedcoat and expose the pale interior, then sow in the usual way.

Some species, such as the South African *Gladiolus*, which originate in areas where bush fires are common, require smoke treatment to initiate growth. They do not need heating, but should simply be placed in a seed tray in a large fish smoker, or in a sieve suspended in a small, smoky fire, for 15–20 minutes. The chemicals in the smoke destroy the germination inhibitors.

*Cyclamen* seed is best sown as soon as ripe, but if sowing dry seed, soak for 24 hours in tepid water to which a single drop of household detergent has been added, then rinse before sowing.

### Sowing and growing on

A loam-based seed medium mixed with one-quarter part by volume of sand or seed-grade perlite is a suitable medium in which to germinate most bulb seeds. When sowing the seed of woodland bulbs, add an additional quarter part of sieved peat moss.

## Offsets .................................................

Many bulbs produce numerous small offsets around the mother bulb, and removing these to grow on individually is one of the easiest methods of propagation.

◄ **1** When the bulbs or corms of *Narcissus*, *Crocus*, or *Colchicum* become overcrowded, lift the clumps during the dormant period. Separate the bulbs from the soil or potting medium and then gently pull offset bulbs from the mother bulbs. Rub off the old dry tunics and dried remains of root. Select only plump bulbs and offsets with no sign of damage or disease. Mother bulbs and larger offsets may be re-planted directly into fresh soil or in a new site.

► **2** Grow on smaller offsets in pots or in a nursery bed, for one or two growing seasons. Plant them at twice their own depth, at least one bulb width apart.

## Bulbils and bulblets ...........................

Collect leaf axil bulbils as they ripen during summer, before the parent plant dies back. They will be dark brown and will come away easily when fully ripe. To collect stem bulblets, lift the parent bulb when the flowering stem has died back after flowering or cut the dry stem from the parent bulb. Pick off the bulblets carefully, to avoid damage to any small roots that may already have formed.

◄ **1** Set bulbils on the surface of a prepared pot of medium, pushing them into the surface to no more than one-quarter of their depth. Cover with sand, so that it barely covers the nose of the bulbil. Grow on in a cold frame. The following fall, pot up individual bulbs separately, or if growth has been vigorous, set the whole pot out in the flowering site.

◄ **2** Plant bulblets at twice their own depth in a prepared pot of medium, and mulch with a layer of sand. After car is as for bulbils (*see above*); both will take 2–5 years to flower.

### Alternative techniques for seed

A few genera and species germinate more reliably using other techniques. Seeds of many genera in the family Amaryllidaceae have flat seeds of a papery texture that rot easily in moist soil before sprouting. Float them on the surface of water in a saucer, or sow onto damp kitchen paper. Keep out of direct sun, at 61–77°F (16–25°C). Do not allow to dry out, even for a short time. Viable seed should sprout within 4–10 days. When the leaf is 2–3in (5–8cm) long, transfer the seedlings to an open, pasteurized medium to grow on.

In some genera, like *Nerine* and *Crinum*, the seeds may germinate while still attached to the flower stem and immediately develop small bulbs that produce first roots and then leaves. These may be removed and treated as small plants to be grown on in the same soil medium as used for the parent plant.

### DIVISION

When grown in favorable conditions, many bulbs produce numerous offsets or daughter bulbs which, through overcrowding and competition for nutrients, eventually deteriorate and become non-flowering. So, as well as increasing your stock, division is also necessary to maintain vigor and flowering potential.

There are several slight variations in technique when dividing different bulbs. Bulbs like *Crinum*, *Nerine*, and *Amaryllis belladonna* form substantial clumps with a mass of fleshy roots and often produce several offset daughter bulbs from the base of large mother bulbs.

*Anemones* have different types of rootstock. Divide the finger or twig-like rhizomes of *Anemone nemorosa* and its cultivars during the dormant season, by simply breaking them apart at the natural joints. The knobbly tubers of *A. blanda*, *A. coronaria*, and the De Caen and St. Brigid cultivars may be cut or broken into small pieces not less than ½in (1cm) in diameter, prior to re-planting in fall.

### BULBILS AND BULBLETS

A number of bulbous plants, including *Tritonia* and, especially, *Lilium* species, produce stem bulbils in the leaf axils on the aerial parts of the stem, or bulblets on the underground rooted portions of the stem. Both provide an easy and convenient means of propagation, and the offspring will be identical to the parent plant.

## Scaling

Scaling is a useful method of propagation for bulbs that produce numerous and/or clearly defined scales that can be separated easily from the parent bulb. *Lilium* is the prime example, and, as the bulbs can be expensive to buy, it is well worth propagating your own during the dormant season from plants already growing in the garden.

◄ **1** Between late summer and fall, select a sound, healthy bulb with plump scales. Pull the outer scales away from the parent bulb's basal plate. Reject any that are soft and flabby, or dried and wrinkled. Place the scales in a plastic bag of fungicide powder and revolve it gently to coat them thoroughly. Empty out the scales and dust off any surplus fungicide.

◄ **2** Transfer the scales to a clean plastic bag with about four times their own volume of a mix of equal parts peat and sand, vermiculite, or perlite. Shake gently to ensure even distribution. Inflate the bag, seal with a wire tie, and put in a dark place, at about 61–77°F (16–25°C). Inspect periodically, and remove any scales that show signs of rotting.

◄ **3** When small bulbs appear at the base of the scales, after about 4–12 weeks, separate them from the medium. Remove the bulblets from any scales that have become soft, but leave firm scales in place. Pot each scale in a small pot of well-drained potting medium, and top-dress with a ¼in (5mm) layer of sand, so that the scale tip is just at the surface. Grow on in a shaded coldframe.

◄ **4** At the end of the first growing season, separate out the bulblets, pot them on individually, and return them to the coldframe. The following spring, if large enough, they can be set out in their permanent positions in the garden. Alternatively, grow the bulblets on in pots for a further year before planting them out.

## Chipping

This method of propagation is useful for increasing stocks of *Hippeastrum* and other bulbs that are not easy to divide and/or do not set seed. Cutting the bulb into chips stimulates it to produce bulblets in response to the injury. Success depends upon meticulous hygiene at all stages to prevent disease penetrating the cut surfaces.

◄ **1** Choose a healthy, mature bulb and cut off the roots and nose using a sharp knife, which has first been sterilized using alcohol. The roots should be sliced off immediately below the basal plate. The length of the section of nose removed must be no more than one-quarter of the overall length of the complete bulb.

◄ **2** Place the bulb nose-end down on the work surface and slice in half vertically, cutting down through the basal plate. Repeat until you have 16 chips, ensuring that each has a segment of basal plate attached (smaller bulbs will yield four to eight chips). Soak the chips in fungicide for 15 minutes, stirring occasionally to ensure good penetration.

◄ **3** Place the chips in a plastic bag containing moistened vermiculite (1 part water to 2 parts vermiculite by volume is sufficient). Seal the bag and place it in a warm, dark place at about 68–77°F (20–25°C), for around 12 weeks. Young bulblets will develop from the basal plate between the chips.

◄ **4** Once the bulblets are large enough to handle, or have begun to form roots, you can separate them and pot them up individually into small pots of potting mix. Otherwise, plant the chip and bulblets in their entirety so that each bulblet nose is just below the surface of the potting mix.

## Twin-scaling

As with chipping, good hygiene is essential throughout the twin-scaling process to prevent disease from entering the cut surfaces of the bulb segments.

◄ **1** The procedure for twin-scaling is essentially the same as for chipping, except that each of the segment is further divided into pairs of scales, each with a tiny section of basal plate.

► **2** The bulblets appear from between the pairs of scales. Chipped or twin-scaled bulbs take about 2–3 years to reach flowering size.

## CUTTAGE

There are several propagation techniques that fall under the general heading of cuttage, all of which cause intentional damage to the bulb, corm, or tuber in order to encourage the production of small offsets from dormant buds. These are all slightly more specialist techniques, that demand greater care, attention to detail and good hygiene.

### Scaling

*Lilium* species and cultivars are often increased by this method. It can also be used on bulbs of *Fritillaria* species, which generally have only a few scales, but these are usually clearly defined and are easily separated by making a vertical cut between them with a sharp knife.

After separating from the parent bulb and dusting with fungicide, the scales may be planted directly into trays of propagating medium. Bury them to half their depth with the tips uppermost, and cover with sphagnum moss or a plastic sheet. Keep the scales moist and shaded, at about 70°F (21°C) for 4–12 weeks, then transfer to a coldframe to grow on.

### Chipping and twin-scaling

Chipping is often used to increase cultivars of *Hippeastrum* and it is equally suitable for increasing other true bulbs, especially those that do not divide readily, or that do not set seed. Twin-scaling, often used to increase *Narcissus* and *Galanthus* (snowdrops), is similar to chipping, but involves reducing the cut sections to pairs of scales. Since both of these methods involve cutting through the flesh of the bulbs, good hygiene is essential at all stages to minimize the risk of contamination of cut surfaces by disease. All work surfaces and cutting blades should be sterilized by swabbing them with alcohol.

## Scooping

Scooping is a good method for propagating hyacinth cultivars, which produce few offsets and do not come true from seed. Only the simplest of tools – an old spoon with a specially sharpened edge – is required.

◄ **1** Hold the bulb upside down and scoop out the central section of the basal plate with the sharpened edge of a spoon. Take good care not to damage the outer rim of the basal plate. The depth of the scoop is critical: aim to remove only as much basal plate tissue as is necessary to expose the lowest point of the scale leaves. Dust the cut surface with fungicide.

◄ **3** Small bulblets will form over the cut surface of the mother bulb after about 12–14 weeks. They should be left in place for the next stage of the process.

◄ **2** Set the prepared bulbs in a tray of sand, with the cut surface uppermost. Place the tray in a warm, dark place, at 77°F (25°C), for about a week. Then moisten the sand in the tray to prevent desiccation of the bulbs.

◄ **4** Plant the parent bulb upside down in a pot of sandy, sharply drained potting medium with the noses of the bulblets just below the surface. The bulblets will produce leafy growth in spring. In late summer, when the leaves die down, lift, separate, and pot up individually. The plants will flower in 3–4 years.

## Scoring

▲ Similar to scooping, scoring is a relatively simple procedure. Make two V-shaped incisions at right angles to each other across the basal plate of the bulb. Dust with fungicide and treat as for scooped bulbs. The bulblets will take 2–3 years to reach flowering size.

### Scooping and scoring

Scooping and scoring are techniques that involve injuring the basal plate to promote the production of small bulblets. They are most commonly used for bulking up hyacinth cultivars that are slow to form offsets and would not come true from seed, even if any were produced. Scooping is generally the more productive method. Commercially, a special curved tool is used, but an old spoon with a sharpened edge will perform equally well. In late summer, select sound healthy bulbs about 3–4in (7.5–10cm) in diameter. Scoring produces fewer, slightly larger bulblets.

## CORMELS

A number of genera, particularly *Gladiolus*, produce numerous cormels, usually around the junction of the old and new corms, which are insurance against the death of the parent. These do not usually grow to flowering size unless separated off. Lift the parent corm complete with its stem, in late summer or early fall as it dies back after flowering. Store in a dry frost-free place over winter. In late winter or very early spring, separate the small cormels from the parent.

## Cormels

◄ The parent corm will produce numerous cormels of varying size. Plant them in trays half-filled with potting mix, 1in (2.5cm) apart and with their growing points uppermost. Then fill the trays so that the growing points are just below the medium surface. Grow on in a frost-free bulb frame or greenhouse. They will take about 2–3 years to reach flowering size.

## Dividing *Dahlia* tubers

The tubers of *Dahlia* species and cultivars can be divided to produce flowering plants in the same year. Start the tubers into growth before division to ascertain which tuber or tubers have viable growths.

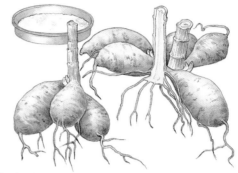

▲ 1 Set the tubers on the surface of a tray of medium in warmth, mist over, and keep warm and slightly moist. The new growth emerges just where the top of the tuber meets the old dried stem. When the "eyes" are clearly visible, section the parent tuber using a sharp knife, so that each section has one eye or shoot, with one or more, plump healthy tubers attached. Dust the cut surfaces with fungicide.

► 2 Set each division in a pot of potting medium, with the base of the shoot just at surface level and the growing tip above the surface. Grow on and plant out in the flowering site when all danger of frost has passed.

## CUTTINGS AND SETTING FROM TUBERS

Several genera that have tubers, such as *Dahlia*, *Tropaeolum*, and *Begonia*, may be propagated by basal cuttings, using much the same technique as for herbaceous perennials (*see pp. 30–31*).

The process begins in late winter or very early spring. Bring the tubers into growth, in warmth, and, when the shoots have reached about 2–4in (5–10cm) in length, trim them from the tuber together with a heel of basal tissue. Dip the cut surface in hormone rooting powder and then pot up the shoots into an open, moist medium. Enclose both the pot and the plant in a plastic bag and keep them in a warm, light place until growth indicates that the shoots have rooted, when the small plants can be potted up or planted out individually. This technique will produce plants of flowering size the same year.

# DIRECTORY OF
# BULBS, CORMS, AND TUBERS

*Below are key tips on a selection of bulbs, corms, and tubers that are particularly suitable for propagation. Unless otherwise specified, follow the detailed instructions given under the propagation section on pages 58–63.*

## Key

| | |
|---|---|
| H | Height |
| S | Spread |
| PD | Planting depth for bulbs |
| B | By bulbils |
| ❦ | By seed |
| ✎ | By division |
| ❦ | Germination |
| ☀ | Time to maturity |

## Allium (Onion)　　　　Liliaceae

Some 700 species of hardy, bulbous or rhizomatous, spring, summer, or fall-flowering perennials, ranging from dwarf species, 2in (5cm) tall, to robust species 6ft (1.8m) in height. Flowers are bell-, star-, or cup-shaped, borne in spherical or hemispherical umbels, above clumps of basal, narrow or strap-shaped leaves. Large species are best purchased as bulbs initially and then increased from garden-collected seed. They require an open, sunny situation.

### Large species

***A. aflatunense.*** Dense, spherical umbels, 4in (10cm) across, of pink-purple flowers, in summer. H 3ft (1m), S 5–6in (12–15cm), PD 2–4in (5–10cm).

*Allium aflatunense*

***A. cristophii,*** syn. *A. albopilosum, A. christophii.* Gray-green leaves and open umbels to 8in (20cm) across, of metallic, pink-purple flowers in early summer. H to 24in (60cm), S 8in (20cm), PD 2–4in (5–10cm).
***A. giganteum.*** Produces star-shaped, lilac-pink flowers in dense umbels to 4in (10cm) across, in summer. H to 6ft (2m), S 8in (20cm), PD 2–4in (5–10cm).

### Small species

The following species increase easily and soon form clumps:
***A. beesianum.*** Bears open, 1in (2.5cm) wide umbels of bell-shaped, blue or white flowers in late summer or early fall. H to 8in (20cm), S 3in (7.5cm), PD 2–4in (5–10cm).
***A. moly*** (Golden garlic). Produces dense rounded umbels, 2in (5cm) wide, of golden-yellow, star-shaped flowers in summer. H to 8in (20cm), S 3in (7.5cm), PD 2–4in (5–10cm).

### PROPAGATION

❦ as soon as ripe, in an open frame. ❦ 6–12 weeks ☀ 1–3 years for small species; 4–6 years for large species. ✎ early fall. Divide established clumps of small species. Separate offset bulblets for larger species.

*Anemone coronaria*

## Amaryllis　　　　Amaryllidaceae

One species of bulb.
***A. belladonna.*** Funnel-shaped, pink flowers in fall, followed by long strap-shaped evergreen leaves. Needs well-drained soil at the base of a warm wall. Mulch in winter where temperatures fall below 23°F (-5°C). H 24in (60cm), S 4in (10cm), PD nose at soil level.

### PROPAGATION

✎ midsummer. Divide established clumps or separate offsets. ☀ 1–2 years. Cuttage – late summer with mature bulbs. Scoop, or section into quarters. ☀ 2–3 years.

## Anemone (Windflower)　　　Ranunculaceae

About 120 species of hardy or half-hardy, tuberous, rhizomatous, or fibrous-rooted perennials, usually with finely cut foliage and cup- or bowl-shaped flowers in spring, summer, or fall. (*See also Herbaceous Perennials directory, p. 35.*)
***A. blanda.*** Knobbly, woody tubers bearing slightly cupped, many-petaled, blue, pink, or white flowers in spring. Hardy. H and S 6in (15cm), PD 2–3in (5–8cm).
***A. coronaria.*** Knobbly tubers producing erect plants with shallowly cup-shaped, white, blue, or red flowers, in spring. Includes the **St Brigid Series**, with double flowers, and the single-flowered **De Caen Series**, both with flowers in a wide range of colors. Hardy. H 12in (30cm) or more, S 6in (15cm), PD 2–3in (5–7.5cm).
***A. nemorosa*** (Wood anemone). Short, slender, finger-like creeping rhizomes that in spring produce delicate, cup-shaped, white flowers. Dark green ferny leaves. Variants bearing lavender-blue flowers are also available. Hardy. H to 6in (15cm), S 12in (30cm) or more.

*Arum italicum*

*Begonia pendula*

*Colchicum speciosum*

## PROPAGATION

🌱 as soon as ripe, in an open frame. Cultivars do not come true.

🌱 slow, unpredictable, and erratic.

☀ 2 years.

🔧 summer, when dormant. Break finger-like rhizomes at the joints. Cut tubers into sections not less than ½in (1cm) in diameter. Do not allow to dry out.

## Arum (Cuckoo pint)　Araceae

Some 26 species of tuberous, herbaceous perennials, with large attractive leaves and flowering spathes followed by spikes of fleshy berries.

*A. italicum.* Arrow-shaped, deep green leaves are veined white, and pale green spathes appear in late spring or early summer. They are followed by poisonous orange-red berries. **'Marmoratum'** has leaves marbled with a mixture of cream and pale green. Hardy. H 12in (30cm), S 8in (20cm), PD 6–10in (15–25cm).

## PROPAGATION

🌱 when ripe, in a coldframe. Extract seed from the berries wearing gloves, as the flesh of the berries can irritate skin.

🌱 2–10 weeks. ☀ 4–7 years to flower; good foliage within 1–2 years.

🔧 mid- to late summer. ☀ 2–3 years.

## Begonia　Begoniaceae

About 900 species of mainly frost-tender annuals, perennials, and shrubs, with tubers, rhizomes, or fibrous roots. Begonias are grown for foliage and flowers, both in a range of colors, for bedding, as pot plants, and in hanging baskets. A complex genus, divided into seven horticultural groups, of

which only the most important tuberous or rhizomatous types are described here. All need a fertile, humus-rich medium and some shade from hot sun.

**Rex-cultorum Group.** These are usually rhizomatous and mainly evergreen plants. They are grown as houseplants for their colorful, often beautifully textured foliage. Optimum temperature: 70–75°F (21–24°C).

**Tuberhybrida** and **Multiflora Group.** These grow from tubers and are dormant in winter. They are cultivated as houseplants for their usually brilliantly colored flowers. Some are also suitable for using as summer bedding. Start into growth at 61–64°F (16–18°C).

**Pendula Group.** Tuberous with a pendent habit that makes them ideal for hanging baskets. Best at 61–64°F (16–18°C).

## PROPAGATION

🌱 late winter to early spring at 61–68°F (16–20°C). Mix with fine sand and sow on the surface.

Tubers – late winter. Section tubers of Multifloras, each piece with a growth bud. Basal cuttings – spring, all types 2½–3in (6–8cm) long. Start stock plants into growth at 61°F (16°C) in early spring.

## Chionodoxa (Glory of the snow) Liliaceae

Genus of six species of hardy bulbs, bearing racemes of blue and white, star-shaped flowers in spring. Their leaves are linear and basal. Grow in any well-drained soil in sun or light, dappled shade. They often self-seed.

*C. forbesii,* syn. *C. luciliae* (hort.), *C. seihei, C. tmolusi.* Produces stems bearing white-centered, blue flowers above erect to spreading leaves in spring. H to 8in (20cm), S 1in (2.5cm), PD 3in (7.5cm).

## PROPAGATION

🌱 early fall, as soon as ripe, or mid-winter, in a coldframe.

🌱 spring. ☀ 2–4 years.

🔧 late summer to early fall. Separate offsets from established clumps.

☀ 1–2 years.

## Colchicum (Autumn crocus)　Liliaceae

Some 45 species of mostly hardy corms bearing goblet-shaped flowers, often with attractive chequered markings, in late summer, fall, or winter. They are more closely related to lilies than crocuses. Leaves appear after flowering is over and are often large and coarse, so site colchicums where they will not swamp smaller plants. Those described below thrive in a range of soils, in sun or light dappled shade. They may also be naturalized in grass.

*C. autumnale* (Meadow saffron). Pinkish-lavender flowers in fall; white or double-flowered variants are available. Hardy. H 6in (15cm), S 3in (7.5cm), more when in leaf, PD 4in (10cm).

*C. speciosum.* Produces pale to deep pink-purple flowers in fall. Hardy. H 7–8in (18–20cm), S 3–4in (7.5–10cm), more when in leaf, PD 4in (10cm).

## PROPAGATION

🌱 early fall, in a coldframe. Transfer to a nursery bed at the end of their second year and grow on.

🌱 spring. ☀ 4–6 years.

🔧 mid- to late summer. Divide established clumps every third or fourth year to maintain vigor. Remove offsets at the same time and re-plant immediately.

☀ 1–2 years.

*Crinum × powellii*

*Crocus tommasinianus* 'Whitewell Purple'

*Dahlia* 'Biddenham Sunset'

## Corydalis
Papaveraceae

Some 300 species of hardy annuals, biennials, and rhizomatous or tuberous perennials, a few of which are evergreen. Those featured are herbaceous. They have brittle, finely divided, fern-like leaves, often blue- or gray-green, and produce spurred, tubular flowers in early spring or summer. A few are difficult to grow – even in the shelter of an alpine house – but those described here tolerate sun or dappled shade, and thrive in moist but well-drained, humus-rich soil.

*C. cashmeriana.* Tuberous perennial bearing dense racemes of intense blue flowers, with curved spurs, above bright green foliage in early summer. It prefers cool damp conditions and part-shade. H to 10in (25cm), S 6in (15cm).

*C. fumarifolia.* Tuberous perennial with glaucous leaves and racemes of azure or purple flowers, with triangular spurs, in late spring and early summer. It enjoys similar conditions to *C. cashmeriana.*
H to 6in (15cm), S 4in (10cm).

*C. lutea.* Mound-forming, rhizomatous perennial with fern-like, pale green leaves. It produces a succession of yellow flowers, with blunt spurs, from spring to fall.
H to 16in (40cm), S to 12in (30cm).

**PROPAGATION**

☙ as soon as ripe, in a coldframe. Seed rapidly loses viability if allowed to dry out.
🌱 spring. ☀ 2–3 years.
🔪 after flowering. Tubers divide naturally after flowering and can be lifted and separated when dormant. Tubers of *C. cashmeriana* may be carefully separated when in full growth if immediately re-planted and kept moist ☀ 1–2 years.

## Crinum
Amaryllidaceae

About 130 species of mostly frost-tender bulbous perennials, bearing showy funnel-shaped flowers on leafless stems in spring, summer, or fall. They usually have strap-shaped basal leaves. All need fertile, sandy, but humus-rich soil and sun. Frost-tender species are ideal for containers in the home.

*C. × powellii.* This bears pale or dark pink flowers in late summer. Hardy, given sharp drainage, a sheltered sunny site, and a winter mulch. In very cold areas, grow in large tubs and overwinter in the greenhouse.
H 4–5ft (1.2–1.5m), S 12in (30cm), PD nose at soil level.

**PROPAGATION**

☙ as soon as ripe at 70°F (21°C). Practical only for the very patient gardener.
🌱 spring. ☀ up to 15 years.
🔪 spring. Lift and separate offsets and re-plant immediately. ☀ 2–3 years.
Cuttage – spring. Scoop or section.

## Crocosmia (Montbretia)
Iridaceae

Seven species of cormous perennials forming clumps of narrow, ribbed, sword-shaped leaves and bearing branched spikes of funnel-shaped flowers in late summer. Several named cultivars are available with flowers in shades of yellow to deep scarlet-red. Most cultivars are hardy to at least 23°F (-5°C), *C. × crocosmiiflora* and *C. masoniorum* to about 14 to 15°F (-10–15°C). All need humus-rich, moist but well-drained soil in sun or part-shade, with a winter mulch until well established.

*C. × crocosmiiflora.* Bears slender, branched spikes of orange or orange-yellow flowers in late summer to early fall.

H 24in (60cm), S 3–4in (7.5–10cm), PD 3–4in (7.5–10cm).

*C. masoniorum.* This species bears upright, orange-red flowers in arching spikes, in mid- to late summer. H to 4ft (1.2m), S 3–4in (7.5–10cm), PD 3–4in (7.5–10cm).

**PROPAGATION**

☙ as soon as ripe, or between winter and early spring, in sandy medium in a coldframe. Cultivars do not come true.
🌱 spring. ☀ 3–4 years.
🔪 spring. Separate offsets or divide overgrown clumps, before growth begins.
☀ 1–2 years.

## Crocus
Iridaceae

Some 80 species of cormous perennials that produce a range of goblet-shaped flowers between late fall and late spring. They mainly have linear leaves with a central silver stripe. A wide range of variants is available, with flowers of many sizes and in various colors, including striped or marked with contrasting colors and often with conspicious styles. Easy to grow and propagate. Some, such as cultivars of *C. nudiflorus* and *C. tommasinianus,* may be naturalized in turf. Those most commonly thrive in sun in light, well-drained soil. PD 3–4in (7.5–10cm).

**PROPAGATION**

☙ as soon as ripe, in late summer to fall, in sandy medium in an open frame.
Cultivars do not come true. Some species will self-seed.
🌱 in spring. ☀ 3–4 years.
🔪 in mid- to late summer, when dormant. Separate offsets or divide crowded clumps. *C. nudiflorus* produces stolons that provide a rapid means of increase. ☀ 1–3 years.

*Eranthis hyemalis*

*Erythronium californicum* 'White Beauty'

*Fritillaria meleagris*

## Dahlia     Compositae

Some 30 species of tuberous-rooted herbaceous perennials and many thousands of cultivars, grown for their exuberantly colored flowers in late summer and fall. There are ten recognized flower types that vary greatly in appearance from single to pompon- or waterlily-like. In frost-prone areas, the tuberous roots are lifted when the leaves have turned black after the first frost, and overwintered in frost-free conditions. Bedding dahlias are raised from seed; the remainder from cuttings or divisions. Grow in fertile, humus-rich, moist but well-drained soil in a sunny site.

**PROPAGATION**

🌱 between late winter and early spring, at 64–68°F (18–20°C). Thin out into individual pots when two true leaves have formed, and pot on successionally. Pinch out stem tips when three leaf nodes have formed, to promote branching.

🌱 5–10 days.  ☀ 5–6 months.

✂ mid- to late winter or early spring. Bring stored tubers into growth in gentle warmth and with careful watering. As dormant buds swell, section tubers so that each has at least one growing point.  ☀ 4–5 months. Basal cuttings – late winter or early spring. Bring tubers into growth at 59–64°F (15–18°C). Take cuttings when shoots are 3in (7.5cm) long.  ☀ 4–5 months.

## Eranthis     Ranunculaceae

Some seven species of hardy perennials with knobbly tubers and finely divided leaves, bearing cup-shaped, yellow flowers in winter and very early spring. Grow in moist but well-drained soil in sun or light shade.

*E. hyemalis* (Winter aconite). This bears rich yellow flowers in late winter and early spring, with a ruff of bright green leaves immediately beneath. H 3–4in (7.5–10cm), S 2in (5cm), PD 2in (5cm).

**PROPAGATION**

🌱 as soon as ripe, in late spring, in an open frame. Seed has short viability and is rarely available commercially. Self-sows freely. Cultivars do not come true.

🌱 14–21 days.  ☀ 1–3 years.

✂ when dormant, immediately after flowering. If the tubers are allowed to dry out they will be difficult to re-establish.

☀ 1–2 years.

## Erythronium     Liliaceae

About 22 species of hardy, spring- to early summer-flowering bulbs. They have pendent flowers with reflexed petals, and conspicuous stamens. They need moist but well-drained, leafy soil in a cool site in partial shade. Do not allow bulbs to dry out before planting.

*E. californicum* 'White Beauty'. Has creamy-white flowers each with a red basal ring in spring. H to 14in (35cm), S 4in (10cm), PD 4in (10cm).

*E. dens-canis* (Dog's-tooth violet). White or pale pink-purple flowers in spring, above bronze-marbled leaves. H to 6in (15cm), S 4in (10cm), PD 4in (10cm).

**PROPAGATION**

🌱 in fall, in an open frame. Cultivars do not come true.

🌱 in spring.  ☀ 4–6 years.

✂ mid- to late summer or immediately after flowering. Separate offsets or divide established clumps and re-plant immediately.

☀ 1–2 years.

## Freesia     Iridaceae

Genus of about six species of frost-tender cormous perennials, with hundreds of cultivars bred for the cut-flower market. They usually bear fragrant, funnel-shaped flowers in a wide range of colors in late winter and spring. The leaves are linear and grow in fans at the base. In areas that experience winter frost, they should be grown in a greenhouse, prepared bulbs for summer-flowering can be planted outdoors. Grow in sandy, slightly acid, humus-rich soil. H 8–16in (20–40cm), S 1–2in (2.5–5cm), PD 3in (7.5cm).

**PROPAGATION**

🌱 in fall, at 55–64°F (13–18°C) for species. Sow seed of cultivars in late winter, and grow on at 61–64°F (16–20°C).

🌱 10–14 days.  ☀ 18–24 months for species, 8–10 months for cultivars.

✂ separate offsets when dormant (fall).

☀ 1 year.

## Fritillaria     Liliaceae

About 100 species of mostly hardy bulbous perennials, usually with subtly colored, often checkered, tubular to bell-shaped flowers. Most flower in spring or early summer. Some small species need protection from rain when dormant and are grown in a bulb frame. The fleshy bulbs produce offsets or rice-grain bulbils. Those described here are suitable for the open garden.

*F. imperialis* (Crown Imperial). Bears large bell-shaped flowers in shades of yellow, orange, or red in early summer. Needs well-drained soil and sun. Hardy, but prone to rot in very moist soils. H 4–5ft (1.2–1.5m), S and PD 8in (20cm).

*Galanthus nivalis*

*Gladiolus communis* subsp. *byzantinus*

*Hyacinthus orientalis* 'Blue Magic'

*F. meleagris* (Snake's head fritillary). Pendent, square-shouldered, chequered white, pink, or purple flowers in spring. It needs moist, fertile, heavy soil in sun or dappled shade. Hardy. H 12in (30cm), S 2–3in 5–7.5cm), PD 4–6in (10–15cm).

### PROPAGATION

❦ in fall, or as soon as ripe, in an open frame. Except for *F. imperialis* and other large species, grow on in pots for 2 years.
❦ in spring. ☀ 3–7 years, depending on size.
❦ separate offsets or sow rice-grain bulbils when dormant, in fall. Pot up or line out in a coldframe. ☀ 1–3 years.
Bulb scales – the bulbs of several species are made up of 2–10 fleshy scales. Separate the scales when dormant, and set in sandy medium. Water sparingly and carefully until established. Plant out the following fall.
☀ 1–3 years.

### Galanthus (Snowdrop) Amaryllidaceae

Genus of 15–20 species of hardy bulbous perennials with single or double, pendent, pear-shaped, often green-tipped, white flowers, mostly in late winter to early spring, although a few are fall-flowering. Most are easily grown in light dappled shade, in leafy, humus-rich – even heavy – soil that does not dry out too much in summer. Snowdrops can be difficult to establish from dry bulbs, and are best divided in the green, and re-planted at their original depth.

*G. nivalis* (Common snowdrop). Bears green-tipped, white flowers in late winter to early spring. Several cultivars are available with larger, double, or yellow-marked flowers. H 4–6in (10–15cm), S 4in (10cm).

### PROPAGATION

❦ as soon as ripe, in a shaded frame. Snowdrops hybridize freely, and established clumps will self-seed.
❦ 2–3 weeks. ☀ 3–4 years.
❦ immediately after flowering. ☀ 1 year.
Cuttage – scoop or quarter when dormant.
☀ 1–3 years.

### Gladiolus Iridaceae

About 180 species of cormous perennials with spikes of open, funnel-shaped flowers and fans of linear leaves. Hybrids and cultivars are the most often grown, valued for their exuberant spikes of brilliantly colored flowers in summer. They are ideal for cutting for flower arrangements. Gladioli are classified into three main groups – the *Grandifloras*, *Nanus* and *Primulinus* – and the flower sizes range from giant, at 5½in (14cm) or more across, to the miniatures, with flowers to 2½in (6cm) wide. They need fertile, very well-drained soil, in a warm, sunny site, with shelter from strong winds. Provide support. Plant 4–6in (10–15cm) deep, in spring. Most are half-hardy or frost-tender and need lifting before the first frosts occur in fall, but a few, such as *G. communis* subsp. *byzantinus*, with magenta flowers and *G. italicus*, which has deep purple-pink flowers, are hardy.

### PROPAGATION

❦ spring, in a coldframe for hardy species; at 59°F (15°C) for all half-hardy and frost-tender species.
❦ in spring. ☀ 1–3 years
❦ fall, when lifting. Separate cormlets from the parent corm and store in dry, frost-free conditions for the winter. Plant out in spring in furrows in a nursery bed. Lift again in fall, as for adult corms. They may reach flowering size in their second year; if not, grow on for a third season, lifting before frosts, as before.
☀ 2–3 years.

### Hyacinthus (Hyacinth) Liliaceae

Genus of three species of hardy bulbous perennials. The legion cultivars of hyacinth, derived from the species *H. orientalis*, are available with single or double flowers, in a range of colors from pure white to indigo-blue, through pink, orange, and yellow. They bloom between late winter and spring; specially prepared and forced bulbs bloom in early to midwinter. They are easily grown in any well-drained soil in sun or light shade. H 8–12in (20–30cm), S 3–4in (7.5–10cm), PD 4in (10cm) in open ground, nose at soil level in containers.

### PROPAGATION

❦ as soon as ripe, or in fall in a cold frame. Species only; cultivars rarely set seed.
❦ in spring. ☀ 4–6 years.
Cuttage – bulbs may be scooped or sectioned when dormant. ☀ 2–3 years.

### Iris Iridaceae

Some 300 species of mostly herbaceous bulbs and rhizomatous or fleshy-rooted perennials grown for their flowers, which consist of "falls" and "standards" and are often very impressive. Most are hardy but some require specific conditions in order to do well. Classified botanically into a number of subgenera; the most important bulbous irises are described here. (*For Bearded and Rhizomatous irises, see p. 41.*)

*Iris xiphium* 'Professor Blaauw'

*Leucojum aestivum*

*Lilium regale*

## Xiphium irises

*I. latifolia*, syn. *I. xiphioides* (English iris). Bears white, blue, or violet flowers in early summer. Enjoys moist but well-drained soil. Hardy. Bulbs may be left *in situ* through the winter. H 10–24in (25–60cm), S to 6in (15cm), PD twice bulb's height.

*I. × hollandica* (Dutch iris). This has white, blue, violet, yellow, bronze, or apricot flowers from late spring to midsummer. Hardy, but protect with plastic covers in harsh winters. Lift bulbs as the leaves wither if grown in damp, heavy soils. H 10–24in (25–60cm), S to 6in (15cm), PD twice bulb's height.

*I. xiphium* (Spanish iris). With pale or dark blue (like **'Professor Blaauw'**), occasionally yellow or white, flowers in late spring and early summer. Hardy, but prefers warmer, drier conditions than Dutch or English irises. Bulbs are best lifted after foliage dies; re-plant in fall. H 16–24in (40–60cm), S to 6in (15cm), PD twice bulb's height. Grow Dutch, English, and Spanish irises in neutral to alkaline, well-drained soil in a sunny position.

### PROPAGATION
as soon as ripe, in fall in a coldframe. Off-spring from hybrids and cultivars will not come true.

in spring.  3–5 years.

fall. Lift established clumps and separate offsets. Grow on in a nursery bed.

1–3 years.

### Reticulate irises

*I. reticulata*. This bears violet-blue flowers, with a yellow ridge on the falls, in late winter or early spring. Several excellent cultivars are available, including **'J.S. Dijt'**, **'Harmony'**, and **'Katherine Hodgson'**. All are suitable for a rock garden in well-drained, neutral to alkaline soil in sun. Hardy, and the most easily grown of the Reticulate group, many of which need bulb frame or alpine house protection. H to 6in (15cm), S to 2in (5cm), PD twice bulb's height.

### PROPAGATION
as soon as ripe, or in fall in a coldframe. Seed is rarely available for cultivars; garden-collected seed will not come true.

in spring.  3–5 years.

late summer. Numerous bulblets are produced at the base of the bulb. Separate and grow on in pots or in a nursery bed.
1–3 years.

## *Leucojum* (Snowflake)  Amaryllidaceae

Some 10 species of mostly hardy bulbs, grown for their snowdrop-like flowers in spring or fall. They have narrow, strap-shaped to linear leaves. Plant in moist but well-drained soil in sun; *L. aestivum* and *L. vernum* are best grown in moist soil that does not easily dry out.

*L. aestivum* (Summer snowflake). Bears pendent white flowers with green tips, in spring. Hardy. H to 24in (60cm), S 4in (10cm), PD 3–4in (7.5–10cm).

*L. autumnale*. Produces bell-shaped white flowers in summer and fall. Hardy. H to 6in (15cm), S 3in (7.5cm), PD 3–4in (7.5–10cm).

*L. vernum* (Spring snowflake). Green-tipped, white flowers in early spring. Hardy. H to 12in (30cm), S 3in (7.5cm), PD 3–4in (7.5–10cm).

### PROPAGATION
in fall, in an open frame.

in spring,  2–4 years.

early summer. Lift established clumps and separate offsets.  1–2 years.

## *Lilium* (Lily)  Liliaceae

Genus of about 100 hardy and half-hardy bulbous species and many hybrids. They produce beautiful trumpet- or funnel-shaped, bowl-shaped, or turkscap flowers, in early, mid-, or late summer, and fall usually on long slender stems that are clad with short, linear leaves, sometimes in distinct whorls. Many are highly fragrant. Most lilies prefer moist but well-drained, neutral to acid soil in sun, with the roots in cool shade. They grow extremely well in containers. Lily bulbs have fleshy, overlapping scales; a few also produce rhizomes and several bear bulbils in the leaf axils or on the stem. Those described below are hardy.

*L. bulbiferum*. Bears umbels of upright, bowl-shaped, orange flowers, which are spotted maroon or black. Early to midsummer. Produces leaf axil bulbils. Tolerant of alkaline soils. H to 5ft (1.5m), S 12in (30cm), PD 2–3 times bulb's height.

*L. candidum* (Madonna lily). Produces sweetly scented, trumpet-shaped white flowers in midsummer. Needs neutral to alkaline soil. H to 6ft (1.8m), S 12in (30cm), PD just beneath soil surface

*L. lancifolium*, syn. *L. tigrinum* (Tiger lily). This has nodding, rich orange-red turkscap flowers, with dark spots and papillae at the throat, in late summer and early fall. It produces bulbils in the leaf axils, and on the rooting stem. H to 5ft (1.5m), S 12in (30cm), PD 2–3 times bulb's height.

*L. regale* (Regal lily). Bears scented, trumpet-shaped white flowers flushed with purple. It blooms in midsummer and prefers full sun. H to 6ft (1.8m), S 12in (30cm), PD 2–3 times bulb's height.

69

*Muscari armeniacum*

*Narcissus* 'February Gold'

*Ornithogalum umbellatum*

**PROPAGATION**

For all vegetative methods of propagation, use only virus-free stock.

❦ as soon as ripe in a coldframe. Hybrids and cultivars do not come true. Lilies show two distinct types of germination. Some, like *L. regale*, produce growth above the medium surface within 2–4 weeks (known as epigeal germination), others like *L. bulbiferum*, first produce a small bulblet below ground before top growth emerges (hypogeal germination), usually after a cold period. Retain pots, undisturbed, for at least 18 months if germination is not immediate. ⚥ variable from 2–4 weeks to 18 months. ☀ 2–4 years.

Leaf axil bulbils – late summer. Detach and sow immediately. ☀ 1–3 years.

Stem bulblets – separate as the leaves and stem wither after flowering. Plant bulblets immediately. ☀ 1–3 years.

Scaling – between late summer and early spring, when dormant. Suitable for all species and hybrids. ☀ 3 years.

✄ established clumps between fall and spring. Some lilies increase readily, producing several new bulbs each year. Lift and divide every third or fourth year to relieve overcrowding. ☀ 1–2 years.

## Muscari (Grape hyacinth) Liliaceae

About 30 species of mostly hardy, mainly spring-flowering bulbs with spikes of blue flowers on leafless stems. The leaves are linear and fleshy, and can grow surprisingly long. They need well-drained soil in sun.

**M. armeniacum.** Bright blue flowers with white-rimmed mouths in spring. Hardy. H to 8in (20cm), S 2in (5cm), PD 4in (10cm).

**M. comosum 'Plumosum'**, syn. 'Monstrosum'. Violet flower spikes, comprising tassel-like sterile flowers. Frost hardy. H to 12in (30cm), S 3–4in (7.5–10cm), PD 4in (10cm).

**PROPAGATION**

❦ as soon as ripe, or in fall in an open frame. Many species self-seed freely.

⚥ in spring. ☀ 3–5 years.

✄ late summer to fall. Most readily produce numerous offsets. Lift and divide established clumps regularly to relieve overcrowding and promote vigor. ☀ 1 year.

Cuttage – bulbs may be scooped or sectioned when dormant. The most effective method of increasing sterile forms rapidly. ☀ 2–3 years.

## Narcissus (Daffodil) Liliaceae

Some 50 species and thousands of cultivars of mostly hardy bulbous perennials bearing characteristic trumpet-shaped flowers in shades of yellow, with ruffs often of a different shade. The myriad cultivars of *Narcissus* are among the most rewarding and easily grown of spring-flowering bulbs and, with careful selection, blooms may be had from late winter with *N.* 'February Gold', almost until early summer, with *N. poeticus* and its cultivars. A few less common species flower in fall. *Narcissus* is divided into 12 horticultural divisions, ranging in form from the trumpet and large-cupped daffodils to the small-cupped and split-corona types, and those – often fragrant – narcissus that bear heads of several flowers. Nearly all are excellent cut flowers. Most thrive in sun or light shade, in any moderately fertile soil that remains moist during the growing season.

Smaller species generally need more sharply drained soils. *Narcissus* range in height from about 4in (10cm) for the smallest species, like **N. bulbocodium**, to about 20in (50cm) for the most robust species and cultivars. Plant all at 1–3 times the bulb's height.

**PROPAGATION**

❦ as soon as ripe, or in fall in an open frame. Many species self-seed and will hybridize. Seed is the most practical method for small species, some of which rarely produce offset bulbs.

⚥ in spring, ☀ 3–5 years.

✄ late summer. Separate offsets from parent bulbs. Lift and divide established clumps and re-plant immediately. ☀ 1–2 years.

Cuttage – large bulbs may be scooped or sectioned when they are dormant. Also by twin-scaling. ☀ 2–3 years.

## Ornithogalum (Star of Bethlehem) Liliaceae

About 80 species of mostly hardy, spring- or summer-flowering bulbous perennials with star-shaped, usually white flowers, often marked or striped green outside. Most are low-growing and are suitable for a position in the front of the border, or in a rock garden. Grow in well-drained soil in full sun but avoid very hot, dry spots; *O. umbellatum* tolerates light shade.

**O. nutans.** Nodding, one-sided racemes of green-striped, silver-white flowers in spring. Hardy. H to 24in (60cm), S 2in (5cm), PD 4in (10cm).

**O. umbellatum.** Bears star-shaped white flowers in broad racemes in early summer. Increases rapidly and can be invasive. Hardy. H 8in (20cm), S 2in (5cm), PD 4in (10cm).

Scilla siberica

Sternbergia lutea

Tulipa 'Ad Rem'

## PROPAGATION

🌱 fall or spring, in a cold frame. Grow seedlings on for 2 years, do not allow to go dormant in the first season.
🌷 in spring, ☀ 3–4 years.
🔨 late summer. ☀ 1–2 years.
Cuttage – scoop or section bulbs of larger species when dormant. ☀ 2–3 years.

## Scilla (Spring squill)    Liliaceae

About 90 species of mainly hardy, spring- or summer-flowering bulbs grown for their star-shaped, usually blue, occasionally white or pink, flowers. The linear to elliptic leaves are sometimes channeled. They need well-drained soil and sun or light dappled shade.
**S. siberica** (Siberian squill). This produces pendent, bright blue, bell-shaped flowers in spring. **'Spring Beauty'** has deep blue flowers. Hardy. H to 8in (20cm), S 2in (5cm), PD 3in (7.5cm).

## PROPAGATION

🌱 fall, or as soon as ripe, in an open frame.
🌷 in spring, ☀ 3–5 years.
🔨 fall. Lift established clumps, separate offsets, and re-plant immediately.
☀ 1–2 years.
Cuttage – bulbs may be scooped or sectioned when dormant. ☀ 2–3 years.

## Sternbergia    Amaryllidaceae

Genus of eight species of mainly fall-flowering bulbous perennials, most of which are cold-hardy but dislike winter wet. They are grown for their goblet-shaped, crocus-like flowers that usually emerge before the leaves. The leaves are linear to strap-shaped. They need light well-drained soil in a warm,

sunny sheltered site, preferably at the base of a wall or similar. Mediterranean plants, they flower best if provided with a warm, dry summer dormancy.
**S. lutea.** Shining yellow goblet-shaped flowers in fall. They appear at the same time as the leaves, which elongate after flowering. Frost-hardy. H 6in (15cm), S 3in (7.5cm), PD 4–6in (10–15cm).

## PROPAGATION

🌱 fall, in a cold frame.
🌷 in spring. ☀ 3–5 years.
🔨 late summer. Lift established clumps, divide, and re-plant immediately.
☀ 1–2 years.
Cuttage – mature bulbs may be scooped or quartered when dormant. ☀ 2–3 years.

## Tigridia    Iridaceae

Genus of 23 species of frost-tender cormous perennials with bell-shaped, or more usually iris-like, flowers, mainly in shades of white, yellow, pink, orange, and red, and often with marks or mottling in contrasting colors. The leaves are narrowly lance-shaped to sword-like, mostly in a basal cluster. They need well-drained, preferably sandy soil, in a sunny position. In cold areas, lift before the first frosts.
**T. pavonia** (Peacock flower, Tiger flower). Bears orange, yellow, pink, or white flowers, with contrasting, darker central marks, in mid- to late summer. H 20–24in (50–60cm), S 4in (10cm), PD 4in (10cm).

## PROPAGATION

🌱 fall, or spring, at 55–64°F (13–18°C). Careful watering and feeding, without over-heating, will ensure large bulbs that may be flowered in their second year.

🌷 in spring, ☀ 2 years.
🔨 late summer to early fall. Separate offsets and re-plant immediately, or store in dry frost-free conditions and re-plant in spring. ☀ 1–2 years.

## Tulipa (Tulip)    Liliaceae

Some 100 species and thousands of cultivars of mostly hardy bulbous perennials with linear to broad leaves that are green to blue-green and may have darker stripes. The many cultivars of *Tulipa* are grown in bedding, in borders and containers, and for cut flowers. They bloom between early spring and early summer. The genus *Tulipa* is divided into 15 horticultural divisions, with flowers ranging in shape from cup-shaped, bowl-shaped or goblet-shaped, to the classic ovoid flowers of the **Darwin hybrids,** and the frilled and dissected blooms of the **Parrot tulips**. Flowers are available in a huge range of colors from almost black to white. Most thrive in sun in any well-drained, fertile soil. Aside from the true species and most of the **Kaufmanniana hybrids,** tulip bulbs are best lifted as the leaves wither after flowering, and stored dry until re-planting in fall. Plant all at a depth of 4–6in (10–15cm).

## PROPAGATION

🌱 as soon as ripe, or in fall in an open frame. Suitable for species only; the cultivars and hybrids seldom set seed, and do not come true.
🌷 in spring. ☀ 3–7 years.
🔨 summer. Separate offsets from the parent bulbs after lifting. Grow on in a nursery bed. Some species tulips produce bulblets on the ends of long stolons. Lift and separate them from the stolons in fall. ☀ 1–2 years.

71

# ALPINES AND ROCK GARDEN PLANTS

A precise definition is elusive, but in the popular imagination alpines are dwarf plants, usually free-flowering and tolerant of low winter temperatures. Gardeners often convert to growing them after witnessing a riotously colorful Swiss mountain meadow in early summer, or tiny plants decorating apparently barren rock slides.

Many alpines are easy to grow; others present a challenge. Regular propagation ensures that you have ample numbers of easy alpines to give to friends, while maintaining your stock of those that are more challenging to grow.

◄ *The rock garden provides niches for both true alpines and rock plants, and is kept well stocked by a range of propagating techniques.*

# WHY PROPAGATE
# ALPINES AND ROCK GARDEN PLANTS?

*The range of techniques for increasing alpines is as wide as the plants themselves are diverse. Depending on the available facilities, many alpines can be propagated throughout the year. A mist unit and a heated propagating bench do much to extend the propagating season and certainly speed root development, but are by no means essential.*

**Alpine species may not set seed in cultivation. For these and named cultivars, vegetative methods must be used.**

Many alpines have growth habits adapted to their specialized habitats, and these give clues to the propagator. For example, some Himalayan androsaces have runners that produce new plantlets like a strawberry plant (peg down the tips and detach the rooted rosettes); some, like soldanellas, occur in alpine turf, forming mats of surface roots that are easily divided. Some mat- and cushion-formers produce new shoots with vestigial roots that develop further only when moisture and a suitable substrate are present, but otherwise shrivel away.

If a plant is happy with its garden siting, self-sown colonies may appear without help from the gardener. Such natural effects can be captivating, but one species often prospers at the expense of its more delicate neighbors – a good reason to rescue less vigorous plants by propagating fresh stocks to re-plant elsewhere. Not all plants set viable seed in cultivation, however, and for these, and for named hybrids and cultivars,

vegetative methods must be used. A few alpines can only be propagated at specific points in their growth cycle (*see the plant directory, pp. 80–85*).

## MULTIPLYING FROM SEED

Raising alpines from seed is by far the most important means of propagating these plants. It is also usually the only practical method of increasing rare or newly introduced species.

Seedlings vary in their vigor or adaptability to certain conditions. Furthermore, seed-raised plants are free of virus, which is a particularly important consideration with susceptible genera such as *Primula*. Some species will hybridize freely (saxifrages, for instance), so there is always the potential to produce exciting new cultivars.

## DIVISION OF CLUMPS

Division is an appropriate technique for almost any plant that forms multiple rootstocks or a clump of basal shoots, or that spreads by underground stolons. Success with this form of propagation is almost sure if divisions are cared for until they are re-established.

## TAKING CUTTINGS

Shrubby alpines can be increased by a range of different types of cutting. Perennial alpines that are unsuitable for division may be propagated by stem-tip cuttings. Most rosette-forming species can be increased by detaching individual rosettes from around the edge of the plant.

### *Alpines in Tufa*

*Crevice-dwellers often grow more compactly and flower more freely in tufa – a porous limestone deposit – than in soil. Recently propagated material is essential – established root systems are too bulky to introduce successfully. Hose down the tufa to remove the outer shell of sludge-like material. Drill a planting hole, 4in (10cm) deep. Shake the roots free of soil and insert them using a funnel of stiff card. Add tufa debris around the roots. Water in well and cap with tufa fragments around the plant collar.*

This compact, crevice-dwelling saxifrage thrives and flowers freely when planted in tufa.

For those with thicker roots, propagation by root cuttings is possible. A few alpines can be increased by leaf cuttings. Relatively few alpines are increased by leaf cuttings but, using healthy, young mature material, the technique is successful with the somewhat fleshy leaves of some sedums and with members of the family Gesneriaceae, such as *Haberlea* and *Ramonda*.

**(1)**

**Hypericum olympicum**
A neat shrublet that may be increased by seed, although species hybridize freely. Stem-tip cuttings result in offspring identical to the parent plant, which is useful if you have a good color form.

**(2)**

**Iberis sempervirens**
This evergreen subshrub can be increased by softwood or semi-ripe cuttings, but since they develop an extensive root system, they may need to be repotted several times before they can be planted out.

**(3)**

**Helianthemum cultivar**
Helianthemums are free-flowering subshrubs but are not always reliably hardy. Insure your stock by propagating regularly from softwood or semi-ripe cuttings, and overwinter young plants in frost-free conditions under glass.

**(4)**

**Aubrieta deltoides cultivar**
Aubrietas are very free-flowering and often self-seed, but self-sown seedlings usually produce inferior flowers to selected cultivars. Propagate named forms by softwood or semi-ripe cuttings, or by division.

**(5)**

**Phlox douglasii cultivar**
The evergreen mats of foliage eventually become less floriferous at the centers; keep them compact by trimming after flowering and take cuttings from the resulting new shoots to replenish your stock.

**(6)**

**Aurinia saxatile 'Citrinum'**
The species can be raised from seed, but if you find its chrome-yellow too bright, grow this paler-flowered cultivar and keep it healthy and floriferous by propagating regularly from cuttings.

# ALPINES AND ROCK GARDEN PLANTS

*The techniques for propagating alpines and rock garden plants are essentially the same as those for other plant groups, except that, in most cases, the material used is much smaller in scale. The other most important modification to ensure success is to use sharply drained propagating mediums to suit the special needs of alpine plants.*

## ALPINES FROM SEED

Generally speaking, sow seed as soon as it is ripe or on receipt from a seed distribution; Ranunculaceae especially (*Anemone, Pulsatilla, Ranunculus*) come up quickly if sown fresh but erratically if dried. Fresh seed is critical in some cases (notably *Celmisia*) – conversely, with plants from low rainfall areas, it matters much less, though seeds can germinate irregularly over several years. If possible, sow stored seed before late winter or germination may be delayed a further year. With rare species, save seedpots for three years before discarding, if space allows, since germination can be erratic.

One of three pre-treatments may be necessary. In all cases, sow immediately after treatment.

### Stratification

The the seed is exposed under moist conditions to temperatures around freezing to break its dormancy. Either mix the seed with moistened sand in a container or sow it as normal (*see right*): in both cases, place in the bottom of the refrigerator for 1–3 months. In areas with cold winters, sowing the seed in late fall and leaving the pots outdoors in an open frame is a simpler option.

### Soaking

This is used for older seeds that are large enough to be handled individually. As well as hydrating dried seed, chemical inhibitors in the seedcoat may be leached out. Quarter-fill a tumbler with tepid water, add the seeds and a drop of household detergent (to help the seed to sink). Change the water after 12 hours, leave for a further 12 hours, and then sow as normal. *Sorbus* seeds should be soaked in vinegar; *Lupinus* need to be immersed in hot – but not boiling – water, and then allowed to cool.

## Alpines from seed . . . . . . . . . . . . . . . . . . . . . . . . . . . .

For most purposes, sowing seed in a 3½in (9cm) pot will provide you with sufficient seedlings for planting out in the rock garden and for giving away to gardening friends.

◄ **1** Fill pots to within ⅓in (1cm) of the rim, tap firmly to settle the medium, and firm gently with a flat board.

► **2** Sow the seed evenly over the surface. Cover the seed with a ⅛–¼in (3–5mm) layer of washed coarse sand. Water overhead or place the pots in a shallow container filled with 2in (5cm) of water for 30 minutes, allowing the medium to become moist by capillary action. Fine seed is best sown onto the surface of the coarse sand – watering or rainfall will wash it down to the correct level.

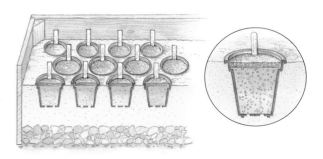

▲ **3** Place the pots outdoors in an open frame; genera sensitive to overhead watering (*Dionysia* especially) need the shelter of an alpine house or a covered frame, while tiny seeds of those alpines that suffer if allowed to dry out (such as those of the Ericaceae) should be kept humid in a closed propagator.

### Scarification

Some seeds have hard seedcoats that need to be broken down before germination can take place. Large seeds can be rubbed with sandpaper; smaller ones can be mixed with sand and shaken in a screw-top jar.

Most seed is best sown in 3½in (9cm) pots. Deep-rooted plants, however, especially those that form underground tubers, benefit from being left in the same pot until they go dormant after a second season of growth; the seed of these should be sown in deep pots (these are available from specialist suppliers). Prepare a seed germination medium of two parts pasteurized loam, one part ¹⁄₁₆–⅛in (2–4mm) lime-free sand, and one part horticultural grade vermiculite or perlite, all parts by volume. (Substitute ericaceous medium for the loam for lime-hating plants.)

Most alpines germinate from late winter as temperatures rise and days lengthen. Good light is needed once the seedlings emerge. Keep the medium moist and protect the seedlings from strong sun, particularly under glass. As a general rule, thin seedlings into 2½in (6cm) pots once the first true leaves are apparent, using the same potting mixture (*for exceptions, see plant directory, pp. 80–85*). Once the seedlings are well established, apply a dilute liquid fertilizer.

## DIVISION

Division is one of the simplest and most reliable techniques, and is used for clump-forming perennials. Most can be divided in either early spring or early fall. Semideciduous, summer-flowering genera such as *Campanula* are best divided in early spring, while *Primula* and others that experience fresh growth in autumn are best divided in early fall. Alpines with a summer dormancy habit are usually best divided before new root growth is underway, in late summer.

Lift large fibrous-rooted plants grown in the open garden, insert two handforks back-to-back and prise the plant apart. On brittle plants with very matted roots, cut the crown into sections, wash off old soil and dip the wet roots in sand before replanting. Small plants may be pulled apart, but well-rooted sections are best cut cleanly with a knife. Most divisions can be planted out immediately at the same depth as previously. Pot up smaller divisions and grow on in a cold frame; re-plant only when substantial new root growth is apparent.

For plants in containers, always re-plant in fresh medium, protecting them from hot or windy weather using shade or a closed coldframe. The divisions may appear etiolated once separated from the surrounding clump. To counteract this, plant them deeper than their previous level. For *Hepatica* and some suckering species of *Cassiope* that resent disturbance, a two-stage process is sometimes used: after flowering, make a deep cut into the crown of the plant with a spade, insert a thin divider into the cut, then leave until the following spring before lifting the division. This stimulates the production of new roots before the new plant is lifted and separated, and therefore enables the division to become established more quickly.

### Irishman's or pre-rooted cuttings

This technique is for alpines with a creeping habit but which form a woody rootstock that does not lend itself to conventional division. It can also be used for plants spread by runners or that produce offsets. The best results from pre-rooted cuttings are achieved when root growth is active – between early spring and mid-fall.

**Irishman's or pre-rooted cuttings** . . . . . . . . . . . . . .
Irishman's cuttings are, in effect, an easy short cut to producing small pre-rooted plants that require the minimum of aftercare to guarantee success.

▲ **1** Lift a section from the edge of the plant, cutting it free either immediately below the developing root mass or along any runner that connects it to the core of the plant. Trim away any dead, damaged, or excess foliage to leave a small crown of shoots (*see right*).

◄ **2** Dip in a fungicide solution, then pot up in a growing medium, with the lowest leaves just at the medium surface. Mulch with sand.

# CUTTINGS

Cuttings are usually taken from mid-spring through to mid-fall, but can be rooted earlier using a soil-warming cable or heated propagator. Softwood cuttings are taken in spring, semi-ripe cuttings in summer, and ripewood or hardwood cuttings from late summer to mid-fall. With rare species it is worth taking cuttings at any time if the parent plant shows signs of distress.

When taking cuttings of alpines, use sharp pruners or a razor blade to avoid crushing plant tissues. In most cases, coarse, lime-free horticultural sand, sterilized by pouring over boiling water through muslin, is an adequate rooting medium, packed into 3½in (9cm) pots. Allow the sand to cool completely before inserting the cuttings. Add a little slow-release fertilizer to the base of the container, so that the cutting may be kept in the container for a month or so after rooting. Some alpines show superior rooting rates in a less free-draining medium of equal parts of perlite and fine composted bark.

## Softwood cuttings

Softwood and stem-tip cuttings are taken from new growth, produced in the first half of spring, after flowering. They are liable to wilt if not used immediately. The optimum time for taking them is early morning or on cool, overcast days when plant tissue is most turgid. To produce cuttings later in the season, cut back a portion of the plant to promote new shoots for use any time up to late summer.

## Semi-ripe and hardwood cuttings

Semi-ripe and hardwood cuttings are taken from midsummer to late fall. Semi-ripe shoots are firm to the touch at their bases but still in active growth. Preferably, use short side shoots, 2–4in (5–10cm) long. Pull them away from the main shoot with a heel, then trim the heel neatly. Hardwood cuttings are used when shoot growth for the year is completed, often indicated by the formation of an apical bud. Take a cutting 2–4in (5–10cm) long, with a heel. Alternatively, cut just above a node, trim the cutting below its lowest node and cut off the lower leaves. Wound the base of the cutting by removing a sliver of bark about ¼in (5mm) long.

Insert the cuttings in the rooting medium, then place them in a closed propagator or cover them with plastic bags supported by stakes. Species intolerant of high humidity are best if left uncovered, but under cold greenhouse protection.

Dipping the cutting base in a rooting compound can help rooting. Most compounds also contain fungicide, but you must still check weekly to remove any cuttings that show signs of rotting or fungal attack. Softwood cuttings will root in 2–6 weeks; semi-ripe cuttings may take two months. Hardwood cuttings will probably not form roots until the spring after they were taken.

Lift out the rooted cuttings with care, pot up using the same growing medium as the parent plant, and, in warm weather particularly, grow them under glass for a

## Softwood cuttings ..............................

Softwood cuttings root more readily than other types of cutting, but need more aftercare. Do not allow them to wilt at any stage, and root them in gentle heat and good light.

► 1 Take cuttings about 1–3in (2.5–7.5cm) long from the tip of non-flowering shoots with a straight cut just below a node.

◄ 2 Trim off the lower leaves to about a half or one-third of the overall length of the cutting.

► 3 Insert to the depth of the lowest leaves into the rooting medium. Firm in gently. Ensure that the leaves do not touch each other.

◄ 4 Seal in a plastic bag, keeping the plastic clear of the leaves with stakes or wire hoops.

week before hardening them off in a semi-shaded area. Rather than disturb recently rooted material just before winter, leave all but the earliest to root in their pots, and overwinter with minimal heat until spring.

### Using individual rosettes as cuttings

Numerous alpines form neat domes or cushions comprising many small rosettes. These can be propagated by snipping off tiny shoots – as little as ⅛in (3mm) in diameter – from the periphery of the clump about a month after flowering (in earliest summer for many alpines). Only the new growth is required, and all persistent dead leaves along the length of the stem must be meticulously trimmed away as far as possible, leaving a small shoot core (the rosette) at the end of a stem ¼–½in (5–10mm) long. Insert in sand up to the base of the rosette. Rooting normally follows after three weeks. When potting on, make a small hole in the growing medium, drop the young plantlet into it, and backfill with sand. When watering, avoid wetting the foliage.

## LEAF CUTTINGS

This method is principally applicable to *Ramonda myconi* and its relatives (e.g. *Haberlea, Jankaea,* and *Lewisia*). Remove mature leaves from the parent plant in early spring or (preferably) late summer: it is not necessary to tear away the base of the leaf stalk (petiole) as is sometimes recommended. Insert the cuttings upright in moist sand in a propagator or a pot enclosed in a plastic bag and keep them shaded. Rooting should occur in 4–6 weeks, even without bottom heat. Pot up plants 2–6 weeks after a small rosette has appeared at the base of the parent leaf.

## ROOT CUTTINGS

Alpines that have long, thickened roots can be increased by root cuttings. Lift the plant soon after the onset of dormancy – generally in midsummer (*Dodecatheon, Pulsatilla*) or fall (*Phlox, Verbascum*). Either shake the roots free of soil or potting medium or wash them clean. Select younger, usually paler, thicker roots and cut them into sections 2in (5cm) long – those nearest the crown will have the greatest potential to form new plants. For the continued health of the parent plant, be sure to remove no more than one-third of the roots for propagation.

## Material for the different types of cuttings .....

The figures below indicate the type of material that is used for various types of cutting. The dotted lines represent the depth of insertion for each type.

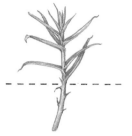

▲ **1** Take stem-tip cuttings, 1–3in (2.5–7.5cm) long, from herbaceous plants in spring.

▲ **2** Take rosette cuttings, up to ½in (10mm) long, in early to midsummer.

▲ **3** Take leaf cuttings from entire, healthy, young leaves in spring or late summer.

▲ **4** Take softwood cuttings, 1–3in (2.5–7.5cm) long, in spring.

▲ **5** Take semi-ripe cuttings, 2–4in (5–10cm) long, in midsummer.

▲ **6** Take ripewood (hardwood) cuttings, 2–4in (5–10cm) long, in late summer or fall.

Trim the cuttings with a straight cut at the top (nearest the crown) and an angled cut at the bases, to ensure correct insertion, then trim off side roots, treat the cut surfaces with fungicide and insert vertically in sandy medium, covering with a ⅜in (1cm) layer of coarse sand. Place in a closed coldframe or under the greenhouse staging until topgrowth is well underway, usually at least a month later. Topgrowth often appears before a new root system is well established, so check that the cuttings are well rooted before potting up.

# DIRECTORY OF
# ALPINES & ROCK GARDEN PLANTS

*Below are key tips on a selection of alpine and rock garden plants that are particularly suitable for propagation. Unless otherwise specified, follow the detailed instructions given under the propagation section on pages 74–79.*

## Key

H    Height
S    Spread
🌰    By seed
🔧    By division
🌱    Germination
☀    Time to maturity

## *Aethionema* (Stone cress)    Cruciferae

About 40 species of hardy, evergreen and semi-evergreen subshrubs, and woody-based perennials with prolific flowers in spring and early summer. As the roots dislike being disturbed, division of plants is rarely a successful method of increase.

*A. grandiflorum*, syn. *A. pulchellum*. Evergreen subshrub with fragrant pink flowers. H to 12in (30cm), S to 12in (30cm).

### PROPAGATION

🌰 as soon as ripe in midsummer or when available. Self-seeds freely.

🌱 1–6 months.   ☀ the following spring.
Softwood cuttings – summer. Clip over after flowering and use the new shoots that arise as a result. When side shoots appear, cut back the main stem by one-third; this will also promote bushiness.

## *Androsace* (Rock jasmine)    Primulaceae

Over 100 species, most of which are mat- or cushion-forming, hardy, evergreen perennials, valued for their tiny flowers. Some thrive in the open garden; others, like the high-alpine Aretian cushion-formers, either demand scree conditions or rock crevices with protection from excess winter rainfall, or are grown in an alpine house. Seed generally needs stratification and is the principal method of propagation.

*A. sempervivoides*. Stoloniferous mat-former with leathery, glossy leaves that bears fragrant pink flowers in spring. H to 2in (5cm), S to 8in (20cm).

*A. vandellii*. Aretian cushion-former with silver-gray, hairy leaves. It bears white flowers in spring. Some plants flower so freely that the foliage is obscured, and the size of the flowers varies; select accordingly. H to 2in (5cm), S to 5in (12.5cm).

### PROPAGATION

🌰 as soon as ripe or before late winter. Prick out seedlings as soon as the first true leaves appear.

🌱 3–6 months.   ☀ 2–3 years.
Rosette cuttings – early to midsummer. Root single rosettes and avoid wetting the foliage by watering from below.   ☀ 1 year.
Stolons – after flowering. Stoloniferous species usually produce 4–5 stolons which root on contact with the mulch. They take about 1–2 months to root well.
☀ 1 year.

## *Armeria* (Thrift)    Plumbaginaceae

Some 80 species of cushion- or mound-forming, evergreen perennials with hardy rosettes of linear leaves, grown for their slender-stemmed, spherical flowerheads in spring or summer. They need well-drained, not too fertile soil, in sun. They hybridize freely, so vegetative propagation is the preferred method of increase to ensure that the new plants are true to type.

*A. juniperifolia*, syn. *A. caespitosa*. Neat hummock-former with gray-green leaves and white, pink, or pink-purple flowers in spring. H to 3in (7.5cm), S 6in (15cm).

*A. maritima* (Sea thrift). Forms clumps of dark green leaves, and bears white, pink, or dark red-purple flowers, in spring and summer. H to 8in (20cm), S to 12in (30cm).

### PROPAGATION

Cuttings – summer. Root single rosettes or branchlets bearing 3–5 rosettes, taken with a heel of older wood at the base.
☀ 9 months.
🔧 early spring.   ☀ 6 months.

## *Aubrieta*    Cruciferae

Some 12 species of evergreen, mat- or mound-forming, hardy perennials, grown for their 4-petaled, cross-shaped flowers, which are borne in profusion in spring. They are easily grown in any well-drained soil in a sunny position. Usually it is the numerous, free-flowering cultivars that are found in cultivation. Recommended cultivars include: **'Dr Mules'** with violet flowers; **'Joy'** with

*Androsace vandellii*

*Aubrieta deltoides* 'Dr Mules'

*Aurinia saxatilis*

*Campanula carpatica* 'Blue Clips'

*Cyclamen hederifolium*

mauve double flowers; and **'Red Carpet'** with deep red flowers. Cut back after the main spring flowering – they will often flower again in summer. H 2in (5cm), S to 24in (60cm).

**PROPAGATION**
Cuttings – softwood or semi-ripe, in summer or early fall. Avoid any woody material and keep part-shaded and uncovered.
☀ the following spring.
✂ early fall after re-growth following cutting back. Split with a spade and re-plant immediately, watering well until established.
☀ 6–8 months.

## Aurinia
Cruciferae

Some seven species of hardy biennials and woody-based, evergreen perennials of which *A. saxatilis* and its cultivars are the most commonly cultivated. They are valued for their heads of small yellow flowers, borne above the foliage in mid-spring. They grow in any well-drained soil in sun.
*A. saxatilis*, syn. *Alyssum saxatile*. Produces masses of bright yellow flowers in spring and has mounds of hairy, gray-green leaves. **'Citrinum'** produces soft lemon-yellow flowers. **'Dudley Nevill'** has attractive buff-yellow blooms. **'Plenum'** has double flowers. All are best sheared back after flowering to maintain a compact habit. H 8in (20cm), S 12in (30cm) or more.

**PROPAGATION**
☙ late summer. Species only.
Ⴤ 10–14 days. ☀ 8–12 months.
Cuttings – summer. Softwood cuttings of new sideshoots formed after cutting back. Line out, uncovered, in a cuttings frame.
☀ 10 months.

## Campanula
(Bellflower) Campanulaceae

About 300 species of annuals, biennials, and perennials with bell-shaped flowers, including many dwarf species and cultivars that are suitable for a rock garden or alpine house. *(See also Herbaceous Perennials directory, p. 36.)* Some are short-lived or monocarpic and a few self-seed or sucker too freely for all but the largest rock garden. The most robust rock garden species thrive in sun or dappled shade, in moist but well-drained soil.
*C. carpatica*. Hardy, mat-former with bell-shaped, white, blue, or violet flowers during summer. **'Blue Clips'** and **'White Clips'**, with blue and white flowers respectively, are smaller and more compact than the species. H to 12in (30cm), S 12in (30cm) or more.
*C.* **'Joe Elliott'**. Dwarf in habit, this hardy plant bears large mid-blue, bell-shaped blooms in abundance during summer; the very generous display means that cuttings are hard to find during its long flowering period, so it is best to divide specimens frequently or take cuttings in mid-spring. H 3in (7.5cm), S to 8in (20cm).

**PROPAGATION**
✂ early spring, before flowers form.
☀ 3–6 months.
Softwood cuttings – early to mid-spring, before flowers form. ☀ 6–8 months.
☙ as soon as ripe or in fall.
*C.* 'White Clips' comes almost true from seed.
Ⴤ spring. ☀ 6–8 months.

## Cyclamen
Primulaceae

About 19 species of tuberous perennials, a few of which are well suited to growing in a rock garden. Selected forms are sometimes increased by sectioning a tuber with multiple growing points, although this technique is specialized and rather risky. Those described need well-drained, not too fertile, humus-rich soil, in dappled shade. Establish colonies beneath shrubs or trees to provide the dry conditions needed during summer dormancy.
*C. coum*. Hardy. Bears small, white, pink to carmine flowers with rounded, dark green, often silver-marked leaves, in late winter or early spring. It exhibits numerous leaf forms, some breeding largely true, e.g. the **Pewter Group**, which have heavily silvered leaves. H to 3in (7.5cm), S 4in (10cm).
*C. hederifolium*, syn. *C. neapolitanum*. Fall-flowering hardy species, with flowers in white and many shades of pink, and leaves that are dark green, often heavily silver-patterned, and flushed purple beneath. May self-seed. H 5in (12.5cm), S 6in (15cm).

**PROPAGATION**
☙ as soon as ripe in late summer, in a soilless medium. Fresh seed is distributed by some specialist suppliers within weeks of harvesting. Stored seed should be soaked and well rinsed. Seedlings emerge when the species concerned comes into leaf and produce only one leaf in the first year. Keep part-shaded and moist after sowing. Shelter seedlings from strong wind and insulate the frame in severe winter weather.
☀ 18–24 months.

## Daphne
Thymelaeaceae

Over 50 species (and numerous hybrids and cultivars) of slow-growing, deciduous and evergreen shrubs, valued for their often very fragrant flowers. Some alpine species seldom set seed (which are poisonous) in lowland

*Daphne petraea* 'Grandiflora'

*Dianthus deltoides*

*Dryas octopetala*

gardens, but can be propagated – depending on species – by layering in spring or early summer (see p. 108), by semi-ripe cuttings in mid- or late summer, (some slow to root and grow on), or by apical-wedge grafting in winter or summer (see p. 111). The choice of rootstock is critical if grafted plants are not to become over-vigorous. Grow in moist but freely drained, approximately neutral, humus-rich soil, in sun.

*D. blagayana*. Prostrate, hardy evergreen with trailing branches bearing fragrant creamy-white flowers in spring. It is easily layered. H to 15in (38cm), S to 36in (90cm).

*D. petraea* 'Grandiflora'. Compact, slow-growing, hardy evergreen with scented, rosy-pink flowers in spring. It is usually grafted (*D. tangutica* and *D. longilobata* are preferred rootstocks) in late winter or high summer. If, however, this plant is to be grown in tufa, the bulkiness of the rootstock makes grafting unsuitable so semi-ripe cuttings are necessary. They are slow to grow,, often not rooting properly for 6 months.
H 4in (10cm) or more, S 10in (25cm).

**PROPAGATION**
Layering – prostrate species and cultivars. Layer new growth, about 4–8 weeks after flowering. Leave for a year before severing from the parent plant. ☀ 1 year.
Semi-ripe cuttings – midsummer.
☀ 12 months
Grafting – *D. petraea* 'Grandiflora'. Choose a seedling rootstock with a stem diameter of ¼in (5mm) and cut off below the cotyledons. Insert a (¾in) 2cm scion, with a wedge-shaped base ½in (1cm) long. Spray with fungicide, bandage firmly with grafting tape, and place in a closed propagator at

64°F (18°C). Midsummer grafts will not need artificial heat. Union takes around a month, after which the plants are hardened off.
☀ 9–15 months.

## *Dianthus* (Carnation, Pink) Caryophyllaceae

Over 300 species, including a number that are suitable in scale for a rock garden, raised bed or trough. *(See also Herbaceous Perennials directory, p.37 and Annuals and Biennials directory, p.18.)* Most rock garden species, some of which are short-lived, are gray-leaved, evergreen perennials with a neat mat-, cushion-, or mound-forming habit, bearing diminutive flowers in spring and early summer. Many named cultivars are available and must be increased by cuttings. Most need sandy, sharply drained, neutral to alkaline soil in sun.

*D. alpinus*. Short-lived cushion-former, bearing pink to crimson flowers in summer. H 3in (7.5cm), S 4in (10cm).

*D. deltoides* (Maiden pink). Short-lived, mat-former with often dark-eyed, white, pink, or red flowers with toothed petals, in summer. Fresh or stored seed germinates abundantly. H to 8in (20cm), S 12in (30cm).

*D. pavonius*, syn. *D. neglectus*. Neat, free-flowering mat-former with pale to deep-pink flowers in early summer. It needs acid soil. H 3in (7.5cm), S 8in (20cm).

**PROPAGATION**
❦ fall to early spring, in an open frame. Expect the seedlings to vary and select accordingly. Compact, large-flowered seedlings can then be increased by cuttings.
❧ spring. ☀ 12 months.
Cuttings or pipings – after flowering.
☀ 8–10 months.

## *Dryas* (Mountain avens)     Rosaceae

Just three species of hardy, prostrate, evergreen subshrubs with leathery, wrinkled, dark green leaves and cup- to bell-shaped flowers. *D. octopetala* is the most commonly cultivated. All are easily grown in sandy, humus-rich soil in sun.

*D. octopetala*. Mat-forming subshrub with oak-like foliage, and up-turned, cup-shaped creamy-white flowers in early summer, followed by feathery seed heads. H 4in (10cm), S 36in (90cm) or more.

**PROPAGATION**
❦ as soon as ripe. Germinates poorly and uncertainly unless sown fresh.
❧ Spring. ☀ 2 years.
Ripewood cuttings – late summer, 2–3in (5–7.5cm) long. ☀ 10 months.
Irishman's cuttings – at any time, preferably in the growing season. ☀ 6–8 months.
Layering – spring. Strip away the dead winter foliage before layering. Detach rooted layers in fall.
☀ 6 months, flowering the following year.

## *Edraianthus* (Grassy bells) Campanulaceae

Over 20 species of tuft-forming, herbaceous and evergreen perennials, generally short-lived. They have grass-like leaves and bell-shaped flowers in summer. They need sun in sharply drained, sandy, alkaline soil.

*E. pumilio*, syn. *Wahlenbergia pumilio*. Cushion-forming perennial, with delicate, bell-shaped flowers in early summer that can be almost any shade of blue when raised from seed. H 1in (2.5cm), S 4in (10cm).

**PROPAGATION**
❦ as soon as ripe or in late winter. Dislikes root disturbance apart from at the seedling

*Gentiana verna*

*Iberis sempervirens* 'Weisser Zwerg'

*Lewisia cotyledon*

stage so place plants in their permanent position as early as possible.

♈ unlikely before mid-spring.

☀ 15 months minimum.

Softwood cuttings – from basal shoots either before the flowers form (early spring) or more readily after flowering. (Disbud spring cuttings if flowers form in same year.)

☀ 12 months.

## Gentiana (Gentian)  Gentianaceae

About 400 species of hardy annuals, biennials, and perennials, the last group including many alpines, with trumpet- or five-pointed flowers in intense shades of blue (*see also Herbaceous Perennials directory, p.39*). They need humus-rich, moist but well-drained soils; the fall-flowering species almost invariably prefer acid soils. Except in climates with cool, damp summers, most need protection from hot sun.

*G. acaulis*, syn. *G. excisa*, *G. kochiana* (Trumpet gentian). Bears blue flowers late spring to early summer. Fowers more reliably in fertile, loamy soil. H 3in (7.5cm), 10in (25cm).

*G. sino-ornata*. Bears blue flowers, striped blue and white outside, in fall. The foliage dies down in winter to clusters of shoots with plump roots (thongs), which are easily divided. H 3in (7.5cm), S to 12in (30cm).

*G. verna* (Spring gentian). Short-lived, forming mats of evergreen rosettes that are clothed in spring with tubular, blue flowers. **subsp. balcanica** is usually longer-lived and more free-flowering, blooming from its second year onwards. H 2in (5cm), 4in (10cm).

### PROPAGATION

☘ as soon as ripe or no later than midwinter or germination will be delayed a year. Sow

*in situ* or prick out the seedlings early to avoid root damage.

♈ spring.  ☀ 12–24 months.

🔧 after flowering or in early fall. Divide *G. sino-ornata* just before growth restarts in early spring.  ☀ 12 months.

Cuttings – late spring. Take basal cuttings of non-flowering shoots ¾–1¼in (2–3cm) long, severed close to the crown and detached cleanly just below a node. Rooting occurs fastest at 50–59°F (10–15°C).

☀ 12 months.

## Helianthemum (Sun rose)  Cistaceae

Some 110 species of small, evergreen and semi-evergreen subshrubs and shrubs, grown for their often vividly colored saucer-shaped flowers and silver or gray-green foliage. A few are not reliably hardy in cold, very wet winters, and all prefer light, well-drained soil in a warm sunny site. Several named cultivars, about 8–12in (20–30cm) tall and across, are available with single or double flowers, in white, yellow, orange ('Ben-Nevis') pink ('Rhodanthe Carneum'), and carmine ('Ben Hope'). Trim the plants back after blooming.

*H. oelandicum* subsp. *alpestre*. Hardy, neat mat-forming species that has downy, gray-green leaves and produces yellow flowers from spring to summer. H 4in (10cm), S 8in (20cm).

### PROPAGATION

Softwood or semi-ripe cuttings – early, or mid- to late summer respectively. Encourage bushiness and cutting material by pinching out stem tips in spring. Overwinter cuttings in frost-free conditions in their first year.

☀ 10–12 months.

## Iberis (Perennial candytuft)  Cruciferae

About 40 species of annuals, perennials, and evergreen subshrubs, several of which are ideal for rock gardens. They are grown for their heads of cross-shaped flowers in late spring or early summer. Cultivate in not-too-fertile, well-drained soil in sun. Cut back once the flowers begin to fall.

*I. sempervirens*. Evergreen, mat- or mound-forming species with small, dark green leaves. It bears heads of small white flowers in spring. **'Weisser Zwerg'**, syn. 'Little Gem', is much more compact and well suited to the smaller rock garden. H 12in (30cm), S 12in (30cm).

### PROPAGATION

Softwood or semi-ripe cuttings – early and midsummer respectively. The root system is very extensive, so pot on frequently or plant out before the second re-potting.

☀ the following spring.

## Lewisia  Portulacaceae

Some 20 species of evergreen and deciduous, rosette-forming perennials, grown for their cup-shaped, often vividly colored flowers, in spring or summer. They need sharply drained soil, and protection from rain and snow in winter – plant them on their sides in crevices in a rock garden or on a wall, or grow in an alpine house.

*L. cotyledon*. Evergreen perennial with flowers of white, yellow, orange, and pink-purple. **Cotyledon hybrids** are robust with flowers in very vivid colors. They are variable from seed, but produce many attractive shades. Good color forms can be increased by offsets. H 6–12in (15–30cm), S 10–16in (25–40cm).

*Lithodora diffusa* 'Heavenly Blue'

*Phlox subulata* 'Tamaongalei'

*Pulsatilla vulgaris*

### PROPAGATION

🌱 as soon as ripe, or in fall/early winter. Use a peat-based medium mixed with 50% coarse sand and some perlite. Keep seedlings frost-free. As they are brittle, thin out carefully either singly or into seed trays – the seed tray seedlings will need moving up into 3in (7.5cm) pots by late spring.
☀ 9 months.

Offsets – midsummer. Detach with ½in (1cm) of stem, treat with rooting hormone, and keep under shaded glass in a peat-based potting medium.
☀ 9 months.

*L. rediviva* (Bitter root). Deciduous perennial that is summer dormant. It bears funnel-shaped, white or pink flowers in early spring and summer and has linear to spoon-shaped dark green leaves. H 2in (5cm), S 4in (10cm).

### PROPAGATION

🌱 early fall. Space-sow seeds under glass; they will germinate over winter. Keep just moist until the seedlings die down naturally. No further water is necessary until re-growth is apparent after about 4 months. Leave in the original pot for a further year, giving occasional liquid feeds during the second spring. Pot the carrot-like crowns individually into 3in (7cm) pots in mid-fall.
🌱 3–6 months. ☀ 2–3 years.

### Lithodora                    Boraginaceae

Some seven species of evergreen shrubs and subshrubs that produce funnel-shaped, usually intense blue flowers, in late spring and early summer. The acid-loving *L. diffusa*, especially its cultivar 'Heavenly Blue', is the most commonly grown. It needs well-drained, humus-rich soil.

*L. diffusa*, syn. *Lithospermum diffusum*. Prostrate evergreen, producing deep blue flowers in late spring and summer. The flowers of **'Heavenly Blue'** are profuse, and deep azure. H 6in (15cm), S 24–30in (60–75cm).

### PROPAGATION

Semi-ripe cuttings – mid- to late summer. Use an ericaceous rooting medium. A mist unit will increase rooting percentage.
☀ 7–8 months.

### Phlox                    Polemoniaceae

Some 67 species of annuals and mainly perennials, including a number of small, mat- or cushion-forming species for the rock garden *(see also Herbaceous Perennials directory p. 43)*. There are numerous named cultivars that will brighten the rock garden in late spring. They need well-drained, not-too-fertile soil, in sun. Those described are hardy.

*P. bifida* (Sand phlox). Neat, evergreen mound-former, that is clothed in lacy-petalled, usually white or lavender flowers. H 6–8in (15–20cm), S 6in (15cm).

*P. douglasii*. Low, mound-forming evergreen with white, blue, or pink flowers in spring. **'Crackerjack'** has crimson flowers, while those of **'Iceberg'** are palest lilac. H 6–8in (15–20cm), S 12in (30cm).

*P. subulata*. Forms evergreen mats or cushions studded with star-shaped flowers, in a range of colors, usually red or purple, but also violet, pink, or white. **'Tamaongalei'** is a striking white and bright pink cultivar. H to 6in (15cm), S to 24in (60cm).

### PROPAGATION

Softwood cuttings – after flowering the mats benefit from a light cutting back and the resultant clippings should contain new shoots, ¾–1¼in (2–3cm) long, which root easily. *P. bifida* has firmer, larger shoots that root best in equal parts peat and sand.
☀ 8–12 months.

### Pulsatilla                    Ranunculaceae

Around 30 species of hardy perennials forming clumps of finely divided leaves, and bearing bell- or cup-shaped – usually very silky-hairy – flowers, in spring or early summer, followed by silky seed heads. Often long-lived but resentful of transplanting except during dormancy. Grow in well-drained soil in sun.

*P. halleri*. Bears lavender or purple, silky-haired, bell-shaped flowers in late spring. It is best with very good drainage and protection from winter rains and snow. H 6–14in (15–35cm), S 6–8in (15–20cm).

*P. vulgaris* (Pasque flower). More robust than *P. halleri* and bears nodding, bell-shaped, silky flowers in white or shades of red, pink, or purple. H and S 10in (25cm).

### PROPAGATION

🌱 as soon as ripe in early summer, or store in the bottom of the refrigerator; old seed germinates erratically. Thin out in early spring, and pot into deep pots to reduce chances of disturbance when planting out the seedlings.
🌱 4–8 weeks if sown fresh.
☀ 2–3 years.

Root cuttings – fall. Root cuttings present some risk to the parent. Cut thick roots into 1¼in (3cm) sections. Keep at about 50°F (10°C). Adventitious buds develop after 2 months, but leave plantlets *in situ* until spring. ☀ usually 2 years.

*Saxifraga burseriana 'Gloria'*

*Sedum spathulifolium 'Cape Blanco'*

*Sempervivum arachnoideum*

## Ramonda
Gesneriaceae

Just three species of hardy, evergreen perennials with rosettes of leaves and flowers like those of the related African violets, in late spring or early summer. Grow in an alpine house or in vertical rock crevices in partial shade. They dislike exposure to winter wet which will make the rosettes rot if it is allowed to accumulate in them.

*R. myconi*, syn. *R. pyrenaica*. This produces lavender, pink, or white flowers in late spring. The leaves are dark green. It is the easiest species to grow. H 4in (10cm), S 8in (20cm).

**PROPAGATION**

Leaf cuttings – early fall. Often used to increase selected color forms.
☀ 12–24 months.

✎ early fall. Wash the dense root system free of medium, then root individual rosettes.
☀ 12 months.

🌰 as soon as ripe. Surface sow the dust-like seed onto the surface of milled sphagnum, then place the pot in a plastic bag to maintain high humidity. Keep out of direct light. Water from below. Prick out only when the seedlings have developed several leaves and are about ½–¾in (1–2cm) across.
🌱 2 weeks.  ☀ 3–5 years.

## Saxifraga (Rockfoil)
Saxifragaceae

Complex genus of about 350 species of mostly rosette-forming, evergreen or semi-evergreen perennials, and many hybrids often with complicated parentage, but sometimes easier to grow than the species. Plants in cultivation range from the easily grown Mossy saxifrages, with mats of soft mossy, bright green leaves, to high-alpine rock, scree, and cliff dwellers that form tight,

firm rosettes of often lime-encrusted leaves (the 'Silvers'). Division is best reserved for the vigorous Mossy hybrids; many of the smaller rosette-formers have a single 'neck' or are difficult to re-establish.

*S. burseriana* '**Gloria**'. Rosettes of gray-green leaves and large white flowers at the end of winter. H to 2in (5cm), S 6in (15cm).

*S. cochlearis* '**Minor**' Lime-encrusted with red-spotted white flowers. Neat in a trough or raised bed. H 4in (10cm), S 6in (15cm).

*S.* '**Tumbling Waters**' Cascades of small white flowers in spring. As with several others, the rosette dies after blooming, but is replaced with daughter rosettes. H 18in (45cm), S 12in (30cm).

**PROPAGATION**

Rosette cuttings – 4–10 weeks after flowering. Root in sand at 50–59°F (10–15°C). If taken too early, the fragile rosettes may fall apart during preparation.
☀ 12 months.

🌰 fall, in an open frame. Hybrids do not come true, although many are fertile and unexpected color forms occur in the resulting seedlings.
🌱 spring.  ☀ 2–3 years.

## Sedum (Stonecrop)
Crassulacea

Around 400 mostly succulent species, not all hardy or long-lived, but with many suited to the rock garden in scale and hardiness. Most bear tiny star-shaped flowers in summer or fall. They need well-drained soil and sun.

*S. spathulifolium* '**Cape Blanco**', syn. *S. s.* '**Cappa Blanca**'. Forms a wide, low mound of bloomed, gray leaves, spangled with heads of star-shaped yellow flowers in summer. H 4in (10cm), S 18–24in (45–60cm).

**PROPAGATION**

Cuttings of non-flowering shoots – early to midsummer. Exceptionally easy to root, even rosettes of a species knocked accidentally onto the surrounding soil may form new plants.
☀ 12 months.

✎ spring or late summer.  ☀ 6 months.

## Sempervivum (Houseleek)
Crassulaceae

Some 40 species, and hundreds of cultivars, of hardy, fleshy, rosette-forming perennials valued mainly for their foliage, although many are also very handsome when bearing their flat-topped heads of star-shaped flowers in summer. The flowered rosette dies after flowering finishes, but surrounding rosettes erupt with short stolons that soon fill the gap with satellite rosettes. Grow in sun in sharply drained, sandy soil of low fertility.

*S. arachnoideum* (Cobweb houseleek). Rosettes of leaves cobwebbed with white hair and bears flowering columns topped by starry pink flowers. H 2–3in (5–7.5cm), S to 12in (30cm) or more.

*S. tectorum* (Common houseleek). Produces large rosettes of very fleshy, purple-tipped, blue-green leaves and has red-purple starry flowers on hairy stems. H 4–6in (10–15cm), S 18–20in (45–50cm).

**PROPAGATION**

Offsets – spring or early summer. Once the stolons have developed new rosettes, sometimes with vestigial roots, these can be severed and grown on separately. Those with congested rosettes can have groupings of rosettes from the edge of the cushion pulled off and rooted in a sandy medium.
☀ 12 months.

# CONIFERS AND HEATHERS

Conifers and heathers are interesting and versatile plants that deserve a place in every garden. They provide form, structure, and continuity while giving a year-round display of colorful flowers, foliage, and cones. There are species and cultivars in every form and for every situation: majestic sequoias to dominate the skyline, exquisite dwarfs for the rock garden, and carpeting heathers to brighten the darkest winter days.

Propagating heathers is easy, rewarding, and economical using only the simplest equipment and the most basic techniques. Conifers can be as easy as heathers, or they can be challenging, a test for the best propagator. Here are all the techniques you will ever need.

◄ *Despite their almost infinite variety in form and color, conifers and heathers are increased by a range of mostly very simple propagating techniques.*

# WHY PROPAGATE
# CONIFERS AND HEATHERS?

*Highly prized for their year-round impact, conifers and heathers deserve a place in every garden. Choice specimens can be expensive yet are easy to propagate at home, so bulk up one or two store-bought plants and grow your own hedge, exchange unusual varieties with friends, or simply expand your border display.*

**Most ornamental cultivars of conifers and heathers do not come true from seed; increase relies on vegetative methods.**

It is well worth while producing a reserve stock of both conifers and heathers. Although most conifers are hardy plants adapted to harsh conditions, casualties can occur in all gardens. Tall heathers, such as *Erica arborea* and *E. australis*, are similarly vulnerable, and mat-forming types, which tend to become bare at the base after a few years, are best replaced regularly with fresh stock. With conifers there is also the attraction of raising your own hedges and saving a lot of money in the process.

## EXCITING POSSIBILITIES

Seed is the key to variety. Some conifer species have unstable genes that are capable of producing plants of widely differing size and habit. Hence, there is always the chance that a new form will emerge.

Most garden heathers, however, are cultivars and can only be reproduced vegetatively. Fortunately, the techniques involved are so easy that propagation from seed is rarely practiced except to produce new hybrids.

# SHORTCUTS

Cuttings are the simplest way of reproducing conifer cultivars though they must be taken from young plants. Heathers are equally easy, but a heated propagator is necessary for those taken early and for tender varieties.

Pines, cedars, firs, and spruces are almost impossible to strike from cuttings and so are usually grafted. Exceptions are the prostrate and dwarf firs and spruces which do root, albeit slowly, from softwood cuttings.

# LOW-MAINTENANCE OPTION

Layering and dropping are two useful techniques for increasing heathers, provided the plants have an area of bare soil around their bases. Dropping is also an ideal way to rejuvenate heathers that have become leggy and bare at the base. Both methods are more reliable and easier than cuttings, since the young plants are rooted while still attached to the parent

### Overcoming Seed Dormancy

*The cones of most conifers ripen in their first year, opening to release the seed before winter. If the seed germinated immediately, the seedlings would face the rigors of winter, but a natural dormancy period – broken by a period of cold – prevents the seed from germinating until the spring, when the young plants have a full season to establish themselves. Storing seeds of conifers such as the pines,* Abies *and* Picea, *at 36–39°F (2–4°C) in the refrigerator for four weeks mimics this process and assists even germination.*

*Abies koreana, the Korean fir, is a species of conifer that can be grown from seed, as discussed above.*

plant and so need much less aftercare; although young plants should always be kept watered and weed-free as a matter of course.

Many dwarf conifers, especially prostrate or arching variants, can can also be layered successfully, provided that young or juvenile shoots are selected. They invariably root quicker than older wood.

 **Taxus baccata 'Hibernica'**
Yews are slow to propagate from seed, which can take 2 years to germinate. Heeled semi-ripe cuttings root in 2–3 months, making this a productive method for both species and cultivars.

 **Abies balsamea, dwarf cultivar**
Prostrate and dwarf cultivars of *Abies* root reasonably easily from softwood cuttings; larger cultivars are propagated by side-veneer grafting onto *A. grandis* or *A. procera*. Only species are increased by seed.

 **Thuja occidentalis cultivar**
Cultivars of *T. occidentalis* include golden-leaved, dwarf, upright, and weeping forms; all are increased by semi-ripe or ripewood cuttings. The species makes good hedging, and ample numbers can be grown cheaply from seed.

 **Chamaecyparis lawsoniana cultivar**
*Chamaecyparis lawsoniana* often mutates and has given rise to over 500 mostly seed-raised variants. Once a sport or mutant has arisen, however, it must be increased vegetatively.

 **Chamaecyparis lawsoniana 'Elwoodii'**
This upright cultivar makes a good screen in the garden. It is one of the easiest of variants to root either from semi-ripe or from ripewood cuttings, so raising sufficient numbers yourself is an economical proposition.

# PROPAGATING TECHNIQUES
# CONIFERS AND HEATHERS

*Although most conifers are relatively straightforward to raise from the seed that they hold
in their woody cones, there are also a number of other techniques available to the home
propagator. Cuttings of both conifers and heathers will strike quite readily,
provided they are taken from young stock at the correct time of year.*

There are three main methods used to propagate conifers: seed, cuttings, and grafting. Each method has its benefits and each requires a different set of techniques and skills. Heathers are usually grown from cuttings or from various types of layers; seed is generally used only to produce hybrids.

## SEED

The majority of conifers hold their seed in woody cones. Most ripen in their first fall, opening to release the seed before the cone falls. Conifers from dry, fire-prone areas such as California (many pines and cypress species) retain the cones unopened on the tree for up to 25 years as part of a natural survival strategy.

Junipers and yews have fleshy cones (commonly referred to as berries) that are either blue–black or red respectively when ripe. Seed from these can be slow to

germinate, taking up to three years, and both genera are more quickly propagated from cuttings. Some conifers (*Taxus, Juniperus, Araucaria,* and members of Podocarpaceae) have male and female cones on separate plants, so that both need to be present before seed will be produced.

Cones are green when immature and change to a brown or purple color when ripe. Collect them just before they start to open, from late summer onward. The seed can be extracted by placing the cones in a paper bag and keeping them in a warm, dry place such as an airing cupboard until they open. Tap out the seed onto a clean piece of paper. If the cones refuse to open, soak them overnight in water and then repeat the procedure. The seed found in the lower part of the cone is the most likely to germinate.

Soak the seed overnight in clean water; any still floating on the surface by morning should be discarded as it is probably not viable.

To extract the seed from yews and junipers, squash the ripe berries between thumb and forefinger. Wash the seed in tepid water and dry it on kitchen paper.

Conifer seed can be stored dry at 39°F (4°C) in airtight containers in the bottom of the refrigerator until the following early spring, and then sown.

### Stratification

The majority of conifer seed germinates best after 3–4 weeks' cold storage at 32–41°F (0–5°C) in the refrigerator before being sown. This mimics conditions found in the wild and overcomes dormancy. Mix the seed with a little damp peat, vermiculite, or perlite in a plastic bag. Alternatively, first sow the seed as described below, and then place the pots in the bottom of the refrigerator.

**Choice of cutting material** . . . . . . . . . . . . . . . . . . . . .
Always take cuttings from shoots with a distinct growing point. Bun-shaped forms may be produced if lateral or feather shoots, especially from yellow-leaved cultivars, are selected.

◀ If the shape of the parent plant is to be maintained, select the cuttings from the leading shoot or from epicormic growth (where this is present) on the main stem or branches. If horizontal side shoots are used, prostrate plants will result and a leading shoot may form only after a number of years – if at all.

Fill 5in (12.5cm) pots with standard seed germination medium. Place the seed on the surface, then cover it with its own depth of medium. Mix the fine seed of species such as *Sequoia* and some cypresses with a little fine sand and surface sow. For the large seed of conifers such as *Araucaria*, sow one seed per 3in (7.5cm) pot.

Place the pots in an open frame outdoors and pot up the seedlings when they are large enough to handle.

## CUTTINGS

Many conifers can be propagated from cuttings. *Juniperus, Taxus, Thuja, Cryptomerias,* and some other genera listed in the directory can be rooted in a coldframe or in a pot tented with a plastic bag.

It is important to use young healthy specimens as a source for cuttings: over six years of age, *Picea* and *Pinus* become increasingly difficult to root. All conifers root faster if the stock plants are young, disease-free, and vigorous. Junipers and cypresses root more easily from cuttings with juvenile foliage than from those that have adult leaves.

Heather cuttings are taken in the same way as conifer cuttings, but you do not need to wound the base; it is sufficient to remove the lowest leaves.

### Timing and types of cuttings

Conifer cuttings are usually taken when the current season's growth begins to turn from green to brown.

Most are best from semi-ripe cuttings taken between late summer and early fall (most genera and cultivars). Early summer (softwood cuttings) may be more successful for *Abies* and *Picea*; late winter (fully ripe cuttings) is suitable for junipers and *Metasequoia*.

Early cuttings of *Abies* and *Picea* (softwood) need a closed, heated propagator to root successfully; cuttings taken in the late winter or early summer also need the protection either of a greenhouse or of an unheated propagator. Watch the cuttings throughout the summer, when they will be more at risk of drying out or being scorched by the sun.

Heather cuttings can be taken at any time in the growing season. All cuttings from midsummer onward are best taken with a "heel" – a small piece of bark from the main stem of the parent plant.

The later in the season you take the cutting (and the harder the wood), the slower the rooting process will be. With bottom heat, conifer and heather cuttings taken early in the season (softwood) should root within six weeks. Cuttings taken around midsummer (semi-ripe) should root by the next spring; late-summer and fall cuttings (fully ripe) by the next fall.

Prepare a well-drained, pasteurized propagation medium of equal parts peat (or alternative) and coarse sand, sharp sand or perlite, or vermiculite. A layer of coarse sand or perlite on the top helps prevent fungal infection and the growth of liverworts and mosses.

### Semi-ripe and fully ripe cuttings ...........................................................................

Whether propagating directly into a coldframe or placing a cutting-filled container in a coldframe, always make sure that the environment is scrupulously clean, in order to minimize the risk of disease while the cuttings produce roots. Protect the cuttings in winter by insulating the coldframe, and in summer by airing and shading it.

▲ **1** If possible, take the cuttings with a "heel" – a small section of wood from the main branch. Grasp the stem near its point of origin, and pull it sharply away, tearing off a small piece of bark.

▲ **2** Trim this with a sharp knife. Strip off the lower leaves, and remove the tip of the cutting if this is very soft.

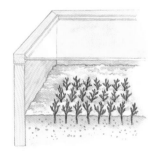

▲ **3** Insert the cuttings in the cuttings medium, either directly or in a coldframe or in a pot, ensuring that the leaves do not touch each other. Place the pot in the cold frame, then shut the lid.

▲ **4** Once the cutting has rooted well, replant it in a 3–4in (7.5–10cm) pot, while it becomes further established, or transplant it to its final position in the garden.

### Semi-ripe conifer cuttings with a heel .........

A cutting taken with a heel has a firm base to protect it against rots. It also has a high capacity to produce roots from the exposed, swollen base of the current season's growth.

◄ **1** Pull a vigorous shoot away from the parent stem, with a heel.

▼ **2** Trim off any leaves on the bottom third of the cutting.

For most conifers, cuttings should be 3–5in (7.5– 12.5cm) long; for heather cuttings and some dwarf conifers, cuttings can be 2in (5cm) long, or even less if suitable material is not available.

### Softwood cuttings

These are taken early in the season, but only when the terminal bud is clearly visible – cuttings taken earlier do not have sufficient food reserves to survive until rooting occurs. On conifers, trim the cuttings at the base just below a node and strip off the lower leaves.

Dip both conifer and heather cuttings in fungicide and insert the bases in hormone rooting powder (or dip the base of the cutting in a hormone solution), then insert them up to one-third of their length in the propagation medium. Spray the cuttings with fungicide and place them in a closed propagator at 70°F (21°C). Inspect them regularly for any signs of fungal diseases and remove any affected cuttings; spray the remainder with fungicide to prevent the spread of disease.

## GRAFTING

Grafting is a more challenging technique used for propagating conifer cultivars and species that are very difficult to strike from cuttings or raise from seed. *Ginkgo* and large-growing cultivars of *Abies*, *Cedrus*, *Picea*, and *Pinus* are often propagated by grafting. The best time for the grafting is during the late winter, before active growth starts. Side-veneer grafting is the technique that is most commonly used (*see also p. 113*).

### Side-veneer grafting

Rootstocks for grafting should be 2- or 3-year-old pot-grown seedlings of the same or a closely related species. All pines are mutually compatible, provided the needles are borne in bundles of the same number.

Coax the rootstock out of dormancy by bringing it into a cool greenhouse about three weeks before grafting. Keep it on the dry side: you do not want to encourage active growth that would result in too great an upsurge of sap.

Scions should be collected from either the terminal shoots or from strong upright shoots of the plant to be propagated. They should be 3–5in (7.5–12.5cm) long and from the previous season's growth. Keep the stems in a plastic bag in the refrigerator until ready to graft.

Cut back the rootstock to within 3in (7.5cm) of the base. Remove a sliver of wood about 1in (2.5cm) long from near the base of the rootstock. Cut the scion just below a pair of buds, then remove a 1in (2.5cm) sliver of wood from its base to match that of the rootstock. Bring the exposed sections of the pith together and bind them with graftingtape. Ideally, the scion and rootstock will be of equal thickness. Where the rootstock is thicker, align the rootstock and scion, bringing into contact the cambium layers (the layer of cells just beneath the bark). Seal with grafting wax.

Keep the grafted conifer humid at 59°F (15°C) until the scion produces fresh growth, normally within four weeks. Pinch out any shoots that arise on the rootstock below the grafting joint. Cut back the rootstock to just above the graft union when the scion is growing well.

## LAYERING HEATHERS

Layering is a successful method of propagating heathers and dwarf conifers, provided the surrounding soil is fairly friable. The layers take about a year to develop good roots, after which they can be severed from the parent plant and planted out or potted up. Layering is best done in fall. Heathers can also be increased by mound layering in spring, but both methods will only be successful on plants that have strong, sturdy stems.

In all cases, the principle is the same: to bring the stems into contact with soil and so encourage rooting at that point. The stems of most other shrubs need wounding in some way to encourage root production, but this is not necessary with heathers.

## Heathers from cuttings . . . . . . . . . . . . . . . . . . . . . .

The best time to propagate heathers from cuttings is when there are plenty of vigorous, non-flowering shoots available. This is most likely to be during the summer.

◄ **1** Take semi-ripe cuttings about 1–1½in (2.5–4cm) long from the top of a non-flowering shoot. Make a straight cut at the base of each cutting.

► **2** Strip off the leaves from the lower half of the stem.

◄ **3** Fill the tray with propagation medium. Insert the cuttings to the depth of the lowest leaves. Place the tray in a closed coldframe.

In the fall, dig a shallow trench around the perimeter of the plant. Mix the soil excavated from the trench with sharp sand and peat, or with potting soil, to make a friable mixture. Spread out the outer stems of the heather and sprinkle in the soil mixture. Hold the stems in position with a length of wire bent into a U–shape.

### Mound layering

Mound layering is a modification of the simple layering described above. Prepare a sandy, friable mixture of garden soil, sand, and peat, or planting medium. Cut out any dead or woody stems, then spread the remainder apart with your fingers and mound the soil around them. Cover the stems to within about 1–2in (2.5–5cm) of their tips. Keep the mound well watered during dry spells in summer, and replenish the mound if heavy rainfall washes some of the soil away. The shoots should root by late summer, when they can be separated from the parent and potted up or planted out.

### Dropping

Heathers that have been allowed to become woody at the base are best increased by dropping. Dropping is done in the dormant season, since it involves lifting the plant, a procedure that normally checks growth. The whole plant is buried in a planting hole that is large enough to accommodate the rootball, with the stem tips 2.5–5cm (1–2in) above soil level.

## Dropping heathers . . . . . . . . . . . . . . . . . . . . . . . . . . . . . . . . . . . . . . . . . . . . . . . . . . . . . . . . . . . . . . . . . . . . . . .

Unless heathers are trimmed regularly after flowering, they tend to become sparse and leggy with age, producing unattractive tufts of foliage at the ends of bare stems that is particularly ineffective as groundcover. In this case, dropping is a simple and productive way of renewing your stock.

◄ **1** In late winter to early spring, lift the entire plant, keeping the rootball intact. Cut back any straggly or congested stems.

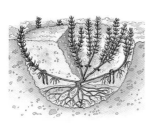

◄ **2** Sink the plant in a hole with 1–2in (2.5–5cm) of the stem tips above soil level. Mix the excavated soil with sand and peat and work it around each stem so that each is separated from its neighbors. Roots will develop by the following fall. Pot them up or plant them out.

◄ **3** The shoots can be arranged in the planting hole in one of two ways: so that they form a straight line, which is easy to keep weed-free; or in a bowl shape.

◄ **4** Heathers arranged in a bowl shape should have their stems evenly spread around the circumference of the planting hole, if possible.

# DIRECTORY OF
# CONIFERS AND HEATHERS

*Below are instructions and special propagating tips on selected conifers and heathers.*
*Unless otherwise specified, follow the detailed instructions given under*
*the propagation section on pages 90–93.*

## Key

H    Height
S    Spread
☙    By seed

## Conifers

### Abies (Firs)        Pinaceae

About 50 species of hardy, evergreen conifers with a pyramidal habit and single, linear, flattened leaves, spirally arranged or in two ranks and often with distinctive pale silvery-blue bands beneath. Mature cones are upright, resinous, usually purple, and break up at the end of the first year. Useful as specimen trees or for windbreaks. Grow in sun in neutral to acid, well-drained soil. **A. balsamea f. hudsonia**. Makes a dense, conical mound. H 24in (60cm), S 3ft (1m). **A. koreana** (Korean fir). Bears purple cones prolifically at a young age; suitable for small gardens. H to 30ft (10m), S to 20ft (6m). **A. nordmanniana** (Caucasian fir). Cultivars include dwarf and prostrate forms. **'Golden Spreader' is a dwarf form.** Prostrate and golden-leaved. H 3ft (1m), S 5ft (1.5m).

**A. pinsapo 'Glauca'**. Makes a fine specimen tree with dark glaucous foliage. H 80ft (25m), S to 25ft (8m).
**A. procera 'Compacta'**. One of the best blue-foliaged dwarfs. H 3ft (1m), S 5ft (1.5m).

#### PROPAGATION

☙ species only. Collect in fall before cones start to break up. Sow immediately outdoors or in an open frame. Seed does not store well.
Grafting – larger cultivars only. Using a side-veneer graft in late winter, graft onto 2-year-old rootstocks of *A. grandis* or *A. balsamea*.
Softwood cuttings – for dwarf and prostrate cultivars. Collect cutting material in early summer, treat with rooting hormone and root in a propagating case.

### Araucaria        Araucariaceae

Genus of 20 species of evergreen conifers, only one of which is reliably hardy. Grow in well-drained soil with shelter from cold wind. **A. araucana** (Monkey puzzle, Chile pine). One of the most characteristic of conifers with an unusual branching habit, symmetrical dome, and sharp-pointed, leathery, dark green leaves. Good as a specimen or in a group. H to 80ft (25m), S to 30ft (10m).

#### PROPAGATION

☙ collect the large seeds from beneath female trees in fall. Sow immediately in individual pots and keep in a coldframe or greenhouse. Lay the seed horizontally on the surface – do not cover, and keep moist.

### Calocedrus        Cupressaceae

About three species of evergreen conifers. **C. decurrens** (Incense cedar). Fairly common in cultivation and hardy. It is an upright, fast-growing tree with tough, scale-like, aromatic dark green leaves. It is resistant to root rots and is not very fussy about soil or site. **'Berrima Gold'** has sprays of yellow foliage. H to 130ft (40m), S to 28ft (9m).

#### PROPAGATION

☙ collect cones when they turn red-brown in fall, dry, and separate the seed. Store seed dry at 37°F (3°C) for 21 days, then sow outdoors in a seedbed or in pots in an open frame. Species only.
Cuttings – early fall. Semi-ripe cuttings root slowly in a coldframe or propagator.

### Cedrus (Cedar)        Pinaceae

Genus of 4 species of hardy, evergreen conifers with slender, needle-like foliage that varies from green to blue; there are also golden and deep blue cultivars. Young trees are pyramidal or conical, and fastigiate and weeping forms are available. All species make excellent specimens for large gardens. Grow in an open site in well-drained soil. **C. libani** **'Sargentii'** is a good weeping form while **'Comte de Dijon'** is a slow-growing dwarf. **C. atlantica 'Glauca'** has deep blue-gray foliage, and **C. deodara 'Aurea'** has foliage that is golden in spring and summer. Mature cones are barrel-shaped, and upright; they ripen in their second year and break up on the tree. H to 130ft (40ft), S to 30ft (10m).

*Abies koreana*

*Cedrus libani*

*Chamaecyparis lawsoniana* 'Ellwoodii'

*Cryptomeria japonica* 'Globosa Nana'

*Ginkgo biloba*

## PROPAGATION

🌰 species only. Collect cones as they start to open and extract the seed by drying. Store dry seed at 37°F (3°C) until early spring and then sow in seedbeds or in pots in an open frame. Seedlings are variable as the species hybridize, but good forms can be selected for foliage color.
Grafting – cultivars are grafted onto rootstocks of the same species or seedlings of *C. deodara*, using a side-graft in late winter.

## *Chamaecyparis* (False cypress)

Cupressaceae

Genus of five species and usually three recognized varieties of hardy, evergreen conifers with aromatic, paired, scale-like leaves in flattened sprays, small rounded cones, and characteristic nodding tips. The species and their cultivars are widely used as specimens, hedges, or screens. Best in neutral to slightly acid, well-drained soil in sun.
*C. lawsoniana* (Lawson's cypress). Has over 500 cultivars – most originating as chance seedlings. There is an almost limitless choice of dwarfs, groundcovers, and weeping and upright forms, offering blue, golden, or variegated foliage. **'Ellwoodii'** is dense and conical. H 10ft (3m). **'Ellwood's Gold'** has gold-green juvenile foliage and forms a small compact tree. H to 10ft (3m). **'Kilmacurragh'** has bright green foliage and a narrow habit. H to 50ft (15m), S 6–15ft (1.8–5m).
*C. obtusa* **'Crippsii'**. Dense sprays of golden foliage and makes a fine specimen. H 50ft (15m), S 25ft (8m).
*C. pisifera* **'Boulevard'**. Makes a small conical shrub when young and has soft blue juvenile foliage. H 30ft (10m), S 15ft (5m).

## PROPAGATION

🌰 collect in fall before the cones open. Sow outdoors in seedbeds or in an open frame. Germination is erratic. Species only.
Cuttings – fall to spring. The best way to propagate all the species and cultivars. Shoots with juvenile foliage root most readily, as semi-ripe or ripewood cuttings, in a coldframe. Most species can be rooted in the same way. Cuttings of *C. nootkatensis*, *C. obtusa*, and their cultivars should be taken in late winter to early spring.

## *Cryptomeria*

Taxodiaceae

One or two hardy evergreen species that tolerate most soils and sites.
*C. japonica* (Japanese cedar). Fast-growing tree with awl-shaped leaves and rounded cones. H to 80ft (25m), S to 20ft (6m).
**'Cristata'** has flattened, crested growths.
**'Elegans'** is a large bushy shrub with bronze-purple foliage in winter. H 25ft (8m), S 12ft (4m).

## PROPAGATION

Cuttings – early fall to spring. Take heeled semi-ripe cuttings, wound lightly at the base, and root in a coldframe or propagator.

## X *Cupressocyparis*

Cupressaceae

Single hardy hybrid between *Chamaecyparis* and *Cupressus*. Grow in deep fertile soil in sun or partial shade.
X *C. leylandii* (Leyland cypress). Has long flat sprays of foliage, nodding leading shoots, vigorous growth, and a columnar habit. Cultivars are among the most widely planted and fastest-growing conifers and are often used for hedging and screens. **'Haggerston Grey'** and **'Green Spire'** are commonly

grown, while **'Robinson's Gold'** is one of the best gold-leaved cultivars. All are propagated by cuttings. H to 120ft (35m), S to 15ft (5m).

## PROPAGATION

Cuttings – collect at almost any time between late summer and early spring. Both heeled and nodal cuttings give good results.

## *Cupressus* (Cypress)

Cupressaceae

About 20 species of evergreen conifers. Most are conical, with scale-like foliage in three-dimensional sprays. Persistent cones often retain the seed. Grow in any well-drained soil in full sun.
*C. arizonica* var. *glabra* **'Pyramidalis'**. Has bluish-green new growth. H to 50ft (15m), S to 15ft (5m).
*C. macrocarpa* **'Donard Gold'** and **'Goldcrest'**. Both good golden forms. H to 15ft (5m), S to 8ft (2.5m).
*C. sempervirens* **'Stricta'**. Fastigiate form of the Italian cypress. H 70ft (20m), S 10ft (3m).

## PROPAGATION

🌰 species only. *C. sempervirens* is variable from seed. Fastigiate forms are best grown from cuttings. The cones are ripe when they turn dark brown and open if dried in gentle heat. Sow in pots in a coldframe or greenhouse after storing for 4 weeks at 37°F (3°C).
Cuttings – early to late fall. Take heeled, semi-ripe cuttings, treat them with rooting hormone, and then root in a coldframe or greenhouse.

## *Ginkgo*

Ginkgoaceae

A single species of hardy, deciduous conifer. Grow in fertile, well-drained soil in sun.
*G. biloba* (Maidenhair tree). Fan-shaped,

*Juniperus communis*

*Metasequoia glyptostroboides*

*Picea glauca* var. *albertiana* 'Conica'

broad, bright green leaves. Fall color is a clear buttery yellow. Male and female cones are borne on separate trees – the plum-like female fruits have an unpleasant odor. H to 100ft (30m), S to 25ft (8m).

### PROPAGATION
Cuttings – midsummer. Collect semi-ripe cuttings, treat with rooting hormone and root in a propagator. Those taken from the shorter lateral shoots will form leaders in their second year.

### Juniperus (Juniper)                    Cupressaceae
Some 45–60 species of hardy to frost-hardy, evergreen conifers. Grow in well-drained soil in sun or light shade. Junipers are variable and versatile. Foliage can be adult (scale-like) or juvenile (sharp, needle-like); sometimes both types appear on the same plant. Habits of both cultivars and species vary.
*J. procumbens* is completely prostrate: H to 30in (75cm); *J. x media* 'Pfitzeriana' is an open vase-shaped shrub: H 4ft (1.2m); while some cultivars of *J. communis* and *J. chinensis* are narrow and conical: H to 70ft (20m), S to 20ft (6m), but very varied. The foliage color in the cultivars also varies with almost every hue of blue, green, or gray available.
*J. squamata* 'Blue Carpet' forms a prostrate blue mat. H 12in (30cm), S to 10ft (3cm).
*J. virginiana* 'Grey Owl' has silvery-gray foliage and a spreading vase-shaped habit. H 6–10ft (1.8–3m), S 10–12ft (3–4m). The seed is in a fleshy fruit and can take 3 years to ripen. Seed germinates erratically over a long period, so most junipers are propagated by cuttings. Prostrate species and cultivars sometimes layer themselves; these can be lifted before new growth begins in spring.

### PROPAGATION
Cuttings – juvenile forms are the easiest but those with adult foliage can also be rooted successfully. Cuttings can be taken as ripewood cuttings in midwinter to early spring, or as semi-ripe cuttings from late summer to early fall. *J. communis*, its cultivars, and other prostrate species are best taken late in the year, while the more upright species and cultivars are best taken in early spring.

### Larix (Larch)                              Pinaceae
Genus of 14 species of hardy, fast-growing, deciduous conifers, with needle-like leaves, clustered or spirally arranged. Grow in deep well-drained soil in sun. Larches are often used in shelter belts or woodland gardens to provide protection for more tender plants; in gardens, they are also valued for their foliage color. They turn shades of old-gold in fall and the spring foliage is the brightest sea-green. Cones persist on the tree for many years although the seed is shed in the first year. Cuttings are very difficult to root and seed is the best method of increase. H 3–100ft (1–30m), S 3–20ft (1–6m).

### PROPAGATION
🌰 collect cones before they open in fall, dry, and extract the seed. Store dry seed at 37°F (3°C) until early spring, then sow outside in a seedbed or open frame.

### Metasequoia                           Taxodiaceae
Genus of a single hardy species. Provide humus-rich, well-drained soil in a sunny site. *M. glyptostroboides* (Dawn redwood). A fine deciduous conifer, with red-brown bark and opposite pairs of linear, soft green leaves

that turn pink or gold in fall. The tight pyramidal habit and the buttresses that form on the older trees make this a fine specimen for medium- and large-sized gardens. Cones ripen in one year and open on the tree. H to 130ft (40m), S to 15ft (5m).

### PROPAGATION
Cuttings – hardwood, in mid- to late winter. Root in a frame, preferably with bottom heat. They root within 4 months. Softwood cuttings collected in early summer and rooted in a propagator are also successful. Collect all cuttings from either terminal or upright shoots.
🌰 collect seed as soon as cones start to open and sow immediately in a coldframe or greenhouse. Seed usually germinates within 2 weeks.

### Picea (Spruce)                            Pinaceae
About 35 species of hardy to frost-hardy, evergreen conifers, distinguished from firs by their narrow, single, usually forward-pointing, spirally arranged needles and their pendent cones. Grow in sun in any deep, moist but well-drained soil, preferably neutral to acid. The foliage is usually dark green, but there are blue cultivars such as *P. pungens* 'Koster' and golden-leaved cultivars like *P. orientalis* 'Aurea'. Some, such as *P. glauca* var. *albertiana* 'Conica' form conical shapes, others make low mounds or "buns"; there are also weeping, fastigiate, or prostrate habits. Sizes vary from H 6in (15cm) to 130ft (40m), S 20in (50cm) to 40ft (12m). Cones ripen and shed their seed in the first year.

### PROPAGATION
🌰 for species. Collect cones when they turn brown, dry, and extract the seed. Sow in

*Pinus mugo* 'Winter Gold'

*Podocarpus salignus*

*Sequoia sempervirens* 'Adpressa'

seedbeds outdoors or in a coldframe.
Grafting – late winter. Larger cultivars are
grafted using a side-veneer graft with *P. abies*
as a rootstock.
Cuttings – dwarf cultivars, such as *P. mariana*
'Nana', *P. abies* 'Little Gem'. Take softwood
cuttings in midsummer, avoiding strong
growth, and root in a propagator.

## *Pinus* (Pine)                    Pinaceae

About 110 species of hardy to frost-hardy,
evergreen conifers, easily recognized by their
bundles of needles. Most have whorled
branches and are fast-growing, conical when
young and dome-shaped at maturity. They
make good individual specimens and smaller
species and some cultivars are suitable for
rock and heather gardens. Grow in sun in
well-drained soil. Cones sometimes remain
closed on the tree for many years. Cuttings
are almost impossible to strike unless really
young plants are used.
*P. mugo* (Mountain pine, Swiss mountain
pine, Dwarf pine). Hardy, round to spreading
tree. **'Winter Gold'** has golden winter
foliage. H to 11ft (3.5m), S to 15ft (5m).
*P. sylvestris* **'Aurea'** (Scots pine). Hardy,
golden conifer; color stronger in fall and
winter. H to 50ft (15m), S to 28ft (9m).

### PROPAGATION
❦ collect cones in fall when they are brown,
dry, and extract the seed. Most species
benefit from 12 weeks' stratification at 39°F
(4°C). Sow into seedbeds outdoors or in
pots in a coldframe in early spring.
Grafting – cultivars are grafted onto rootstocks of the same species or other species
with the same number of needles. Use a
side-veneer graft in late winter.

## *Podocarpus* (Podocarps)     Podocarpaceae

About 100 species of evergreen, coniferous
shrubs or trees, but with only a few hardy
species in cultivation. Tolerant of many soils,
but prefer sun and shelter from cold wind.
*P. salignus*. Drooping branches and long,
glossy, narrowly lanceolate leaves. H to 70ft
(20m), S to 28ft (9m).

### PROPAGATION
Cuttings – late summer to early fall.
Heeled and semi-ripe, the lower half lightly
wounded. Treat with rooting hormone and
root in a coldframe or propagator. They
root slowly.

## *Prumnopitys*                 Podocarpaceae

Genus of 10 species of evergreen, yew-like
coniferous shrubs or small trees with long,
arching branches and linear, glossy green
leaves that are bluish underneath.
*P. andina*, syn. *Podocarpus andinus* (Plum yew).
Fast-growing and hardy, can resemble *Taxus
baccata*. H to 70ft (20m), S to 25ft (8m).

### PROPAGATION
Cuttings – late summer to early fall. Take
heeled semi-ripe cuttings, with the lower half
lightly wounded, treat with rooting hormone
and root in a coldframe or propagator. They
are slow to root.

## *Sequoia*                        Taxodiaceae

One hardy, evergreen conifer species. Grow
in well-drained soil in sun or light shade.
*S. sempervirens* (Redwood, Coast redwood).
Has red-brown bark and broad, scale-like
leaves. Suitable only for larger gardens. H to
365ft (112m), S 20ft (9m). **'Adpressa'** can
be grown as a dwarf shrub. H to 28ft (9m),
S to 20ft (6m). Cones ripen in the first year.

### PROPAGATION
❦ collect cones when they begin to turn
yellow, dry, and extract the seed. Store
at 37°F (3°C) for 3 weeks, then sow in a
coldframe. Keep seedlings moist.
Cuttings – midwinter. Take ripewood cuttings
and root in a coldframe. Leaders will form
after 2 years.

## *Sequoiadendron*

(Sierra redwood, Big tree)      Taxodiaceae

One hardy, evergreen conifer species. Plant
in sun or light shade in well-drained soil.
*S. giganteum*. Has furrowed, red-brown bark
and scale-like to awl-shaped leaves. Suitable
for larger gardens. Cones ripen in second
year. H to 260ft (80m), S to 30ft (10m).

### PROPAGATION
❦ collect in fall and sow immediately
in a coldframe, seedbed, or pot. Seed
germinates readily without pre-treatment.
Cuttings – early fall. Small cuttings of
epicormic shoots (those that arise directly
from the trunk) root easily in a coldframe.

## *Taxodium*                      Taxodiaceae

Genus of three species of hardy, deciduous
conifers, similar to *Metasequoia*, but with
alternate leaves. Grow in moist or wet soil,
preferably slightly acid, in sun or part shade.
*T. distichum* (Bald cypress, Swamp cypress).
Foliage turns brick-red in fall. Cuttings are
difficult to strike, so seed is the best
method to use. H to 130ft (40m),
S 28ft (9m).

### PROPAGATION
❦ collect cones as they turn purple in fall,
extract seed and sow immediately in a
seedbed outdoors, or in a coldframe.

97

*Thuja orientalis* 'Elegantissima'

*Thujopsis dolabrata* 'Aurea'

*Tsuga canadensis* 'Jeddeloh'

## Taxus
Taxaceae

Between 3 and 10 hardy species of long-lived, evergreen conifers, recognized by their flat, linear, glossy, dark green leaves and the fleshy, bright red fruits. Tolerant of a wide range of soils and situations, they also respond well to clipping and are extensively planted as hedges and in formal situations. *T. baccata* (Yew). Has many cultivars ranging from the dark green, columnar **'Fastigiata'** (Irish yew) to the wide-spreading, golden-leaved **'Washingtonii'**. Yews are difficult to propagate from seed (although bird-sown seedlings can sometimes be found) but cuttings strike readily. H 30–70ft (10–20m), S 25–30ft (8–10m).

**PROPAGATION**
Cuttings – fall. Take heeled, semi-ripe cuttings with a small portion of 2-year-old wood at the base. Wound lightly at the base, treat with rooting hormone and root in a coldframe. They will root within 2–3 months. Cuttings taken from side shoots do not form leaders, but they can be used as shrubs.

## Thuja
Cupressaceae

Genus of five species of hardy, fast-growing, conical, evergreen coniferous trees. Grow in well-drained soil in full sun; protect from cold winds. The small, scale-like, usually dark, glossy green leaves are in flattened sprays and very strongly scented. Small, woody cones ripen in their first year and release the seed on the tree. Cultivars include dwarfs, weeping and fastigiate forms with colors from yellow and gold, to green and variegated. *T. occidentalis* **'Rheingold'** (H 3ft/1m) and **'Golden Globe'** H 3ft (1m) are both good golden forms.

Thujas are easy to propagate from seed and from cuttings.
*T. orientalis* **'Elegantissima'**. Conical bush with foliage that turns from yellow to green then bronze over winter. H 15ft (5m), S 20ft (6m).

**PROPAGATION**
 species. Collect cones in early fall as they turn brown, then dry and extract the seed. Sow immediately in a seedbed outdoors, or in a coldframe.
Cuttings – fall to early winter. Take heeled, semi-ripe cuttings and root in a coldframe. They will root over winter.

## Thujopsis
Cupressaceae

One species of hardy, evergreen conifer, similar to *Thuja*, but distinguished by rounded cones and fleshier, flattened sprays of waxy leaves, that are bright green above and marked with white beneath. Seedling offspring are variable and slow-growing, so cuttings are the preferred means of increase. *T. dolabrata*. Conical, dark green tree. H to 70ft (20m), S to 30ft (10m). **'Aurea'** has golden foliage. H to 45ft (15m). **'Nana'** is a dwarf form. H 3ft (1m), S 32in (80cm).

**PROPAGATION**
Cuttings – midwinter. Root ripewood cuttings, collected from erect shoots, in a coldframe.

## Tsuga (Hemlock)
Pinaceae

Up to 11 species of hardy, evergreen conifers, with flattened, narrow, usually linear leaves, arranged radially or in two ranks, and with silver-white bands beneath. They are related to *Picea*, but have softer foliage, and the leading shoot remains pendent while the tree is still growing. *Tsuga* has tiny, almost

spherical male cones, and small ovoid to almost spherical female cones that become pendent at maturity. They resemble the cones of the spruces (*Picea*), but with fewer scales, and turn from shades of purple to brown when ripe, usually in late fall. Grow in humus-rich, moist but well-drained soil in sun or light shade.
*T. canadensis* (Eastern hemlock, Canadian hemlock). Linear leaves, fresh green at first, later dark green, with two broad white bands beneath. H to 80ft (25m), S 30ft (10m). It has given rise to several cultivars that are useful for smaller gardens.
**'Aurea'** is a slow-growing, compact cultivar with golden-yellow young leaves that become darker through the season. H to 25ft (8m) eventually. **'Jeddeloh'** forms a hemispherical specimen. H to 5ft (1.5m), S slightly broader. **f. pendula**, syn. 'Pendula', is slow-growing and mound-forming with pendent branches. H to 12ft (4m). The dwarf cultivars are sometimes used for bonsai. They are propagated by cuttings, since they become too vigorous if grafted onto seedling rootstock.
*T. heterophylla* (Western hemlock). Aromatic, very glossy, dark green leaves, of varying sizes, arranged on each side of the shoot. It is a fine specimen tree for larger gardens and makes excellent hedging. H to 130ft (40m), S to 30ft (10m).

**PROPAGATION**
 for species. Collect brown cones, dry, and extract the seed. Sow in spring, in seedbeds outdoors or in a coldframe.
Cuttings – take semi-ripe cuttings in late summer or fall, treat with rooting hormone, and root in a propagator.

*Calluna vulgaris* 'Foxhollow Wanderer'

*Daboecia cantabrica* 'Bicolor'

*Erica vagans* 'Valerie Proudley'

# Heathers

## *Calluna* (Heather)                    Ericaceae

A single species of hardy, evergreen shrub, *C. vulgaris*, with thousands of cultivars. All need humus-rich, acid soil and are best in open sites in full sun. Heathers have tiny flowers, concealed by colored sepals, and dense, scale-like leaves. H to 20in (50cm), S to 30in (75cm). Only a few of the cultivars can be described here. **'Gold Haze'** has yellow foliage and white flowers. **'Robert Chapman'** has bronze-yellow foliage turning orange in winter, and bears pink flowers in late fall. **'Silver Queen'** has gray foliage and deep mauve-pink flowers. **'Foxii Nana'** is dwarf: H to 3in (7.5cm). **'H. E. Beale'** has long spikes of double pink flowers in late summer to early fall. **'Kinlochreul'** has double white flowers.

### PROPAGATION

Cuttings – late summer. Take 1–2in (2.5–5cm) heeled, semi-ripe cuttings from the lateral shoots below the main shoots. Root in a coldframe or plastic-covered pot. Cuttings can also be taken in early winter, after flowering, from new growth. Root in a greenhouse or propagator, with bottom heat.

## *Daboecia* (Irish heath)                    Ericaceae

Genus of two species of hardy to half-hardy, evergreen shrubs. All grow best in acid soils in full sun but will tolerate neutral soils and partial shade.

*D. cantabrica*, syn. *D. polifolia* (Irish heath or St. Dabeoc's heath) is a hardy species and, along with its cultivars and hybrids, is the most common in cultivation. All have short racemes of relatively large, urn-shaped flowers at the tips of stems bearing lanceolate, lustrous, dark green leaves. **'Bicolor'** has white, magenta, and pink flowers, some striped, often with all these colors on the same raceme. **'Jack Drake'** bears bright purple flowers, and **'Waley's Red'** has deep magenta blooms. H 10–16in (25–40cm), S to 26in (65cm).

### PROPAGATION

Cuttings – mid- to late summer. Take heeled semi-ripe cuttings, 1–2in (2.5–5cm) long, from lateral shoots that arise below the main shoots. Root in a coldframe or plastic-covered pot.

## *Erica* (Heaths)                    Ericaceae

Some 700 or more evergreen shrubs with a few hardy species. They have bell-shaped flowers, with color in the petals, rather than sepals, and whorls of mostly linear, needle-like foliage. Species and cultivars can be selected to give a brilliant display of color almost throughout the year. All grow best in well-drained, acid soil in full sun, although *E. carnea*, *E. vagans* and their cultivars tolerate some lime.

*E. arborea* (Tree heath). Has needle-like, dark green leaves, and bears pyramidal racemes of honey-scented white flowers in spring. **'Aurea'** has golden foliage. H to 6ft (2m), S 5ft (1.5m), often more.

*E. carnea*, syn. *E. herbacea* (Winter heath). Has linear, dark green leaves and bears slender, urn-shaped, pink-purple flowers in late winter and early spring. Most cultivars bloom between winter and mid-spring. **'Foxhollow'** has soft pink flowers and foliage that is a bronze-tipped, pale yellow to deep orange, in winter. **'Myretoun Ruby'** has deep crimson flowers in mid- to late spring. **'Springwood White'** has white flowers and brilliant green foliage. H to 16in (40cm), S to 22in (55cm).

*E. cinerea* (Bell heather). Has linear, dark green leaves and produces spikes of urn-shaped, white, pink, or purple flowers throughout summer. **'C. D. Eason'** has magenta flowers. **'Fiddler's Gold'** has golden then red foliage and pink flowers. **'Hookstone White'** bears white flowers. H to 24in (60cm), S 32in (80cm).

*E. x darleyensis* (Darley Dale heath). Has comparatively broad, lance-shaped green leaves and produces spikes of urn-shaped or cylindrical, white or pink flowers, from late winter to early spring. **'Darley Dale'** has cream-tipped foliage in spring, and bears soft pink flowers; **'J. W. Porter'** has green leaves, tipped red and cream, and bears deep pink-purple flowers. **'White Perfection'** has white flowers and green leaves. H 12in (30cm), S 24in (60cm).

*E. vagans* (Cornish heath). A spreading shrub with dark green, linear leaves. It bears pink or white bell-shaped flowers from midsummer to fall. **'Birch Glow'** has deep pink flowers. **'Valerie Proudley'** has white flowers and yellow foliage. H 16in (40cm) or more, S 30in (75cm).

### PROPAGATION

Cuttings – early to midsummer. Take heeled, lateral cuttings 1–2in (2.5–5cm) long. Cuttings from tree heaths should be 2–3in (5–7.5cm) long. Cuttings of *E. carnea* and its cultivars root better if taken from the tips of the flowered shoots in early summer. Layering – early spring, for dropping or mound layering.

# $\mathscr{S}$HRUBS, TREES, AND CLIMBERS

Trees and shrubs are the backbone of the garden. They are often long-lived plants that make a strong statement or that can provide valued backdrops to others of more transient beauty.

Climbers are woody-stemmed plants that attach themselves either to another plant (as in the wild) or to some kind of support by various means, whether these be twining stems, leaf stalks, or specialized aerial roots.

Such is the diversity of this group of plants that a wide range of propagation techniques can be used for multiplying them. While most can be raised by seed or cuttings, some must be grafted, while for others layering is a possibility.

◄ *Shrubs are a group of long-lived woody plants that do not usually need frequent propagation; when they do, a wide range of techniques can be exploited.*

# WHY PROPAGATE
# SHRUBS, TREES, AND CLIMBERS?

*The backbone of the garden, this important group of plants introduces an element of three-dimensional structure into planting schemes. Specimens draw the eye and thus should always be of the best quality. Propagation allows you to build up a reserve to replace leggy plants or those that have outgrown their allotted space.*

**Many ornamental shrubs and trees are cultivars that must be raised by vegetative means: they seldom come true from seed.**

This group tends to be expensive, since several years are often needed to produce a saleable plant, and many garden centers now stock only the most popular and easily propagated cultivars. Hence gardeners can provide a valuable service by keeping up stocks of those unusual cultivars which may not be viable to propagate on a commercial scale, as well as providing themselves with back-up insurance against winter losses.

In addition, many trees and shrubs are slow-growing but ultimately large plants. Once they outgrow their allotted space it is often better to discard them or pass them on to friends with larger gardens, rather than resort to drastic pruning that may not be successful. Replace them with smaller, newly propagated specimens. Similarly, climbers can often become unmanageable with time and are best replaced.

## PRIMARY METHODS

All species can be propagated by seed and this is often the best method, particularly for trees, assuring healthy, disease-resistant stock. However, seed production and germination can be adversely affected by the climate.

Cuttings are a reliable method of reproducing the characteristics of any given parent plant and are also useful for exploiting growth that "sports" (*see box on opposite page*). However, some difficult genera, such as *Berberis*, require the rather more specialized technique of mallet cuttings.

## ADDITIONAL TECHNIQUES

Subjects which do not strike readily from cuttings can be grafted, a technique which also boosts vigor if a strong-growing rootstock is used. Subjects include selections of *Camellia reticulata*, rhododendrons, plus maples, birches, and beeches. The allied technique of budding is generally used for propagating hybrid roses.

**Betula species**
All birches can be propagated by seed, but cultivated plants are likely to produce variable offspring since birches hybridize so readily. Side-veneer grafts or softwood cuttings are used to increase cultivars.

**Tree peony**
Tree peony species *Paeonia lutea* and *P. delavayi* can be increased by fresh or scarified seed. Named cultivars are increased by apical-wedge grafting in the late summer onto *P. officinalis* or *P. lactea* rootstocks.

**Sambucus nigra 'Guincho Purple'**
Colored or cut-leaved variants of elder are propagated mainly by hardwood cuttings. Cuttings of the hollow stems are taken with a heel, mallet, or just below a node to avoid exposing the pith.

**Cotinus coggygria 'Royal Purple'**
Colored-leaved variants of *Cotinus* can be propagated by softwood cuttings in early summer, but French layering in spring is the most productive method. Layers will be wellrooted by autumn.

Climbers and shrubs that produce long branches near ground level can be layered or, if a branch will not reach the ground, air layered. As a variation, plants that become straggly can be "dropped," whereby the whole plant is almost buried to encourage the stems to root.

Some shrubs produce shoots at ground level which can be divided, almost like an herbaceous perennial. Others have running roots just below the soil surface which have growth buds that can make new plants.

### Sports and Reversion

*Variegated plants usually arise as a sport or mutation on a green-leaved parent species. Other sports may vary in form or habit, or in leaf, flower, or fruit color. They very rarely come true from seed and must be increased vegetatively, usually by cuttings of one sort or another, or by grafting. Sometimes, a branch of the sport reverts back to the normal green type. Since this is almost invariably more vigorous than the mutated form, it is advisable to check regularly*

*and remove reverted shoots as seen. If they are left in place, they will eventually dominate the plant.*

The shrub *Elaeagnus pungens* 'Maculata' must be propagated vegetatively.

**⑤**

### *Euonymus fortunei* 'Silver Queen'

Taking semi-ripe cuttings in summer is one of the easiest and most convenient methods for evergreen *Euonymus*, whether species or cultivars. They root easily, without wounding or hormone rooting treatments.

**⑥**

### *Physocarpus opulifolius* 'Dart's Gold'

Although they do not come true from seed, cultivars can be increased using a range of other methods including division of suckers, and by semi-ripe or hardwood cuttings.

# PROPAGATING TECHNIQUES
# SHRUBS, TREES, AND CLIMBERS

*The most rewarding method of propagating shrubs, trees, and climbers is almost certainly from seed, one of the simplest techniques available to produce something that will have a long-lasting place in the garden or landscape from the very beginning of its life. However, for speedier results there are plenty of other effective ways of multiplying.*

## SEED

The seed of most woody plants can be collected in fall, by which time it has ripened sufficiently for sowing. Rhododendron seed is carried in capsules that release the seed only after frost, so these are best gathered while still green in early fall and stored dry in a paper bag until they open. *Acer* seeds should also be harvested while still green; if left until fully ripe they develop a hard outer coat that inhibits germination. *Hamamelis* "shoots" its seed once mature, so this too is best gathered while the seed capsules are still green.

Squash fleshy berries to release the seed, then wash it in tepid water and dry on kitchen paper. Plants like crabapples that enclose their seeds in firm fruits should be cut open so that the seeds can then be picked out. Sticky seeds that are difficult to handle, such as *Pittosporum,* can be separated by mixing with fine sand.

Nearly all seed is best sown the first spring after gathering. Store in labeled paper bags in a cool, dry place through the winter, or dry and store in airtight containers in the door of a refrigerator. In order to produce viable plants of slow-growing shrubs like rhododendrons, it is necessary to sow seed in the fall so that the seedlings' first growing season lasts as long as possible. This permits the formation of winter buds that will produce the following year's growth.

### Pre-sowing treatments

Some seed benefits from further preparation prior to sowing. In nature, many seeds undergo a period of winter dormancy – a survival mechanism that prevents seeds from germinating until suitable weather conditions for growth return in the spring.

### Scarification . . . . . . . . . . . . . . . . . . . . . . . . . . . . .

For successful growth, it is important to break the dormancy of hard-coated seeds artificially so that water intake is speeded up and germination occurs without hindrance.

### Sowing in containers . . . . . . . . . . . . . . . . . . . . . . . . . . . .

As young seedlings are highly vulnerable to disease, it is vitally important that the container into which they are sown is absolutely clean and that the medium is fresh and sterile.

◄ Nick or shave off part of the hard outer coats of large seeds such as oak (*Quercus*) to allow water to penetrate more easily. It is vital that only the hard outer coat is cut into; any damage to the seed embryo within is likely to prove fatal. Abrade the coat of smaller seeds in a jar lined with glasspaper or rub them between two pieces of glasspaper.

► Fill a seed tray with proprietary seed germination medium and firm gently. Sow the seed thinly and evenly and cover with its own depth of sieved seed germination medium. Top with a ¼in (5mm) layer of coarse sand, water thoroughly, and label. Put the pots either in an open frame or in a well-ventilated cold greenhouse. Keep the medium evenly moist. After germination has taken place, check regularly for any signs of pests and diseases, and spray routinely with a fungicide. Thin out into individual pots once the seedlings are large enough to handle.

## Stratification

With most plants that grow naturally in cool temperate climates, seed needs a period of cold to overcome dormancy before it will germinate. This process of chilling is known as stratification. In areas with cold winters, it can be achieved by sowing the seed as normal in late fall or winter and placing the containers in an open frame, where they will be exposed to winter cold. As soon as the seedlings begin to germinate, bring the containers into a cold greenhouse.

A more reliable method is to mix the seed with moistened peat or perlite in a clear plastic bag and place it in the bottom of a refrigerator (not the freezer). For the majority of species, chilling to 33–41°F (0.5–5°C) is sufficient to break dormancy; if allowed to freeze solid, the seed embryo will almost certainly die.

The length of the chilling period required varies depending on the species. Most hardy deciduous species will need chilling for 6–8 weeks, others may need as little as 3–4 weeks, or as many as 12 weeks. Seeds should be sown as usual as soon as signs of germination are seen, so check them regularly for signs of emergence without opening the plastic bag.

## Scarification

Some species delay germination by forming a hard protective coat around the seeds. This must be breached or broken down so that the embryo can imbibe the water necessary for germination to occur. Seeds of the pea family (Leguminosae) have a hard and shiny coating, particularly after a hot summer, and this is best broken down by covering the seed with hot, but not boiling water, then allowing it to soak over twenty-four hours before sowing as normal.

The hard coat of large seeds can be abraded with a file or nicked with a sharp blade. In the latter case, it is vital that just a small section of the seed coat is shaved away with a shallow cut. A deep cut that penetrates the embryo can be fatal. Smaller seeds can be rubbed gently between two sheets of glasspaper, or shaken gently in a lidded jar lined with glasspaper.

## Sowing

Sow the seed in containers or seed trays containing a proprietary loamless or loam-based seed germination medium or a mixture of equal parts sterilized loam,

## Sowing outdoors in a seedbed ·················

An outdoor seedbed is constructed using a frame of side boards, 8–9in (20–22cm) deep; the bed is 3ft (1m) wide so that it is narrow enough to reach from both sides.

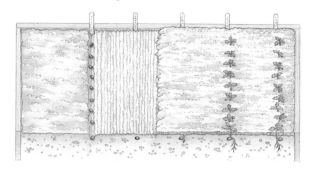

▲ Dig the seedbed thoroughly, then rake to a level surface with a fine tilth. Sow the seeds in furrows or broadcast, sow them in blocks. Cover with sand. By the following spring they should have germinated.

peat (or alternative), and sand to which a slow-release fertilizer has been added at the manufacturer's recommended rate. Firm the medium gently, sow the seed thinly and evenly on top, and cover with its own depth of medium. To offset the growth of mosses and lichens and to prevent the medium surface from capping during this time, mulch it with a layer of coarse sand. Put the pots in an open frame or in a well-ventilated cold greenhouse.

Germination can be erratic, and seedlings may not emerge until the second spring, so keep the medium just moist and do not discard the pots if seedlings do not appear in the first season.

Once the seedlings are large enough to handle, thin into individual pots, taking care to handle seedlings only by their seed leaves to avoid damaging the fragile roots and stems.

## Sowing outdoors in a seedbed

Where large numbers of plants are needed – for example, for hedging – seed may be sown in a specially prepared seedbed outdoors. Seed sown outdoors generally needs less aftercare than that sown in containers, and can be grown on for a further year *in situ* without disturbance.

In the season before sowing, dig over a bed to a spade's depth and incorporate well-rotted organic matter and coarse sand. Just before sowing, rake the bed to a fine tilth and sow the seeds evenly. Cover the seeds

## Semi-ripe cuttings . . . . . . . . . . . . . . . . . . . . . . . . . . . . .

This technique is used to propagate a wide range of deciduous and evergreen shrubs. Each cutting should have a woody base but be soft toward the tip.

▶ 1 Cut a leading shoot just below a node at the base and just above a node at the tip, so it is 4–6in (10–15cm) long. Remove the lowest pair of leaves with a straight cut flush to the stem, then cut the lowest remaining leaves in half with a straight cut across the center of the leaf.

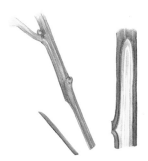

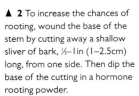

▲ 2 To increase the chances of rooting, wound the base of the stem by cutting away a shallow sliver of bark, ½–1in (1–2.5cm) long, from one side. Then dip the base of the cutting in a hormone rooting powder.

▲ 3 Around the edge of a 5in (12.5cm) pot filled with cuttings medium, insert the cuttings almost to the depth of the lowest leaves. Ensure that the leaves do not touch. Place in a propagator, if bottom heat is needed, or in a coldframe while the cuttings root.

to their own depth with soil, then top with a ⅜in (1cm) layer of coarse sand. Water thoroughly and label each furrow with the plant name and date of sowing.

Keep the seedlings weed-free and water as necessary. Check for and treat any pests and diseases as soon as seen. After a year's growth, if plants are large enough, transplant directly to their permanent site. Otherwise, transfer to a nursery bed and grow on for a further year.

## CUTTINGS

Most woody plants can be propagated by cuttings. For some, cuttings can be taken at any time during the growing season, while for others, timing is critical. Equipment needed depends on the degree of ripeness of the cutting: softwood cuttings need the warmth and humidity of a propagator; hardwood cuttings may be rooted in open ground. Softwood cuttings need more

attention during rooting and establishment but root quickly, and a high percentage generally succeeds.

Use an inert rooting medium of equal parts peat or pasteurized ground bark with sand or perlite. Use pots not less than 5in (12.5cm) across (small pots dry out more quickly and are susceptible to wide temperature fluctuations; both will inhibit rooting).

To ensure a good supply of cutting material, prune a section of a tree, shrub, or climber hard the previous year to stimulate the growth of vigorous new shoots.

The cuttings of some shrubs such as viburnums and azaleas have a tendency to die off during the first winter after rooting. For these, take cuttings as early in the season as possible to provide the longest possible growing season before the first period of dormancy.

### Softwood cuttings

In spring or early summer, in the early morning when plant tissues are at their most turgid, take cuttings about 2–4in (5–10cm) long, cutting just above a node (water the plant copiously the previous evening, if the weather

## Hardwood cuttings . . . . . . . . . . . . . . . . . . . . . . . . . . . .

Hardwood cuttings should be propagated in a site out of full sun and sheltered from drying winds. Cultivate the soil well. Make a trench a spit deep with one vertical side and one sloping one. Line the trench to a depth of 1in (2.5cm) with sand.

◀ 1 Take cuttings of pencil thickness (thinner stems tend to dry out before callusing occurs). Remove any soft tips, cutting just above a leaf joint. Trim the base of the cuttings just below a joint to make cuttings about 6in (15cm) long. (Long, vigorous stems may yield two cuttings.)

▶ 2 Dip the bases of the cuttings in rooting powder, then insert them in the trench up to about two-thirds of their length. If soil heaving occurs during freeze and thaw spells, gently firm in the cuttings with your foot. Hardwood cuttings of some evergreens, such as boxwood, lavender, and *Lonicera nitida*, can also be rooted in the open ground. Remove the lower leaves as for other types of cuttings. It is important not to allow these to dry out, so cover them with plastic covers or root them under polytunnels.

is dry). Put the cuttings in a plastic bag to prevent moisture loss. They must not be allowed to wilt.

Trim the cuttings just below the lowest node and remove the lower leaves to leave three or four at the tip. Clematis stems can simply be cut halfway between the nodes to make nodal cuttings (*see right*); each stem will yield several cuttings. Immerse the cutting in a fungicide solution and then dip the base in hormone rooting powder.

Insert the cuttings in the rooting medium up to half their length so the leaves do not touch each other. Put them into a mist propagation unit or into a closed, heated propagator. If the propagator is large, tent the pots with plastic bags supported by stakes.

Check often for signs of fungal disease. Remove any affected cuttings and spray the remainder with fungicide. The cuttings should root in 2–3 weeks, after which they can be potted up and slowly hardened off. Alternatively, if cuttings are taken later in the season, add some slow-release fertilizer to the rooting medium to avoid having to pot them up immediately. Potting up can then be delayed until the following spring.

### Semi-ripe cuttings

Semi-ripe cuttings are taken in mid- to late summer just as the base of the current year's growth is beginning to harden and turn woody. They are prepared as for softwood cuttings, but are rooted in an unheated propagator or closed coldframe. Or, take the cutting with a "heel" (*see below*). Grasp the stem at its point of origin and pull it sharply away from the stem with a small tail of bark. Trim the bark with a sharp knife.

### Heel cuttings ...................................

A heel cutting is a side shoot that is pulled away from the main stem together with a piece, or "heel", of woody material from the main stem. This protects the cutting from disease.

◄ Holding the main stem firmly in one hand and the side shoot in the other, pull the side shoot down sharply so that the shoot comes away with a "heel" from the main stem. Remove the lowest pair of leaves and neaten the heel, using a sharp knife. Dip the basal cut in a hormone rooting powder. Then insert it in a cuttings medium while it initiates root production.

### Nodal and inter-nodal cuttings ...............

Most woody plants are cut just below a node, that is the point on a stem where a leaf, shoot, or flower arises. A stem cut at a node is less vulnerable to fungal rots than one cut farther up.

▲ **Nodal cutting**
Make a straight basal cut about ⅛in (3mm) below a node on the stem to be propagated. Remove the lowest pair of leaves.

▲ **Inter-nodal cutting**
This type of cutting, with its basal cut made halfway between nodes, is most commonly used when propagating fuchsias and clematis.

### Mallet cuttings

Mallet cuttings are used for shrubs such as *Berberis* with hollow or pithy stems that are reluctant to root if taken as normal. Including a piece of the previous year's growth gives the cutting sufficient food reserves to root.

### Hardwood (fully ripe) cuttings

These are taken at the end of the growing season for evergreens, and for deciduous plants just after leaf drop. Most deciduous cuttings can be rooted in open ground or in a frame; evergreens are best in a shaded coldframe. Add a slow-release fertilizer to the rooting medium, as the cuttings will stay for a year before transplanting.

### Mallet cuttings ...................................

This type of cutting, with its large woody plug from the main stem at the base of the side shoot, is invaluable for cuttings that would otherwise be vulnerable to rotting organisms.

◄ Using a pair of sharp pruning shears cut the main stem straight across just above the side shoot to be propagated. Then make a second, straight cut about ¾in (2cm) below the first cut. Remove any leaves at the bottom of the side shoot, and dip the basal cuts in hormone rooting powder. Then insert the mallet cutting into some cuttings medium.

107

## Leaf-bud cuttings ...............................

To ensure success for this technique, always use young, vigorous stems of the current season's growth, which will have a high capacity to root.

► Cut a semi-ripe stem from the plant and cut it into sections, just above each leaf. Shorten the base to leave a length of stem about 1in (2.5cm) below the leaf. Thereafter, treat them as for semi-ripe cuttings, though the application of hormone rooting powder is not generally necessary.

Prepare a mix of equal parts sand and peat moss or peat moss and perlite. Remove the soil in the frame and replace it with a 6in (15cm) depth of the rooting medium. Alternatively, fill pots with the same mixture.

Take cuttings about 8in (20cm) long of the current season's growth and remove the tips, cutting just above a node. Trim the cuttings at the base just below a node to make a cutting about 6in (15cm) long. For species with pithy stems, take heel cuttings, as described under semi-ripe cuttings (see p. 107). On broadleaf evergreens, remove all but the top three or four leaves. Treat the bases of the cuttings with rooting hormone. Insert the cuttings up to two-thirds of their length in the rooting medium. They will callus over during the winter and begin to root from the following spring. During winter, insulate the frame during cold weather with a thermal blanket. Ventilate during mild periods. During spring and summer, ventilate on warm days and water during dry spells. Lift and pot up rooted cuttings in fall.

### Leaf-bud cuttings

Leaf-bud cuttings are commonly used to propagate camellias and *Mahonia*, which have vegetative buds in the leaf axils. By this method, a number of cuttings can be produced from a stem. Leaf-bud cuttings are best taken in late summer to early fall.

### Eye cuttings

Eye cuttings are similar to leaf-bud cuttings and are used most commonly for grapes (*Vitis*) that "bleed" copiously if cut while the plants are in active growth. Take them in winter when the plants are fully dormant. On plants where the eyes (dormant buds) are widely

spaced, each cutting can be of a single eye as for camellias. Where they are closer together, allow two or three buds per cutting. This method can be used for other deciduous climbers such as *Actinidida deliciosa*.

## LAYERING

Layering has the great advantage that the layered stem is encouraged to root before separating it from the parent plant, which will continue to provide sustenance until the rooted layer is separated. Layering is best done in early spring, when the sap is flowing and stems are pliable, but may also be performed between late fall and early spring. It is vital that the soil around all forms of layers is kept moist and free of weeds during the entire rooting period.

The year before you intend to make a layer, prune back a low branch hard in fall or spring. This will encourage the production of flexible young shoots which root more easily than older material. Just before making a layer, improve the soil around the base of the plant by working well-rotted organic matter and coarse sand into the soil. When layering ericaceous plants, it is important to ensure that the coarse

## Simple layering ...................................

Layers usually take about a year to root and should be ready for transplanting in the fall following layering. The layered stem is cut off just behind the wound, then re-planted.

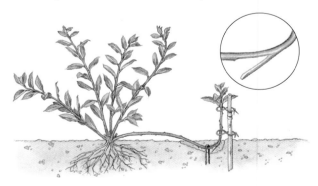

▲ Select a vigorous shoot from low down on the plant and bring into contact with the soil. At the point where the stem meets the soil, wound it by cutting a small tongue of bark in the underside of the stem. Dab hormone rooting powder beneath the tongue with a soft brush. Remove all leaves and side shoots from behind the wound, then peg down the stem with U-shaped piece of wire. Insert a cane next to the layer and tie the rest of the stem to this (the bending of the stem encourages rooting). Bury the pegged-down piece of stem with a little friable soil. Layers take up to a year to root.

## Serpentine layering ........................

This technique is similar to simple layering except that the stem is wounded and pegged down at multiple points near a node, with at least one leaf bud between each wounding site.

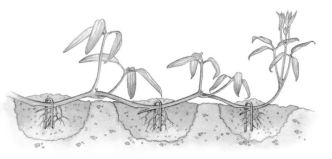

▲ Wound the lower part of the stem by making several angled incisions, 1–1½in (2.5–4cm) long. Peg these to the ground at intervals. The resulting plants, usually rooted by fall, will be smaller than those produced by simple layering and must be potted up and grown on in a coldframe rather than being planted out immediately.

sand and medium are lime-free and acid. There are a number of different types of layering, as follows.

### Simple layering

This, as its name suggests is the simplest form of layering. Layers that are made in early spring should

## Tip layering ........................

The hole for tip layering should be 3–4in (7.5–10cm) deep and have one almost vertical side farthest from the parent plant and a sloping one at the point nearest to the parent plant.

▶ In spring, bring down the tips of long arching shoots to ground level. At the point where they touch the ground, improve the soil at that point, working in potting medium or peat moss and sand. Bury the stem tip. By fall, new roots should have formed and the new plant can be cut from the main stem, potted up, and grown on.

root by fall, but should be left in place until the following spring when they can be separated from the parent plant.

### Serpentine layering

Serpentine layering is a modification of simple layering that can be used to create a number of new plants from the same stem and is suitable for climbers such as clematis and wisteria that produce a quantity of long, flexible stems. The timing is the same as for simple

## French or continuous layering ........................

This technique is carried out over a period of several years. In the dormant season, cut the parent plant back, as for stooling. The stems that emerge are allowed to grow unimpeded during the next growing season. The following spring these stems are reduced to eight or ten in number, and trimmed to equal length.

◀ 1 In spring, once the stems have been trimmed to equal length, spread them out like the spokes of a cartwheel, pegging them to the ground at intervals. This stimulates the production of vertical-growing laterals along each stem.

◀ 3 Continue adding soil as the laterals grow until they are buried up to 4in (10cm) of their length.

◀ 2 When these laterals appear, unpeg each "spoke" stem and cultivate beneath it. Each stem is then pegged down into a long, narrow trench 2in (5cm) deep. When the laterals are 4in (10cm) long, earth up the whole of the each "spoke" stem to a depth of 1in (2.5cm).

◀ 4 In winter, fork away the soil and sever rooted stems from the parent.

## Stooling . . . . . . . . . . . . . . . . . . . . . . . . . . . . . . . . . . . . . . . .

This technique is ideal to raise lilacs, dogwoods, *Amelanchiers*, and *Cotinus coggygria*. The same healthy parent plant can be mound layered each spring for a succession of new plants.

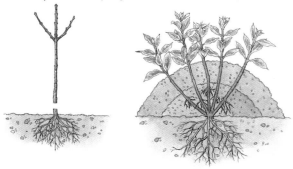

▲ In the dormant season, cut back all stems hard to within 2in (5cm) of the ground. As soon as any new shoots are 6in (15cm) long, earth them up with friable soil. Continue adding soil as they grow until 8in (20cm) of stem is buried. In early winter, gently fork away the mounded-up soil and sever rooted stems below the root system. Healthy plants will produce new shoots the following spring that can be treated in the same way.

layering, with the exception of grapes (*Vitis* spp.) and other genera that "bleed" when sap is actively flowing and it is best that these are done in late winter.

### Tip layering

Tip layering is used primarily on ornamental members of the genus *Rubus*. It is best done in early to midsummer, when the tip growth has begun to harden.

### French layering:

An even more productive method is French or continuous layering, though it is suitable only for shrubs whose stems remain fairly flexible, such as *Cotinus coggygria* or *Cornus alba*. French layering is a modification of stooling (*see above and right*), and also, needs forward planning since the young plant to be used as propagating material needs one complete growing season to establish itself before preparation for layering takes place.

### Air layering

Air layering can be used on shrubs (principally rhododendrons) that do not produce flexible stems near enough to ground level for conventional layering to be an option. In spring, select a sturdy, horizontal stem produced the previous year and remove any leaves and

side shoots along a section about 12in (30cm) in length. Cut a tongue 1in (2.5cm) long into the wood, pulling the knife towards the stem tip. Pack the cut with moistened sphagnum moss to hold it open, then wrap more damp moss around the stem. Cover this with a polyethylene sleeve taped to the stem. The layer should root by fall. Cut the stem just below the rooted section, pot up the new plant and place in a cool greenhouse until well established. (*See also Houseplants, p.137.*)

## MOUND LAYERING OR STOOLING

Stooling is a useful method of producing rootstocks in quantity where propagation by seed would be too slow. It is used commercially to produce rootstocks for grafting fruit trees and, for this purpose, permanent "stool beds" are maintained that are cropped every other year. It is also suitable for ornamental shrubs that respond well to hard pruning, such as *Amelanchier*, *Salix*, *Ribes* and ungrafted specimens of lilac (*Syringa*). It is similar to layering in that the stems are brought into direct contact with the soil, the advantage being that stems within the crown of the plant will also root. It does, however, need planning, since before stooling takes place a young plant must be planted and allowed to establish for a growing season before cutting back in its first dormant season in late winter or early spring. The resulting growth provides material for rooting.

## DROPPING

Dropping is most often used to increase dwarf shrubs, like rhododendrons, heathers, and dwarf cultivars of conifers (*see also Conifers and Heathers, p.95*). In the dormant season, thin out the stems if they are congested to give the remainder sufficient room to produce new roots. In winter or early spring, lift the plant, keeping the rootball intact. Deepen the planting hole and return the plant to it, leaving 1in (2.5cm) of the stems above ground level. Settle the plant in the hole and fill in with friable soil, mixed with potting soil, working it in around each stem with your fingers. Keep well watered during the growing season.

New roots should have formed by fall. Carefully remove the soil from around the stems to check for rooting. Sever rooted stems and pot up into a growing medium, or line out in a nursery bed. Grow on until large enough to be set out in permanent positions.

# GRAFTING

Grafting is a more advanced technique of propagation that is used for plants that do not root easily by other means of vegetative propagation, and which do not come true from seed. It involves the union of a stem from the plant to be propagated (the scion), taken from the most recent season's growth, with a well-established, strong-growing, and compatible rootstock of the same or closely related species, raised by seed or stooling (see opposite). For grafting to be successful, it is essential that the cambium layer (the layer of actively dividing cells just below the bark) of each is brought into close contact. All the cut surfaces must be absolutely clean and must not be allowed to dry out at all, so tools must be kept clean and the graft union made immediately after preparation.

Seedling rootstocks should be 1–3 years old; rootstocks from stooled plants should have been grown on for at least one year after being severed from the parent stock plant. The rootstocks should be either container-grown or containerized during the season before grafting is to be carried out.

With a few exceptions, grafting is best done in late winter, towards the end of the dormant season. Deciduous azaleas and selections of *Acer palmatum*, however, which bleed profusely if cut when the sap is rising rapidly, are more successfully grafted in mid- to late summer, while *Hamamelis* is best suited to late summer. It is essential that the root system of the newly grafted plant does not dry out while the two elements are bonding.

About three weeks prior to grafting, bring the rootstocks into a heated greenhouse to stimulate them gently into growth. Keep the roots fairly dry to encourage root activity and to minimize the speed of upward sap flow. There are a number of variations on the technique of grafting, of which only the simplest and most commonly used are described here.

## Apical-wedge or cleft grafting

Apical-wedge or cleft grafting is a straightforward and relatively simple form of grafting.

Prepare for grafting in midwinter by cutting from the scion healthy, vigorous, 12in (30cm) lengths of the previous season's growth of about pencil thickness. Place them in a labelled plastic bag and store them in the bottom of a domestic refrigerator. At the same time, earmark suitable rootstocks of a compatible species, also with a pencil-thick stem.

Just before grafting, in late winter, lift the rootstock and wash the soil from the roots. Cut the stem back to about 1in (2.5cm) above the roots, with a straight cut. Make a clean vertical incision in the top center of the rootstock, about 1in (2.5cm) deep. Trim back the scions to a length of about 6in (15cm), with a straight cut just above a healthy bud at the top and a straight cut at the base. Trim the base to a wedge-shape by cutting each side with a sloping cut about 1–1½in (2.5–4cm) long. Push the wedge into the cleft in the top of the scion. Ideally, the scion and stock should be of equal thickness, but if they differ slightly, align the cut surfaces so that one edge of the scion is flush with one edge of the stock. Bind the union with grafting tape, pot up into potting medium, and place in a heated propagator

## Apical-wedge grafting . . . . . . . . . . . . . . . . . . . . . . . . . .

The rootstock and scion must be the same size, preferably pencil thick. For the rootstock use a year-old seedling tree, cut down to 1in (2.5cm) above the root collar.

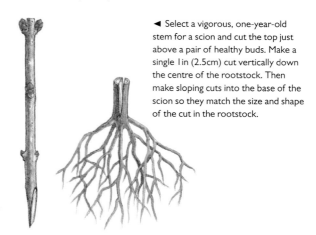

◄ Select a vigorous, one-year-old stem for a scion and cut the top just above a pair of healthy buds. Make a single 1in (2.5cm) cut vertically down the centre of the rootstock. Then make sloping cuts into the base of the scion so they match the size and shape of the cut in the rootstock.

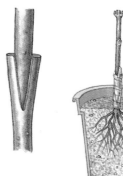

► Insert the prepared scion into the cut in the rootstock, neatly aligning the edges on at least one side so there is maximum connection of cambial tissue. Then bind the union with grafting tape and cover it with wax to protect it against disease. Once the graft union has callused over, remove the grafting tape.

## Saddle grafting ·······························

This technique is a good way to propagate rhododendrons, especially large-flowered hybrids. For the rootstock, use a 1–3 year old seedling, usually of *Rhododendron ponticum*.

◄ **1** In winter, bring the rootstock into the greenhouse about one month before grafting, and induce into growth. Select a vigorous, non-flowering, pencil-thick shoot of the previous year's growth for the scion and in its base cut an inverted V-shape. Then shorten the rootstock to 2in (5cm).

► **2** Starting 1in (2.5cm) from the rootstock base, make an upward, angled cut on either side to match the inverted V-shape on the scion. Push the scion over the rootstock. Secure the union with grafting tape and keep the grafted plant in a closed, humid, heated propagator until the two

at 50–59°F (10–15°C). The graft union will form within about 5–6 weeks. Once the union has taken, remove the grafting tape and harden off gradually. Grow the new plant on for a further year in a coldframe before planting in its permanent position.

### Saddle grafting

This results in strong plants, as a large area of cambium is exposed to form the graft union. It is vital, however, that the scion and stock are about the same thickness. Saddle grafting is most commonly used for evergreen rhododendron hybrids, which are usually grafted onto seedling rootstocks of *Rhododendron ponticum*.

About four weeks before grafting, in late winter or very early spring, bring the potted rootstock into a

greenhouse and bring gently into growth at about 50°F (10°C). For the scions, take 2–5in (5–12.5cm) lengths of healthy, one year old, non–flowering shoots and store in plastic bags in the bottom of the refrigerator until they are ready to graft.

Trim down the rootstock with a straight cut, to 2in (5cm) above the surface of the potting medium. Then make two sloping cuts from the top of the rootstock, to form a V-shape. Prepare the scion by making a cut in its base that matches that in the rootstock in length and angle. Fit the rootstock and scion together and bind with grafting tape. Cut any large leaves in half to minimize water loss. Place in a closed propagator at 50–59°F (10–15°C) and spray weekly with fungicide. The graft union will form within about 4–5 weeks. Once the union has taken, remove the grafting tape and harden off gradually. Grow on for a further year in a coldframe before setting out in the permanent planting site, applying a balanced liquid fertilizer monthly.

### Side-veneer grafting

Side-veneer grafting is the technique that is most often used for trees, and suits a variety of evergreen and deciduous shrubs.

Bring a potted, 1–3 year old seedling rootstock with a pencil-thick stem into the greenhouse three weeks before grafting. Prepare for grafting by cutting back the rootstock stem to 12in (30cm). Starting at about 1in (2.5cm) from the stem base, make a short, inward, downward-sloping cut. Make a second downward cut about 1in (2.5cm) above the first, so that it meets the first cut to form a slight lip. Select a vigorous, one year old, pencil-thick stem from the scion and trim to about 6–10in (15–25cm) long, cutting just above a bud. At its base, make a shallow cut, 1in (2.5cm) long, on one side of the stem, and make a second angled cut across the base. Match scion and rootstock so that the scion fits neatly into the lip cut in the rootstock, with cambiums in close contact. Bind with grafting tape, and brush with grafting wax to seal the union. Place deciduous species in a greenhouse at 50°F (10°C); evergreens in a closed propagator at 59°F (15°C). Union should occur within 4–5 weeks. Harden off the new plant and as the scion starts into growth, gradually cut back the rootstock stem to just above the graft union.

## Side-veneer grafting ...........................

For most hardwood tree species and deciduous shrubs, side-veneer grafting is done just before bud break in mid- to late winter, but *Acer* and *Hamamelis* are propagated in summer.

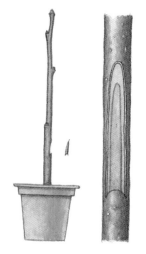

◄ **1** Prepare the receiving notch on the rootstock as follows. About 1in (2.5cm) from the base, make a short, downward cut at an angle of 45°. About 1in (2.5cm) above this, make a straight, downward cut to meet it and so remove a section of rootstock.

► **2** Cut the same length of bark from the base of the scion and trim the base to fit snugly into the V-shaped channel at the bottom of the rootstock cut. Fit rootstock and scion together. Ideally the cambium layers of both will meet on both sides, but if (as is likely) the scion is thinner than the rootstock, line up the cambium layers on one side.

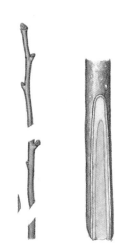

◄ **3** Bind the union with grafting tape; this will expand as the union "takes" and thickens. Wax the graft union. Once the graft has taken, gradually head back the rootstock until it is flush with the top of the graft union.

# BUD-GRAFTING

Chip-budding and T-budding are forms of bud-grafting used to propagate fruit trees, but are also used for ornamental members of the family Rosaceae. Roses are usually propagated by T-budding in the nursery trade. Bud-grafting is done in midsummer, while the plants are still in active growth, onto well-established rootstocks preferably with a stem diameter of at least ½in (1cm). (*See also Fruit, pp.182–183.*)

### T-budding

With roses, cut a strong, ripe, healthy stem from the scion and trim off the leaves. Snap off any thorns cleanly. Holding the stem with its growing point towards you, locate a dormant bud. Place the blade of the knife below it, then draw it toward you, cutting beneath the bud. Pull the knife sharply to tear off a "tail" of bark by which you then handle the bud. Pull off the wood behind the bud, using the knife if necessary. Make a T-shaped cut in the rootstock, cutting no deeper than the bark. Ease back the bark with the tip of the knife and insert the bud, tail uppermost. Trim back the tail level with the top of the "T." Bind the stem with grafting tape. Cut back the rootstock when new scion growth emerges the following spring.

### Chip-budding

The principle of chip-budding is essentially the same as that for T-budding, differing mainly in the form of the cuts made in the stock and the shape of the bud-chip taken from the scion. To remove a bud-chip, make a ¼in (5mm) cut, at an angle of 45°, about ¾in (2cm) below the bud. Make a second cut, about 1½in (4cm) above the first, sloping downward and inward behind the bud to meet the first cut. The cut in the rootstock is made to correspond in shape to that of the bud-chip, with a slight lip at the bottom.

# DIVISION

Suckering shrubs such as *Kerria* that produce quantities of thin, pliable stems can be lifted in spring or fall and divided like herbaceous perennials. Chop the crown into pieces with a spade or knife and replant the divisions immediately. Alternatively, single suckers can be removed from the margins of the clump and potted up or re-planted immediately.

# DIRECTORY OF
# SHRUBS, TREES, AND CLIMBERS

*Below are key tips on a selection of shrubs, trees, and climbers that are particularly suitable for propagation. Unless otherwise specified, follow the detailed instructions given under the propagation section on pages 104–113.*

## Key

| | |
|---|---|
| H | Height |
| S | Spread |
| T | Time for rooting |
| P | Time for planting out |
| U | Time for graft union to appear |
| 🌱 | By seed |
| ✺ | By division |

### Abelia                                    Caprifoliaceae

About 30 species of deciduous and evergreen shrubs with tubular flowers.
*A. x grandiflora.* This hybrid shrub is superior to either of its parents (*A. chinensis* and *A. uniflora*). It has slender, drooping branches with small, glossy, dark green leaves and pink-tinted, white flowers from midsummer to early fall. Evergreen but becomes semi-evergreen in cold winters. Hardy to 15°F (-10°C) or more with shelter from cold winds, it is best against a wall in cold climates. H 6–10ft (2–3m), S 12ft (4m).

**PROPAGATION**
Softwood or greenwood cuttings – early summer, 2½–3½in (6–9cm) long.
T 8 weeks, P 10 months.

Semi-ripe cuttings – mid- to late summer; nodal cuttings, 4–6in (10–15cm).
T up to 6 months, P 12–15 months.
Simple layering – spring. T 6–8 months, P 12–18 months.

### Acer (Maple)                              Aceraceae

About 150 species of evergreen and deciduous trees, ranging in size from forest trees to medium- and small-sized specimens. Valued for their leaf shapes, usually brilliant fall color, and, in some species, such as *A. griseum*, *A. capillipes*, and *A. davidii*, attractively colored or striated bark. The many different colored and sized variants of the hardy Japanese maples, *A. japonicum* and *A. palmatum*, are ideal for smaller gardens. H to 25ft (8m), S to 30ft (10m), but many are considerably smaller. Provide moist but well-drained soil and some shelter from cold winds and very hot sun.

**PROPAGATION**
🌱 if sown as soon as ripe, will germinate the following spring. Collect as the "keys" begin to color. Sow in a seedbed or open frame. Spring-sown seed may not germinate for a further year. T transfer to nursery bed in first spring, P 12–24 months.

Side-veneer grafting – late winter or midsummer, for selected cultivars, especially of *A. japonicum* and *A. palmatum*.
U 6 or 5 weeks respectively, P 2–3 years.
Chip-budding – early to midsummer, for selected cultivars.
U 6 weeks, P 18–24 months.
Softwood cuttings – spring. Requires some skill but can be successful with *A. palmatum* and the *A. palmatum* Dissectum Group.
T 8–10 weeks, P 24 months.

### Actinidia                                 Actinidiaceae

Some 40 species of mainly deciduous climbers grown for foliage, flowers, and fruit. *A. deliciosa*, syn *A. chinensis* (Chinese gooseberry, Kiwi fruit). (*See also Fruit directory p.184.*) Hardy with heart-shaped leaves and clusters of creamy white flowers. 'Blake' is self-fertile. H 30ft (10m).
*A. kolomikta.* Slender and hardy, producing dark green leaves splashed with cream and pink variegation. H 15ft (5m) or more.

**PROPAGATION**
Vine eyes – midwinter.
T 16 weeks, P 18–24 months.
Hardwood cuttings – winter. Take cuttings 8–10in (20–25cm) long.
T 3–4 weeks, P 12–18 months.
Softwood cuttings – early summer. Take basal or nodal cuttings of side shoots, 6in (15cm) long. T 3–4 weeks, P 18 months.
Semi-ripe cuttings – late summer 3–4in (7.5–10cm) long, with bottom heat, in a closed propagator.
T 4–6 weeks, P 12–18 months.

### Aesculus                                  Hippocastanaceae

About 15 species of deciduous trees and shrubs. Most, including *A. x carnea* (Red horse chestnut) and *A. hippocastanum* (Horse chestnut), are large trees, suited

*Abelia x grandiflora* 'Frances Mason'

*Acer palmatum*

*Actinidia chinensis*

*Betula pendula*

*Buddleja davidii* 'Black Knight'

to large gardens. Grow in well-drained soil in sun or part shade. Species can be propagated by ripe seed sown in a seedbed outdoors; cultivars are T-budded in summer.
*A. parviflora* (Bottlebrush buckeye). Spreading, suckering shrub with candles of white flowers that are produced freely in midsummer. The palmate leaves turn yellow in fall. H 10ft (3m), S to 15ft (5m).

**PROPAGATION**
Root cuttings – midwinter.
T 8–12 weeks, P 24 months.
✄ winter; separate suckers when dormant. Re-plant immediately.
Softwood cuttings – spring, 6in (15cm) long, under mist with rooting hormone.

## Andromeda            Ericaceae

Just two species of hardy evergreen shrubs with urn-shaped flowers and narrow gray-green leaves. They need permanently moist, acid soil in sun or light shade.
*A. polifolia*, syn. *A. rosmarinifolia* (Bog rosemary). Dwarf shrub bearing terminal clusters of urn-shaped, pink or white flowers from late spring to early summer. H 16in (40cm), S 24in (60cm).

**PROPAGATION**
Softwood cuttings – early summer, 1½–2⅓in (4–6cm) long.
T 4–5 weeks, P 6–12 months.
Semi-ripe cuttings – late summer, 2½–4in (6–10cm) long.
T 8–10 weeks, P 6–12 months.
Mound-layering – spring.
T 1 year, P 18–24 months.
🌰 as soon as ripe, or in spring. Surface-sow on finely milled moss peat; this is slow, but seedlings come true to type. P 24 months.

## Aucuba            Cornaceae

Three or four species of evergreen, hardy shrubs with large leaves and attractive fruits. They grow in any soil, except waterlogged, in sun or light to full shade.
*A. japonica*. Tough shrub with bold, glossy, ovate leaves and insignificant flowers which give rise to large red berries on female plants. Variegated forms, such as **'Picturata'** and **'Sulphurea Marginata'**(both female), are more attractive but need more light to grow well. The species can be raised from seed sown in a coldframe in fall, but it may take several years to ascertain whether offspring are female, berrying forms.
H and S 10ft (3m).

**PROPAGATION**
Hardwood cuttings – early to late fall. 4–6in (10–15cm) long, with a heel.
T 16–20 weeks, P 30 months.
Semi-ripe cuttings – late summer, 4–6in (10–15cm) long, with a heel.
T 6–8 weeks, P 30 months.

## Berberis (Barberry)            Berberidaceae

Some 200 species of often spiny, deciduous and evergreen shrubs. The hardy barberries are grown for their attractive foliage, which, in deciduous species, often colors well in fall, and for their cup-shaped flowers and the fruits that follow. There are compact species suitable for small gardens and larger spreading species for bigger areas. Pruning keeps them in check. Most thrive in any well-drained soil, in sun or light dappled shade.

**PROPAGATION**
Semi-ripe cuttings – midsummer, under mist or in a plastic tent. Heeled, nodal, or mallet cuttings may all be tried. Treat with rooting

hormone. T 4–6 weeks, P 12 months.
Hardwood cuttings – 3–4in (7.5–10cm) long, early fall in a coldframe. Pre-rooting treatment as above.
T 8–12 weeks, P 18–24 months.
🌰 species only. Separate from the ripe fruit, stratify in boxes of sand over winter, and sow as soon as signs of germination are seen in spring. P 24–30 months.

## Betula (Birch)            Betulaceae

About 60 species of graceful, deciduous trees of light, airy habit that grow in almost any soil, but only reach their maximum size in fertile soils. They have attractive stem colors, from white, in *B. ermanii*, *B. pendula*, and *B. papyrifera*, to pink and shiny orange-brown in *B. albo-sinensis* var. *septentrionalis*; *B. nigra* (River birch) has shaggy, dark brown bark. Many eventually make very large trees up to 50–80ft (15–25m) tall, so choose with care. When grown as multi-stemmed trees, height is reduced and the impact of the stems is increased.

**PROPAGATION**
🌰 sown as soon as ripe, in a seedbed or open frame. Germinates in spring.
P 12–24 months.
Grafting – selected cultivars. Side-veneer grafting in late winter.
U 3–6 weeks, P 8–20 months.

## Buddleja            Loganiaceae

About 100 species of evergreen and deciduous shrubs and trees. Hardy *B. davidii* and its many cultivars are most commonly grown. They are very easy to propagate by hardwood cuttings. Other desirable, early summer-flowering species include

115

*Buxus sempervirens*

*Camellia japonica* 'Flame'

*Clematis* 'Jackmanii Superba'

*B. alternifolia* (hardy), with rounded clusters of lilac-colored flowers, and *B. globosa* (frost-hardy), with rounded panicles of orange blooms. These are best propagated by semi-ripe cuttings. H and S to 15ft (5m). Grow in fertile, well-drained soil in full sun.

*B. davidii* (Butterfly bush). Slender, pyramidal panicles of fragrant white, purple, or lilac flowers, from midsummer, almost into fall. Cultivars have more intensely colored flowers. **'Black Knight'** has deep blue-purple blooms; those of **'Dartmoor'** and **'Royal Red'** are red-purple; and **'Pink Delight'** produces orange-eyed, pink flowers. H 10ft (3m), S to 15ft (5m).

**PROPAGATION**

Softwood cuttings – early summer; nodal cuttings, 4in (10cm) long.
T 3–4 weeks, P 12 months.
Semi-ripe cuttings – midsummer, heeled cuttings, 4–5in (10–12cm) long.
T 4–8 weeks, P 12 months.
Hardwood cuttings – fall, 8–10in (20–25cm) long.
T 16–20 weeks, P 18–24 months.

## *Buxus* (Box) Buxaceae

About 70 species of evergreen shrubs and trees, grown for their foliage. Grow in fertile, well-drained soil in partial shade. All can be propagated in the same way.

*B. microphylla* (Small-leaved box). Dense, hardy shrub. H 30in (75cm), S 5ft (1.5m).
*B. sempervirens* (Common box). Hardy shrub, ideal for hedging and topiary. Variants include some with golden- or silver-variegated leaves. H and S 15ft (5m).
**'Suffruticosa'** is a dwarf form that is ideal for hedging and edging. H to 3ft (1m).

**PROPAGATION**

Semi-ripe cuttings – mid- to late summer, 3–4in (7.5–10cm) long, in a coldframe.
T 12–16 weeks, P 24 months.
Ripewood cuttings – early fall, 3–4in (7.5–10cm) long, in a coldframe.
T 20–24 weeks, P 30 months.

## *Camellia* Theaceae

Genus of some 250 species of evergreen shrubs and trees. The most frequently grown are the named cultivars of *C. japonica* and *C. x williamsii* (*C. japonica* × *C. saluensis*) with flowers in a range of colors from white to deepest crimson, including all shades of pink, and some with splashed and striped petals. Flower forms include single, semi-double, rose, anemone, and peony. They need leafy acid soil, with some shade from the hottest sun, and shelter from cold winds. Most are completely hardy, but the flowers are not – the earliest bloom in late winter or before, with some cultivars still flowering in mid- to late spring, and the flowers are susceptible to frost. Camellias make excellent wall plants, preferably on a shaded, sheltered wall out of early morning sun. Most *C. japonica* cultivars also thrive in tubs in a cold greenhouse.
*C. japonica* cultivars H to 28ft (9m), S to 25ft (8m), but some are smaller.
*C. x williamsii* cultivars H to 15ft (5m), S to 10ft (3m).

**PROPAGATION**

Semi-ripe cuttings – mid- to late summer; nodal cuttings, 3–4in (7.5–10cm) long. Root in 50:50 (v/v) peat and sand, or ericaceous cuttings medium, with gentle bottom heat, in a propagator.
T 7–8 weeks, P 18–24 months.

Leaf-bud cuttings – late summer to early fall, or late winter.
T 10 weeks, P 18–24 months
Simple layering – fall. P 18 months.

## *Chimonanthus* Calycanthaceae

Six species of deciduous and evergreen hardy shrubs with very fragrant flowers. Will grow in any good garden soil, but best planted in full sun in a warm, sheltered site, so that the new growth will ripen fully.
*C. praecox*, syn. *C. fragrans* (Winter sweet). Produces beautifully fragrant, waxy, purple-centered, pale yellow flowers, on leafless branches from mid- to late winter. H 12ft (4m), S 10ft (3m).

**PROPAGATION**

Softwood cuttings – summer. Clones, which are mostly raised from seed, show varying success. T 8–10 weeks, P 24 months.
Simple layering – spring or early fall.
T up to 2 years.
🌰 as soon as ripe, germinates in spring. Seed-raised plants may take 5–7 years or more to reach flowering size.
T in spring, P 3–4 years.

## *Clematis* Ranunculaceae

Some 200 species and over 400 hybrids of mostly deciduous and some evergreen climbers, varying in size and vigor from dwarf to very vigorous species which reach 20ft (6m) or more. Most in cultivation are hardy. There are early-flowering species and their hybrids, flowering in late winter and early spring, and late summer- and fall-flowering species, so with careful selection, you can have clematis in bloom over much of the year. Grow in fertile,

Cotinus coggygria 'Royal Purple'

Cotoneaster x watereri

Daphne x burkwoodii

humus-rich soil with the roots in the shade and growth in the sun. Use "long tom" pots when potting up.

**PROPAGATION**

Softwood internodal cuttings, – early summer; 3–4in (7–10cm) long.
T 4–5 weeks, P 18–20 months.
Semi-ripe internodal cuttings, – midsummer; 4–5in (10–12cm) long, in a propagator, with gentle bottom heat.
T 4–6 weeks, P 18 months.
Serpentine layering – late winter to early spring. T 1 year.
🌰 sown as soon as ripe in a coldframe, germinates in late winter or early spring. Put under glass on germination. Species only. P 9–15 months.

## Cornus            Cornaceae

Some 45 species of mostly deciduous, hardy or frost-hardy, shrubs and trees, including those that are valued for their brilliantly colored winter stems, such as **C. alba** and its cultivars (H and S 10ft/3m), and those that are grown for their elegant habit and flowers, like **C. capitata** (H and S 40ft/12m), **C. kousa** (H 22ft/7m, S 15ft/5m), and **C. nuttalli** (H 40ft/12m, S 25ft/8m).

**PROPAGATION**

Hardwood cuttings – late fall to early winter, 8in (20cm) long, in open ground. C. alba and cultivars. T up to 6 months, P 12 months.
Semi-ripe cuttings – mid- to late summer, 3–4in (7.5–10cm) long, with a heel, with gentle bottom heat in a propagator.
T 6–12 months, P 2–3 years.
🌰 sown as soon as ripe in an open frame or coldframe, may take 18 months to germinate. P 2–3 years.

## Cotinus            Anacardiaceae

Just two species of deciduous trees or shrubs, of which C. coggygria and its cultivars are the more widely grown. It does not require a rich soil, flowering more freely and having richer fall colors on poor soils.
**C. coggygria**, syn. Rhus cotinus (Smoke bush, Venetian sumach). Has rounded green leaves that produce wonderful fall colors. The feathery flower plumes in midsummer give rise to its common name. '**Royal Purple**' is a less vigorous, smaller, red-leaved cultivar. H and S 15ft (5m).

**PROPAGATION**

Softwood cuttings – early summer, 2in (5cm) long. T 6 weeks, P 18 months.
French layering – winter-early spring.
T 9 months, P 12–18 months.

## Cotoneaster            Rosaceae

Over 200 species of evergreen, semi-evergreen, or deciduous shrubs and trees. There is enormous variation in habit from prostrate, ground covering species like **C. dammeri** and C. 'Nan Shan' (H 8in/20cm) to those that make excellent wall shrubs, such as **C. horizontalis**, **C. salicifolius**, and **C. microphyllus** (H 3ft/1m). A number, like **C. frigidus**, **C. diaricatus**, **C. multiflorus**, and **C. x watereri** (H to 30ft/10m), make elegant, upright or arching, large shrubs or small trees. Nearly all have tiny white flowers in spring or summer that give rise to berries in a wide color range. Grow in fairly fertile, well-drained soil in sun or partial shade.

**PROPAGATION**

Softwood cuttings – early summer, 3–4in (7.5–10cm) long (deciduous species).
T 5 weeks, P 18–24 months.

Semi-ripe cuttings – late summer to early fall with a heel, 3–4in (7.5–10cm) long, in a coldframe (evergreens and semi-evergreens).
T 3–6 months, P 18–24 months.
🌰 sown as soon as ripe in an open frame or seedbed, may take up to 18 months to germinate. Remove from the fleshy berries before sowing. P 2–3 years.

## Cytisus            Leguminosae

About 50 species of deciduous and evergreen shrubs and trees, grown for their often fragrant, pea-like flowers. All Cytisus are sun-loving, tolerant of poor, dry soils, and rather short-lived. C. scoparius (Common broom) and its cultivars are commonly grown. Use "long tom" pots when potting up.
**C. scoparius**. Hardy, deciduous, upright shrub with green stems. The species has yellow flowers, but it has produced cultivars of many colors, including bicolors, through salmon-red to deep crimson. H 5ft (1.5m) or more, S 5ft (1.5m).

**PROPAGATION**

Semi-ripe cuttings – mid- to late summer, 3–4in (7.5–10cm) long, in a coldframe.
T 6–8 weeks, P 12–18 months.
🌰 fall or spring. C. scoparius seed produces a mixture of plants with multicolored flowers. Most are suitable for hedging and similar plantings. T 7–8 weeks, P 18–24 months.

## Daphne            Thymelaeaceae

Some 50 species of deciduous, semi-evergreen, and deciduous shrubs noted for the intense fragrance of their flowers. A number are suitable for a rock garden (see p.82), and some have a reputation for being

*Deutzia* x *elegantissima* 'Rosealind'

*Eleagnus pungens* 'Malculata'

*Embothrium coccineum*

difficult to grow. Grow in humus-rich, well-drained but not dry soil. Most prefer slightly acid soil. Position in sun or light shade.

*D.* x *burkwoodii*. Fast-growing, hardy, semi-evergreen with glossy, dark green leaves, and clusters of fragrant, pink-flushed, white flowers in early summer. 'Somerset' has slightly darker flowers and makes an elegant vase-shaped bush. H and S 3–5ft (1–1.5m).

*D. mezereum* (Mezereon). One of the earliest daphnes to flower, producing its fragrant pink to purplish-pink flowers in late winter on leafless stems. *F. alba* has white flowers that show up well on dull winter days. H 4ft (1.2m), S 3ft (1m).

**PROPAGATION**

Softwood cuttings – early summer, nodal cuttings, 2–4in (5–10cm) long.

T 6–8 weeks, P 12–18 months.

Semi-ripe cuttings – midsummer, 2–4in (5–10cm) long, preferably with a heel, under mist or in a coldframe.

T 5 –14 weeks, P 12–18 months.

Simple layering – preferably spring to early summer. Sever rooted layers in fall, but do not lift until the following spring.

T 6 months, P 12–24 months.

🌰 sown as soon as ripe in a coldframe, germinates in the first or second spring. Suitable for species, and the most practical method for *D. mezereum*. P 20–24 months.

## Deutzia        Hydrangeaceae

Some 60 species of mostly deciduous shrubs that bear cup- or star-shaped flowers in spring and summer. They include a number of beautiful hybrids and cultivars like the pink-flowered *D.* 'Mont Rose', and *D.* x *elegantissima* 'Rosealind'. Both of these

are hardy and are suitable for smaller gardens, reaching only 4ft (1.2m) high. Deutzias are easily cultivated, growing well in all fertile soils, preferably in a sunny site.

*D. scabra*. Has erect branches covered with papery bark. The star-shaped white flowers are produced in summer. 'Candidissima' has double flowers. H 10ft (3m), S 6ft (2m).

**PROPAGATION**

Softwood cuttings – early summer, 3–4in (7.5–10cm) long.

T 3 weeks, P 18–24 months.

Hardwood cuttings – fall, 10–12in (25–30cm) long.

T 4–6 months, P 12–24 months.

## Elaeagnus        Elaeagnaceae

Some 45 species of hardy, evergreen and deciduous shrubs and small trees, grown for their attractive leaves and, some, for their tiny, sweetly scented flowers. *E.* x *ebbingei* and *E. pungens* and their cultivars are suitable for growing in smaller gardens. Grow in fertile well-drained soil. The evergreens will tolerate some shade.

*E.* x *ebbingei*. An fall-flowering evergreen with glistening, glossy sea-green leaves, covered in silvery scales beneath. 'Limelight' is more compact, with pale green leaves that are yellow and intensely silvered when young. H and S to 12ft (4m).

*E. pungens*. A spiny-stemmed evergreen with dark green foliage and small, tubular, fragrant white flowers in fall. 'Maculata' has leaves splashed with deep yellow. H to 12ft (4m), S to 15ft (5m).

**PROPAGATION**

Semi-ripe cuttings – late summer, 3–4in (7.5–10cm) long, in a coldframe.

T 4–6 months, P 12–18 months.

Leaf bud cuttings – mid to late summer. Sections with a leaf and one or more buds.

T 7–8 weeks, P 18–24 months.

## Embothrium        Proteaceae

Eight species of evergreen shrubs and trees, only one of which is grown. Frost-hardy, it thrives in shelter and dappled sunlight in moist, acid soil.

*E. coccineum* (Chilean fire bush). Gives a brilliant display of scarlet flowers in late spring and early summer. H to 30ft (10m), S to 15ft (5m).

**PROPAGATION**

Root cuttings – mid- to late winter, in a coldframe. T 4–5 months, P 18–24 months.

🌰 spring, at 55–61°F (13–16°C), in ericaceous seed compost. P 2 years.

⚒ of suckers, in late winter. P immediately.

## Escallonia        Crassulariaceae

Over 50 species of evergreen shrubs. Those most frequently grown in gardens are the named cultivars, often of rather mixed parentage, many of the best raised in Ireland by the Slieve Donard Nursery. They bear clusters of white, pink, or red flowers, above shiny leaves, throughout summer and early fall. They dislike dry winter winds, and although not reliably hardy in inland areas, they thrive in coastal regions where they make good hedges. Escallonias can be grown in any well-drained soil. H and S 5–8ft (1.5–2.5m).

**PROPAGATION**

Softwood cuttings – early summer, 3–4in (7.5–10cm) long. T 8–10 weeks, P 18 months.

Semi-ripe cuttings – mid- to late summer,

*Forsythia* x *intermedia*

*Hamamelis* x *intermedia* 'Pallida'

*Hebe* 'Great Orme'

4–5in (10–12.5cm) long, nodal or with a heel, in a coldframe, or propagator with bottom heat. T 4–5 months, P 18 months.

## Forsythia — Oleaceae

Some seven species of hardy, deciduous or semi-evergreen shrubs valued for their brilliant yellow flowers, produced before the leaves appear in early spring. The most commonly grown are named cultivars, including those of *F.* x *intermedia*, such as **'Lynwood'** and **'Karl Sax'**. H and S 5–8ft (1.5–2.5m). All are easily grown in moist but well-drained soil.

### PROPAGATION
Softwood cuttings – early summer; nodal cuttings, 4in (10cm) long.
T 4–5 weeks, P 18 months.
Hardwood cuttings – fall, 10–12in (25–30cm) long, in a nursery bed or coldframe. T 4–5 months, P 12–18 months.

## Fuchsia — Onagraceae

Around 100 species of evergreen and deciduous shrubs, with some 8000 named cultivars. Of the species, *F. magellanica* is the hardiest, and has been used in the breeding of many modern, more or less hardy cultivars like **'Mrs Popple'** and **'Riccartonii'**. They thrive in both semi-shade and full sun and in any well-drained soil. In cold winters, nearly all the "hardy" fuchsias die back to the base and re-emerge in spring, rather like herbaceous perennials.
*F. magellanica.* Has slender, delicate flowers with a long scarlet tube and sepals, and violet corollas, borne throughout summer.
Var. *molinae*, syn. 'Alba', has shorter-tubed white flowers tinted mauve; **'Versicolor'**

has scarlet and violet flowers, and its leaves are variegated silver-gray and white, flushed pink and purple. H and S to 10ft (3m); less in areas with cold winters.

### PROPAGATION
Most types of cuttings strike freely, but young plants from cuttings late in the season are best overwintered under glass for their first winter, in cold areas.
Softwood cuttings – spring or early summer; nodal or internodal, 3–4in (7.5–10cm) long.
T 2–3 weeks, P 4 months.
Semi-ripe cuttings – mid- to late summer; nodal or internodal, 3–4in (7.5–10cm) long.
T 4–5 weeks, P 16 months.

## Garrya — Garryaceae

Of the 13 species of evergreen shrubs and trees in the genus, *G. elliptica* is the most commonly found in cultivation. It is tolerant of a range of soils and is well suited to growing against a shady wall.
*G. elliptica.* Valued for its glossy, dark gray-green leaves and silver-gray catkins in late winter or early spring. Male and female catkins are borne on separate plants, the males being longer and more decorative. For this reason, it is usually best to propagate it vegetatively from a known male plant. **'James Roof'** has catkins 8in (20cm) or more in length. H and S to 12ft (4m).

### PROPAGATION
Semi-ripe cuttings – late summer, 3–4in (7.5–10cm) long, preferably with a heel, in a coldframe. Roots well but cuttings are often lost when transplanted, so insert in small pots or modules.
T 7–8 weeks, P 18–24 months.

## Hamamelis — Hamamelidaceae

Some six species of hardy deciduous shrubs, of which *H. mollis* and *H.* x *intermedia* and their cultivars form some of the most valuable of winter- and early spring-flowering shrubs. The large leaves turn golden yellow in fall. They need humus-rich, moist, neutral to acid soil and shelter from cold dry winds. Use lime-free compost when propagating.
*H.* x *intermedia.* Cultivars have yellow, orange, or red flowers with crimped petals. **'Arnold Promise'** has bright yellow flowers; **'Diane'** has dark red flowers; **'Jelena'** has copper-tinted flowers; and **'Pallida'** bears soft yellow blooms. H and S to 12ft (4m).
*H. mollis* (Chinese witch hazel). Produces fragrant, spidery, golden yellow flowers on leafless branches in mid- to late winter. H and S to 12ft (4m).

### PROPAGATION
Simple layering – fall or early spring.
T 2 years.
Softwood cuttings – spring or early summer, 4in (10cm) long, under mist.
T 6–8 weeks, P 12–24 months.
🌰 sown as soon as ripe in a coldframe.
Will germinate in first or second winter.
Species only. T 2 years.
Side-grafting – late winter, or budding in late summer, onto seedling understocks of *H. virginiana.* For cultivars.
U 16 weeks, P 2–3 years.

## Hebe — Scrophulariaceae

About 100 species of evergreen shrubs with attractive leaves and spikes of flowers. They range from fully hardy, compact shrubs with small leaves, like *H. pinguifolia* **'Pagei'** (H 12in/30cm), and *H. odora*, syn.

*Hydrangea macrophylla*

*Hypericum 'Rowallane'*

*Kalmia latifolia*

*H. buxifolia* (H 3ft/1m), and "whipcord" types, like the olive-green *H. armstrongii* (H 3ft/1m), to larger species with large, often variegated leaves that are not completely hardy inland but which thrive in coastal gardens. These include the half-hardy *H. x andersonii* '**Variegata**' (H 6ft/2m) and garden hybrids of *H. speciosa*, like '**Tricolor**' (H 4ft/1.2m). Of the many named cultivars, *H.* '**Fall Glory**', with deep purple flowers from midsummer onward, and *H.* '**Great Orme**', with long spikes of pink flowers in summer, have proved to be reasonable hardy and make small shrubs. H and S 3ft (1m).

**PROPAGATION**
Semi-ripe cuttings – late summer to early fall, 2–4in (5–10cm) long; in a propagator, with gentle bottom heat, for less hardy species, or in a coldframe for hardy species.
T 4–8 weeks, P 6–12 months.

## Hydrangea  Hydrangeaceae

Over 80 species of evergreen and deciduous shrubs and climbers, of which the best known are the many cultivars of *H. macrophylla*. Some of these are sold as pot plants. The majority are frost-hardy or fully hardy. They have "lace-cap" or mop-head flowers (the 'Hortensias'), in white, pink, or blue; the blue-flowered cultivars produce true blues only on acid soil, becoming muddy pink in alkaline conditions (and vice versa). H and S most reach about 3–5ft (1–1.5m).
*H. petiolaris*. Deciduous climber that produces creamy white lace-caps in summer. H to 50ft (15m) eventually.

**PROPAGATION**
Softwood cuttings – early summer, nodal or internodal, 3–4in (7.5–10cm) long.

T 3 weeks, P 10–15 months.
Softwood cuttings – late spring to early summer, for *H. petiolaris*. Take cuttings of non-flowering side shoots, 3in (7.5cm) long. T 6 weeks, P 6–12months.

## Hypericum  Hypericaceae

Over 400 species of hardy to frost-tender shrubs, trees, annuals, and perennials, with showy yellow flowers. Those featured are best in well-drained, loamy soil, with some shelter from cold, dry winter winds.
*H. calycinum* (Rose of Sharon). Hardy evergreen or semi-evergreen shrub making valuable groundcover in shade or sun. Rather too invasive for more manicured areas.
H 24in (60cm), S indefinite.
*H.* '**Hidcote**'. Hardy evergreen or semi-evergreen shrub blooming from midsummer to fall. H to 4ft (1.2m), S 5ft (1.5m).
*H.* '**Rowallane**'. Frost-hardy, semi-evergreen shrub bearing cupped, deep golden yellow flowers from late summer to fall.
H to 6ft (1.8m), S to 3ft (1m).

**PROPAGATION**
Semi-ripe cuttings – mid- to late summer, 4–5in (10–12.5cm) long, preferably heeled, in a coldframe. T 6 weeks, P 9–12 months.
✄ fall. Lift small rooted pieces (Irishman's cuttings), remove topgrowth and pot individually; overwinter in a coldframe.
P 6–8 months.

## Ilex (Holly)  Aquifoliaceae

Some 400 species of deciduous and evergreen shrubs and trees, of which the most widely planted are the legion frost-hardy cultivars of *I. aquifolium* (English holly), with prickly leaves, and *I. x altaclerensis*,

which have spine-toothed or entire leaves. These include many with variegated leaves in shades of gold, creamy white, yellow, or silver and with leaf shapes ranging from smooth and camellia-like ('**Camelliifolia**') to the spiny '**Ferox**' (Hedgehog holly). They range in habit from fastigiate to pyramidal or rounded. Hollies bear male and female flowers on different plants; both are needed to produce berries. Only vegetative propagation will produce offspring that are identical to the parent. It is possible to raise species from seed sown in fall in an open frame or seedbed, but they can take 2–3 years to germinate. Grow in moist but well-drained, humus-rich soil in sun or partial shade.
H to 80ft (25m), S to 50ft (15m), but many hollies are smaller.

**PROPAGATION**
Semi-ripe cuttings – mid- to late summer 2–4in (5–10cm) long, nodal or with a heel, in a propagator with gentle bottom heat. Take cuttings before the first frost or they will defoliate. They can also be rooted in early fall in a coldframe. T 7–8 weeks, P 18–24 months.

## Kalmia  Ericaceae

Some seven species of hardy, evergreen shrubs that need similar growing conditions to rhododendrons, in moist but well-drained acid soil. Use ericaceous seed or cuttings mediums. All parts are poisonous.
*K. latifolia* (Mountain laurel). Dense, glossy-leaved evergreen, with distinctive flowers from late spring to midsummer. They open from crimped, deep pink buds, becoming pale to deep pink and cup-shaped.
H and S to 10ft (3m) eventually.

*Kolkwitzia amabilis* 'Pink Cloud'

*Lonicera periclymenum*

*Magnolia stellata*

**PROPAGATION**

🌡 spring, at 45–54°F (6–12°C ). Seed germinates within a few weeks, and is pricked out and grown on in a cool greenhouse until 5cm (2in) high. Harden off and grow on in a coldframe, for the first two winters. P 2–4 years.
Simple layering – late summer to early fall. P 1–2 years.

## Kolkwitzia — Caprifoliaceae

A single hardy species of deciduous shrub.
*K. amabilis* (Beauty bush). Slender branches, covered for weeks in late spring and early summer with foxglove-like, pink flowers with a yellow throat. It grows in any well-drained soil and thrives on chalk. **'Pink Cloud'** has darker flowers. H 10ft (3m), S 12ft (4m).

**PROPAGATION**

Softwood cuttings – spring, 4in (10cm) long, preferably under mist.
T 4 weeks, P 12–18 months.
Greenwood to semi-ripe cuttings – midsummer, 4–6in (10–15cm) long, preferably with a heel, in a coldframe.
T 3 months, P 12–18 months.
Simple layering – late winter to early spring.
T 1–2 years.

## Lavandula — Labiatae

Some 28 species of hardy to frost-hardy, evergreen, aromatic shrubs and subshrubs.
In gardens, *L. angustifolia* and *L. x intermedia* (English lavender), and their cultivars, are favored for low-growing hedges, and in dry, sunny, mixed and herbaceous borders. The flower spikes, from mid- to late summer, vary from darkest violet through pink to white. All

thrive in full sun on well-drained soil.
They range in size from dwarf forms, like **'Nana Alba'** and **'Munstead'** at H 12–18in (30–45cm), S 12–24in (30–60cm), to **'Grappenhall'**, at H and S 3ft (1m).

**PROPAGATION**

Semi-ripe cuttings – from mid- to late summer, 3–4in (7.5–10cm) long, preferably with a heel, in coldframe.
T 4–16 weeks, P 9 months.
🌡 species and some cultivars, like 'Munstead', come true from seed. Sow in spring in a coldframe or cold greenhouse. Stratification in the refrigerator improves germination.
T 6 months.

## Lonicera — Caprifoliaceae

About 180 or more species of hardy to half-hardy deciduous and evergreen shrubs and climbers, many of which are valued for their fragrant flowers; *L. nitida* for its leaves.
*L. fragrantissima*. Hardy, deciduous or semi-evergreen shrub with strongly scented, white flowers in winter and early spring. H 6ft (1.8m), S 10ft (3m).
*L. nitida*. Hardy, evergreen shrub making a dense, low-growing hedge. H and S to 10ft (3m) eventually. **'Baggesen's Gold'** is smaller and has yellow leaves. H to 5ft (1.5m).
*L. periclymenum* (Common honeysuckle, Woodbine). Vigorous, hardy, deciduous climber that has long been valued for its intensely fragrant flowers, borne from mid- to late summer. **'Belgica'** (Early Dutch) blooms from late spring to early summer, and **'Serotina'** (Late Dutch) blooms from midsummer to early fall. H to 22ft (7m).
*L. x purpusii* is similar to *L. fragrantissima*. H to 6ft (1.8m), S 15ft (5m).

**PROPAGATION**
**Shrubs**

Softwood or greenwood cuttings – early to midsummer; nodal cuttings, 4–5in (10–12.5cm) long, for deciduous species
T 6 weeks, P 18–24 months.
Semi-ripe cuttings – mid- to late summer, 4in (10cm) long, in a coldframe, for evergreens.
T 12 weeks, P 12–18 months.
Hardwood cuttings – fall, 10in (25cm) long, in a coldframe, for deciduous species.
T 6–12 months, P 24 months.

**Climbers**

Softwood to greenwood cuttings – summer; nodal cuttings, 4–5in (10–12.5cm) long.
T 6 weeks, P 18–24 months.
Hardwood cuttings – fall, 8–10in (20–25cm) long, in a coldframe.
T 6–12 months, P 12–24 months.
Layering – fall. P 12 months.

## Magnolia — Magnoliaceae

Some 125 species of deciduous and evergreen trees and shrubs with unrivaled flowering displays. Many are best in larger gardens. The following hardy species are among the more suitable for smaller and medium-sized gardens. All early-flowering magnolias need shelter from wind and frost to prevent flower damage. They require moist, well-drained, humus-rich, preferably acid to neutral soil and sun or part shade.
*M. denudata*, syn. *M. hypoleuca* (Yulan). Deciduous shrub or small tree of elegant habit, producing pure white, chalice-shaped flowers, on bare branches in spring.
H and S to 30ft (10m) eventually.
*M. sieboldii* subsp. *sinensis*, syn. *M. sinensis*. One of the smaller deciduous tree

*Mahonia x media 'Charity'*

*Malus 'Profusion'*

*Philadelphus 'Belle Etoile'*

magnolias, bearing lemon-scented, nodding, saucer-shaped white flowers with a center of red stamens in early summer, which in favorable seasons are followed by pendent fruits that display scarlet seeds in winter. H 20ft (6m), S 25ft (8m).

*M. x soulangeana*. Deciduous shrub or tree with large, goblet-shaped, pure white, pale to deep pink, or purple-flushed flowers, borne with or before the leaves in spring. H and S to 20ft (6m) eventually.

*M. stellata* (Star magnolia). Compact, free-flowering, deciduous shrub clothed in slender-petaled, star-shaped white flowers, with or before the leaves in spring. H 10ft (3m), S 12ft (4m) eventually.

**PROPAGATION**
Greenwood to semi-ripe cuttings – mid- to late summer, 3–4in (7.5–10cm) long, with a heel, in a closed propagator, or under mist, with gentle bottom heat.
T 8–12 weeks, P 2–3 years.
Simple layering – early spring. P 2–3 years.
🌢 sown as soon as ripe, in fall, in a seedbed, will usually germinate in spring.
For all species. P 3–4 years.
Softwood cuttings – early summer (*M. x soulangeana* forms). Place stem with one or two leaves and buds under mist or in a propagator with bottom heat.
T 8–10 weeks, P 24 months.

## Mahonia                                    Berberidaceae

About 70 species of hardy, evergreen shrubs, valued for their handsome foliage, and, in most species and cultivars, their fragrant yellow flowers. Grow in fertile, moist but well-drained soil. They prefer shade but will tolerate sun if the soil is not too dry.

*M. x media* 'Charity'. Erect, evergreen shrub, with attractive foliage and racemes of fragrant yellow flowers from late fall to late winter. H to 10ft (3m), 6ft (2m).

**PROPAGATION**
Leaf-bud cuttings – late summer to early fall, in a propagator with gentle bottom heat.
T 12 weeks, P 18–24 months.
Semi-ripe cuttings – late summer to early fall, 3–4in (7.5–10cm) long, in a propagator with gentle bottom heat.
T 12 weeks, P 18–24 months.
🌢 fall, in a seedbed or open frame, germinates in spring. P 1–2 years.

## Malus                                        Rosaceae

Some 35 species, and many cultivars, of deciduous trees and shrubs, that include the flowering crabapples. These are mainly small to medium-sized trees; many are ideal for smaller gardens, with flowers in spring and fruits and foliage colors in fall. The cup-shaped flowers range from white (*M. 'John Downie'*), through pink (*M. floribunda)*, to deepest pink-purple, as in *M. 'Profusion'* and *M. x purpurea)*, both of which also have red-purple foliage. Those notable for their fall fruits, which often persist after leaf fall, include **'Golden Hornet'** and **'John Downie'**. Grow in moderately fertile, well-drained soil in full sun or light shade.
H 22–30ft (7–10m), S 12–25ft (4–8m).

**PROPAGATION**
T-budding or chip-budding – summer, onto clonal rootstocks as used for fruit trees *(see pp 82-83)*. Named cultivars.
U 4–6 weeks, P 12–24 months.
Whip-grafting – winter, as above. U 4–6 weeks, P 12–24 months.

Hardwood cuttings – late fall to early winter. T 6 months, P 12–24 months.
🌢 fall, in a seedbed outdoors, usually germinates in spring. Species only. May take 10 years to reach flowering size. P 12–24 months.

## Paeonia                                     Paeoniaceae

Some 30 species of mostly hardy, mainly herbaceous perennials but with several tree peonies, including *P. lutea*, with yellow flowers, and *P. delavayi* with deep red blooms, in early summer. Many hybrids and cultivars are available, derived from *P. suffruticosa* and *P. lutea*. All have erect, woody stems, divided leaves, and bear flowers of many shades and shapes. Grow in deep, fertile soil in sun or light shade; protect from cold winds and spring frosts. H and S to 5–6ft (1.5m).

**PROPAGATION**
🌢 sown as soon as ripe, in a coldframe, usually germinates in spring, although root growth precedes shoot growth, and this may lead to delayed emergence, especially if seed has been allowed to dry out. P 3–4 years.
Simple layering – spring. P 2–4 years.
Grafting – named cultivars. Usually wedge-grafted in midsummer onto herbaceous peony understock. U 3 months, S 2–3 years.

## Philadelphus (Mock orange)
                                              Hydrangeaceae

About 40 species of mainly deciduous shrubs, the most commonly grown of which are named hybrids and cultivars. They have predominantly white flowers, from early to midsummer, valued for their often intense fragrance. **'Beauclerk'**, **'Belle Etoile'**, **'Sybille'**,

*Potentilla fruticosa*

*Pyracantha* 'London'

*Rhododendron* 'Bashful'

and **'Virginal'** are hardy and among the most reliable and floriferous. Grow in fertile, well-drained soil in full sun or light shade. H 4–10ft (1.2–3m), S 6–8ft (1.8–2.5m).

**PROPAGATION**
Softwood cuttings – early summer, 3–5in (7.5–12.5cm) long.
T 3–6 weeks, P 6–12 months.
Hardwood cuttings – late fall to early winter, 8–12in (20–30cm) long.
T 6 months, P 12 months.

## *Pittosporum*                Pittosporaceae
About 200 species of mostly half-hardy or frost-tender, evergreen shrubs and trees. Grow in fertile, well-drained soil in full sun or light shade; variegation is best in full sun.
*P. tenuifolium.* Hardy to about 23°F (-5°C) in a warm, sheltered site. Evergreen shrub or small tree with chocolate-colored, scented flowers in spring. Cultivars, with variegated or colored leaves, are mostly less vigorous. H 12–30ft (4–10m), S 6–15ft (1.8–5m).

**PROPAGATION**
as soon as ripe in fall, in a coldframe. Rub away the sticky, viscous coating with dry sand. Germination usually occurs in spring. Species only. P 18–24 months.
Semi-ripe cuttings – late summer, for cultivars, 3–4in (7.5–10cm) long, with a heel, in a propagator with gentle bottom heat. Small bushy or branched cuttings of less vigorous cultivars can be rooted successfully. T 10 weeks, P 12–24 months.

## *Potentilla*                Rosaceae
About 500 species of hardy annuals, biennials, perennials, and shrubs. Those most useful in gardens are the many shrubby

cultivars, mostly derived from *P. fruticosa.* These bear saucer-shaped flowers over long periods from early summer until early fall, in a range of colors. Plant in poor to moderate soil in full sun. H to 4ft (1.2m), S 5ft (1.5m).

**PROPAGATION**
Softwood cuttings – early summer; nodal cuttings, 4in (10cm) long, in a propagating case. T 6 weeks, P 6–12 months.
Semi-ripe cuttings – mid- to late summer, 3in (7.5cm) long, in a coldframe.
T 6 months, P 12 months.
Hardwood cuttings – fall, 3–6in (7.5–15cm) long, in a coldframe.
T 6 months, P 12–18 months.

## *Prunus*                Rosaceae
Genus of about 400 species of evergreen and deciduous trees and shrubs. The best known are perhaps the legion cultivars of Japanese cherries, *P. dulcis* (almond), and other spring-flowerers like *P. subhirtella*, *P. cerasifera*, and cultivars, as well as those grown primarily for their bark, *P. serrula* and *P. maackii.* Named cultivars are nearly all chip- or T-budded, as for fruit trees *(see pp. 82-83).* Grow in moist but well-drained, moderately fertile soil in sun; evergreens tolerate partial shade. There is a wide range of sizes: H 3–40ft (1–12m), S 3–20ft (1–6m).

**PROPAGATION**
sown as soon as ripe in fall, in a seedbed or open frame, usually germinates in spring. Species only. P 24 months
Semi-ripe cuttings – midsummer, for smaller trees and shrubs, including evergreens like *P. laurocerasus*, 3–4in (7.5–10cm) long, with a

heel, in a coldframe. Some success with Japanese cherries can be expected with this method. T 12 weeks, P 2 years.
Softwood cuttings – late spring to early summer, 2–3in (5–7.5cm) long, preferably under mist. T 4–6 weeks, P 2 years.

## *Pyracantha*                Rosaceae
Some seven species of thorny, evergreen shrubs, grown mainly as named cultivars and valued for their flowers, foliage, and large, very colorful, yellow, orange, or red fruits in fall. They are grown as free-standing shrubs, or are wall-trained – especially useful for cool, shady walls. Grow in fertile, well-drained soil and protect from cold winds in frost-prone areas. Depending on how they are trained, H and S to 10ft (3m) or more.

**PROPAGATION**
Semi-ripe cuttings – mid- to late summer, 4–6in (10–15cm) long, with a heel, in a propagator, with gentle bottom heat.
T 4–6 weeks, P 12–15 months.

## *Rhododendron*                Ericaceae
Between 700 and 800 species and innumerable hybrids and cultivars, *Rhododendron* includes evergreen and deciduous shrubs that range from prostrate dwarfs with tiny leaves ⅛in (3mm) long, to potentially large trees with leaves that reach 30in (75cm) or more in length. There is a wide range of sizes: H from 12in (30cm) to 22ft (7m), S very similar. The azaleas, both evergreen and deciduous, are a subgroup of the genus. The flowering season begins in late fall or winter and continues into mid- or late summer, for the latest-flowering species and hybrids. The earlier flowers are

123

*Rhus typhina*

*Robinia pseudoacacia 'Frisia'*

*Romneya coulteri*

susceptible to frost damage, so the early-flowering species and hybrids are best planted in sites that escape the early morning sun. All need moist but well-drained, humus-rich, acid soil and the majority need dappled shade, and shelter from cold, dry winds; some of the dwarf alpines and the so-called "iron-clad" hardy hybrids tolerate more open conditions in sun. Use lime-free potting, cuttings, and seed medium for all propagation. As a guide, propagate large-leaved species by seed or layers; hardy hybrids by cuttings or grafting; deciduous azaleas by layers, cuttings, or grafting; evergreen azaleas and alpine (dwarf) species by cuttings; alpines also by seed.

**PROPAGATION**

Semi-ripe cuttings – late summer; nodal, 4in (10cm) long, under mist or in a propagator, with gentle bottom heat. They vary considerably in their speed of rooting, from 10 weeks to 3–4 months. P 3–4 years.
Simple layering – spring or fall. T 2 years.
Side-veneer or saddle-grafting – late winter or late summer. Once the graft union has formed, harden off very slowly.
U about 7 weeks, P 24–30 months.
🌰 sown as soon as ripe, or in spring, at 55–61°F (13–16°C ). Sow seed of hardy alpine species in a coldframe as soon as ripe. Surface-sow on finely milled moss peat. (Seedlings are sensitive to over-fertilizing.) They hybridize freely and garden-collected seed may not come true. P 2–3 years.
Softwood cuttings – early summer (deciduous Knaphill-type azaleas), in a mist propagator or with bottom heat. Well worth trying but success rate varies according to clone. T 8–12 weeks, P 24 months.

## Rhus                                    Anacardiaceae

Some 200 species of hardy to frost-tender, deciduous or evergreen shrubs, trees, and climbers. They are cultivated mainly for their pinnate leaves, which in deciduous species often color well in fall, and their erect, showy fruit clusters. They grow in any moderately fertile well-drained soil, in partial shade or sun, but fall color is best in full sun.
*R. typhina* (Stag's horn sumach). Deciduous, spreading, suckering shrub with finely divided, pinnate leaves that color brilliant red, yellow, and orange in fall. It bears panicles of yellow-green flowers in summer that, on female plants, give rise to conical, red-furry fruit clusters that persist through winter. H 15ft (5m), S 20ft (6m). 'Dissecta', syn. 'Laciniata', is more compact with very elegant, finely cut leaves. H to 6ft (1.8m).

**PROPAGATION**

🌰 species only. As soon as ripe in fall, in a seedbed outdoors. P 2–3 years.
Semi-ripe cuttings – midsummer 4–5in (10–12cm) long, preferably under mist, or root in a propagator, with bottom heat. T 4–5 months, P 16–18 months.
✂ separate suckers during the dormant period, between fall and spring.
Plant out immediately, or pot up and grow on until the roots are well established, then plant out in fall or spring.

## Ribes                                   Grossulariaceae

Some 150 species of mainly deciduous shrubs, of which *R. sanguineum* and its cultivars are the most commonly grown. Reliably hardy and floriferous, it thrives in almost any soil in sun or dappled shade. It

can be raised from seed, and will self-seed, but seedlings often produce flowers of washed-out or undistinguished color.
*R. sanguineum* (Flowering currant). Vigorous deciduous, spreading shrub with clusters of rosy-pink flowers in spring and aromatic lobed leaves, like those of black currants. 'Tydeman's White' and 'White Icicle' have white flower clusters; those of 'Pulborough Scarlet' and 'King Edward VII' are dark red. H and S to 6ft (1.8m).

**PROPAGATION**

Softwood cuttings – early summer, 3–4in (7.5–10cm) long. T 8 weeks, P 18 months.
Hardwood cuttings – fall, 10–12in (25–30cm) long. T 4–5 months, P 18 months.

## Robinia                                 Leguminosae

Genus of between 4 and 20 species of bristly or thorny, deciduous trees and shrubs with finely divided foliage and pea-like flowers in spring or early summer. Most are useful for their tolerance of industrial pollution and poor, dry soils. Prefer sun.
*R. pseudoacacia*. A rapidly growing, suckering tree that is too large for many gardens. H to 50ft (15m), S 15ft (5m) or more. 'Frisia' has golden leaves. It is more suited to smaller gardens. H to 50ft (15m) eventually.
'Umbraculifera', syn. 'Inermis' (Mop-head acacia) has a round-headed habit, and is thornless. H and S to 20ft (6m).

**PROPAGATION**

🌰 species only. As soon as ripe in fall, in a coldframe. Spring-sown seed should be soaked or scarified. P 2–3 years.
Grafting – for cultivars, side-grafting in winter onto seedling understock of the species. U 2–3 months, P 12–18 months.

*Rosa moyesii*

*Rosa 'Gloire de Dijon'*

*Rosa 'Seagull'*

## Romneya                    Papaveraceae

Just two species of frost-hardy, suckering subshrubby perennials or shrubs.

*R. coulteri* (Tree poppy). Upright, deciduous shrub, with divided, intensely glaucous, blue-gray leaves, and large poppy-like white flowers with golden centers, from midsummer to fall. It needs a warm, sunny, sheltered site in sun and may be cut to the ground in hard winters, but usually re-sprouts, especially if given a deep, dry winter mulch. H 3–8ft (1–2.5m), S indefinite.

### PROPAGATION

✄ separate suckers in spring. P immediately.
🌰 early spring, at 55–61°F (13–16°C). Seed will germinate in 10–14 days. Overwinter young plants under glass for their first winter. P 15 months.

Root cuttings – winter, 1in (2.5cm) long, laid horizontally on the medium. They will shoot within 2–3 months, but leave undisturbed until growing vigorously, usually by early summer.

## Rosa                          Rosoceae

Although the genus *Rosa* comprises only some 150 species of evergreen and deciduous shrubs and climbers, the extensive breeding and hybridization that has been carried out – especially over the last 300 or more years – has given rise to many thousands of cultivars. The history of their breeding and parentage is extremely complex, and hybridizing has given rise to such diversity of habit, leaf, and flower form and color, that a system of horticultural classification has been developed to bring some order to their multiplicity. The three main divisions are between species roses, the Old roses, and the Modern roses. The Old roses are held by most authorities to be those bred before the First World War and the major groups include the Albas, Gallicas, Damasks, Centifolias and their mossy sports, the Moss roses, as well as more recent developments like the Teas, Chinas, Bourbons, Portlands, and Hybrid Perpetuals. Many of the Old roses produce only one glorious flush of bloom in midsummer. The best known of the Modern roses, many of which are remontant, or repeat-flowering, include the large-flowered bushes (the Hybrid Teas) and cluster-flowered bushes (the Floribundas), as well as the more recent groups, like groundcover, miniature, and patio roses. Both Old and Modern groups include examples of climbers and ramblers. Roses are available in a huge range of sizes and forms: H and S from 12in (30cm) to 20ft (6m) or more.

Traditionally, nearly all named rose cultivars are increased by budding in mid- to late summer onto various rootstocks that are selected for vigor and suitability to climate and soil type. For example, *R. multiflora* rootstocks prove especially useful where winters are cold, and perform reasonably well on relatively poor soils.

*R. 'Laxa'* is the most widely used rootstock; it is hardy, almost thornless (and therefore easy to handle) and performs reliably in a range of soils and climates. It also has the advantage that it does not sucker freely. On other rootstocks, especially *R. canina*, bud-grafted roses frequently produce suckers from below the graft union which must be removed as soon as seen, if they are not to dominate and eventually out-compete the grafted cultivar. Commercial rose nurseries are geared to the large-scale production of roses by budding. The primary reason commercial growers use this method of propagation is that it produces offspring which are genetically identical to the parent from which the scion material is taken, producing saleable stocks – uniform in habit, vigor, and flowering performance – relatively quickly.

For the amateur grower, where such uniformity is less important, it is possible to propagate many named cultivars using the range of techniques outlined below. It is important, though, to bear in mind that parentage by these means can prove difficult. Parents may be reluctant to root or can give rise to offspring that lack the vigor of their budded parent. It is certainly worthwhile experimenting, however, with any notably vigorous cultivars that you grow in your garden, and with roses that are relatively simple hybrids or cultivars of the species. These would include cultivars of the shrub roses *R. pimpinellifolia* and *R. rugosa*, and rampant ramblers or climbers like 'Seagull', 'Gloire de Dijon', or *R. filipes* 'Kiftsgate'. The propagation of roses from seed – at least by the amateur gardener – is restricted to the true species, like *R. glauca*, *R. moyesii*, or *R. hugonis*. The species will come true from seed, although you can expect a range of natural variation, as with the offspring of any seed-raised species. Commercial growers raise all new cultivars from seed, by the process of hybridization. Given their complex parentage, offspring will be enormously variable and the chances of producing a new cultivar with all the

125

*Sambucus nigra*

*Sorbus aucuparia*

*Styrax japonica*

required virtues of good flower form, color are remote for amateurs. Breeders take note of the chromosome numbers of parents and then undertake controlled cross-pollination by hand. The resulting seedlings are then grown on for two or more years before it becomes clear whether any will produce first-quality blooms. Even then, the hybridization process is only half complete; any promising offspring will be bulked up by budding, and put on trial for several years in order to assess their overall strengths and weaknesses.

**PROPAGATION**

🌰 species only. Collect ripe hips in fall and extract the seed from the flesh. Stratify for 3–6 weeks in a domestic refrigerator and sow in plug trays in a coldframe. They may germinate in the first spring, but can take 12–24 months, so do not discard the containers. P 2–3 years.

🔧 suckering species and cultivars grown on their own roots (e.g. of R. pimpinellifolia, R. rugosa, and R. gallica). Separate suckers when dormant and grow on in a nursery bed. P 1–2 years.

Simple layering – after flowering, in summer. Suitable for many climbers and ramblers, and vigorous cultivars with flexible stems, such as Alba and Damask roses.
T 12 months, P 12–24 months.
Hardwood cuttings – early fall, about 9in (22cm) long. Species and their cultivars, simple hybrids, ramblers, vigorous climbers, and miniature roses. Hybrid Teas and Floribundas may be slow to root by this method. T 12 months, P 2–24 months.
Semi-ripe cuttings – summer after flowering, 6in (15cm) long, in a propagator. Species

and their cultivars, simple hybrids, ramblers, vigorous climbers, and miniature roses. Hybrid Teas and Floribundas may be slow to root by this method.
T 6–8 months, P 12–24 months.

## Rubus                                Rosoceae

Some 250 species of evergreen and deciduous shrubs that include blackberries, hybrid berries, and raspberries, which are grown for their fruit (see p.189). The more ornamental kinds are grown for their flowers and foliage, or for their very attractive winter shoots. Those in cultivation are hardy. They grow in any well-drained, moderately fertile soil, in sun or dappled shade; those grown for winter shoots produce the best results in full sun.
R. 'Benenden'. Smooth-stemmed, deciduous shrub with lobed, dark leaves and large, pure white, cupped flowers in spring and early summer. H and S to 10ft (3m).
R. cockburnianus. Suckering, deciduous shrub whose primary ornamental value is its white-bloomed shoots in winter. It has dark leaves that are white-hairy beneath, and produces small, purple flowers at the stem tips in summer. H and S 8ft (2.5m).
R. thibetanus. Similar to R. cockburnianus, with gray-hairy leaves. H and S 8ft (2.5m).

**PROPAGATION**

Hardwood cuttings – fall to early winter, 8–10in (20–25cm) long.
T 6–9 months, P 12–24 months.
Simple layering – fall, for R. 'Benenden'. T 9 months, P 12 months.
🔧 separate rooted suckers of suckering species in early fall or spring, and re-plant immediately.

## Sambucus                          Caprifoliaceae

Some 25 species of hardy, herbaceous perennials and deciduous shrubs and trees. Grow in humus-rich, moist but well-drained soil in sun or light shade. **S.** *nigra* and **S.** *racemosa* (Red-berried elder) have a number of cultivars with very attractive foliage, that colors well in fall. Among the finest are **S.** *nigra* **'Guincho Purple'**, with deep purple leaves turning red in fall, H and S 20ft (6m); **'Laciniata'**, with finely cut leaflets turning red and gold in fall, is a similar size. **S.** *racemosa* **'Plumosa Aurea'** has very finely cut, golden yellow leaves, that need shade from hot sun; it is ideal for brightening dark corners. H and S 10ft (3m).

**PROPAGATION**

Hardwood cuttings – late fall to winter, 10in (25cm) long, with a heel to avoid exposing the pith. T 1 year, P 15 months.

## Sorbus                                 Rosaceae

Some 100 species of hardy, deciduous trees and shrubs, valued for their habit, fruit, and leaves. Grow in well-drained soil in sun; most prefer acid to neutral soil. The species come true from seed.
S. aucuparia (Mountain ash). Has pinnate dark green leaves often coloring brilliantly in fall. Orange-red berries. H 50ft (15m), S 22ft (7m).
S. cashmiriana. White fruit; dark green pinnate leaves. H 25ft (8m), S 22ft (7m).
S. hupehensis. Pink fruit; blue-green leaflets. H and S 25ft (8m).
S. vilmorinii. Crimson fruit, becoming white; glossy pinnate leaves. H and S 15ft (5m).

**PROPAGATION**

🌰 sown as soon as ripe in a coldframe,

*Vitis* 'Brandt'

*Weigela florida* 'Variegata'

*Wisteria floribunda*

usually germinates in spring. Clean from the
flesh. P 12–24 months, or more.
Cultivars are propagated by T-budding, or
chip-budding, in summer, or by whip-and-
tongue grafting in winter.

## Styrax                                           Styracaceae

About 100 species of deciduous or
evergreen shrubs and small trees. Plant
in moist, humus-rich, neutral to acid soil.
*S. japonica*. Graceful, deciduous tree with
pendent, bell-shaped flowers in early
summer. Hardy, but may suffer damage from
late frosts, so site with some shade from
morning sun. H 30ft (10m), S 25ft (8m).

### PROPAGATION
as soon as ripe. Keep at 59°F (15°C) for
3 months, then transfer to the bottom of
the refrigerator for 3 months. Bring under
glass when signs of germination are seen.
P 18–24 months.
Softwood cuttings – late June to early July,
2½in (6cm) long, in a propagator with gentle
bottom heat. T 6–8 weeks, P 24 months.

## Symphoricarpos                       Caprifoliaceae

About 17 species of hardy, deciduous shrubs
that will grow in any well-drained soil in
sun. They are grown for their fleshy, round
fruit. *S. albus* (Snowberry). Small shrub
with slender, erect shoots forming dense
thickets. Insignificant flowers in summer are
followed by large, globose white berries
which persist through winter. H and S
6ft (2m). *S. x chenaultii* is the most
popular variety.

### PROPAGATION
Hardwood cuttings – fall, 8–10in (20–25cm)
long. T 1 year, P 2 years.

separate suckers when dormant and re-
plant immediately.

## Viburnum                                Caprifoliaceae

About 150 species of hardy to frost-hardy,
evergreen and deciduous shrubs. Grow in
fairly fertile, well-drained soil in sun or partial
shade. This versatile genus has many species
offering sweetly scented flowers, ornamental
fruits, and attractive foliage which, in
deciduous species, often colors beautifully in
fall. The white-flowered *V. farreri* (H 10ft/3m)
and *V. grandiflorum* (H 6ft/1.8m), and the
pink-flowered *V. x bodnantense* (H 10ft/3m),
flower on upright leafless branches from late
fall and in mild spells from winter to spring.
*V. tinus* (Laurustinus) is an evergreen shrub
(H 10ft/3m), useful as a hedge or specimen,
and flowers from late fall to early spring.
Most of the remaining species, including
*V. carlesii*, *V. x carlecephalum*, *V. x burkwoodii*,
*V. dilatatum*, and *V. x juddii*, are spring and
summer flowering. H to 10ft (3m),
S to 8ft (2m).

### PROPAGATION
Softwood cuttings – early summer, 3–4in
(8–10cm) long. T 4 weeks, P 2–3 years.
Semi-ripe cuttings – mid- to late summer,
3–4in (8–10cm) long, with a heel, in a
propagator with gentle bottom heat.
T 4 weeks, P 2–3 years.

## Vitis                                               Vitaceae

About 65 species of hardy, deciduous, mostly
climbing shrubs, some with edible fruits.
Grow in fertile, well-drained soil in sun.
*V. 'Brandt'*. A vigorous, deciduous climber,
with a dual purpose: it produces sweet black
grapes in favorable years, and its green leaves

color red and purple, highlighted with yellow
veins in fall. H 22ft (7m).

### PROPAGATION
Vine-eyes – winter.
Hardwood cuttings – late winter, 10–12in
(25–30cm) long, outdoors or in an open
frame. T 12 months, P 18–24 months.

## Weigela                                    Caprifoliaceae

Some 12 species of deciduous shrubs. They
bear tubular white, pink, or deep ruby-red
flowers in early summer. Grow in well-
drained soil in sun or light shade.
*W. florida*. Has dark pink flowers in mid-
spring, 'Variegata' has white margined leaves.
H and S to 8ft (2.5m).

### PROPAGATION
Softwood cuttings – early summer, 4–5in
(10–12cm) long. T 6 weeks, P 18–24 months.
Hardwood cuttings – fall, 8–10in (20–25cm)
long. P 12 months.

## Wisteria                                   Leguminosae

Genus of 10 species of vigorous, twining,
deciduous climbers. *W. floribunda* (Japanese
wisteria) flowers in early summer, and the
larger *W. sinensis* (Chinese wisteria) blooms
in late spring and early summer. Both bear
long, pendent racemes of violet-blue, pink,
or white flowers, to 12in (30in) or more.
Several cultivars with longer racemes or
stronger-colored flowers are available.
Provide support in fertile, well-drained soil
in sun or part shade. H 28ft (9m) or more.

### PROPAGATION
Serpentine layering – spring. T 1 year.
Softwood cuttings – early summer. Take basal
cuttings of side shoots, 3–4in (7.5–10cm)
long. T 6–7 weeks, P 12–18 months.

# HOUSEPLANTS

Most gardeners want to bring the garden into the home. Fortunately, there are many plants suitable for growing indoors that can be enjoyed when it is too cold or wet to venture outside. Many tolerate low light levels, while others, such as the cacti, thrive in the dry atmosphere of centrally heated rooms, provided they have sufficient light.

Most houseplants originate from the subtropical or tropical zones and make rapid growth given the conditions that suit them. Most of the methods described in this chapter will produce new plants much more quickly than for hardy, cool–climate plants. You can easily increase your stocks of houseplants in the comfort of your home when forays outdoors seem unappealing.

◄ *Even naturally long-lived houseplants occasionally succumb to the less than ideal conditions in the modern home. Propagating them regularly helps ensure your stock.*

# WHY PROPAGATE
# HOUSEPLANTS?

*Although some houseplants can be cherished into a venerable old age, many lose their charms over the years and hence are best replaced from time to time. The satisfaction that comes from propagating those replacements for yourself is tremendous, giving a treasured pot plant the chance of immortality through reincarnation.*

Houseplants include a number of plant groups; some are increased by methods seldom used for outdoor plants.

There are many reasons why you should wish to propagate your own houseplants. First and foremost, since even the best cared-for plants need replacing in time, it is far more economical to create your own stocks of reserve plants than to purchase expensive replacements from commercial growers.

Some houseplants, such as *Ficus elastica*, will eventually outgrow their allotted space in the home and so need to be replaced periodically with smaller specimens. Others can lose their natural habit in time when confined in a pot, becoming gaunt, lop-sided or top-heavy as they stretch towards the light. Shrubby houseplants can also develop bare patches down low as they age, caused by the periods of dormancy that gardeners in cold climates must often impose in order to overwinter them successfully.

In their wild habitats, many of the plants that we grow as houseplants are vigorous and more or less permanently in growth. For this reason, most can be increased extremely rapidly, though some form of supplementary heating is generally required.

## GROWING FROM SEED

A few houseplants are regularly raised from seed, notably *Coleus* and *Cacti*. For some, the treatment is the same as for half-hardy annuals, but others have more specialized requirements. In general, seed will germinate at about 9°F (5°C) higher than the average minimum temperature required for growth.

## VEGETATIVE TECHNIQUES

Many shrubby houseplants, including some of the succulents, can be increased by cuttings as easily as their hardy counterparts, but most need bottom heat as supplied by a soil-warming cable or heated propagator to root successfully. For suitable subjects, large quantities of new plants can also be propagated quickly by leaf cuttings. Indeed, in some cases, a single leaf can be used to produce many offspring.

## SIMPLE DIVISION

Houseplants that develop a crown can be increased by division, rather like herbaceous perennials. This procedure is best carried out when the plant has outgrown its container and is just coming into growth in spring. Many orchids can be divided, but the divisions must be of a certain size, so only a few new plants may be produced. Whereas on other plants the old growth is generally discarded, on orchids the

①

*Dracaena marginata 'Tricolor'*
This houseplant often becomes bare of leaves at the base, but may be cut back and renewed by rooting leafless stem sections in gentle bottom heat in early summer.

②

*Saintpaulia*
The legion cultivars of the African violet do not come true from seed, but are easily increased by single leaf cuttings, which will root within 2–3 weeks in summer.

dormant back-bulbs can also be used for propagation. Some bromeliads and succulents have a central rosette of leaves around which new plantlets form. When the central rosette dies off, the plantlets, or offsets, can be removed and grown on.

## AIR LAYERING

This is a method of propagation for large plants that in the wild would layer themselves in the surrounding soil. This habit can seldom be accommodated in a container, but suitable stems can be encouraged to produce roots by air layering without the need to bring them down to ground level.

### Propagation from Plantlets

*There are several houseplants that produce completely formed young plantlets which grow whilst still attached to the parent plant. These young plantlets are, in effect, ready-made propagules, and increasing stock from them is usually very simple.* Chlorophytum comosum, *the spider plant, is well known for producing cascades of young plants on its flowered stems. They root easily if introduced to a pot of medium and pinned down onto the medium surface.*

A *Chlorophytum comosum* 'Variegatum' with numerous young plantlets hanging from its stems.

**3**

### Chlorophytum comosum

Raising variegated cultivars of the spider plant is so simple that it does not matter that seed is not an option. Increase by division or plantlets at any time of year except winter.

**4**

### Echeveria

An example of one of the rosette-forming succulents that produce ready-made propagules in the form of daughter rosettes They can also be propagated by stem or single leaf cuttings.

**5**

### Aglaonema

This is one of many foliage plants that seldom set seed in cultivation, and rely on vegetative forms of increase. When parent plants become leggy, divide or separate basal suckers in spring.

131

# PROPAGATING TECHNIQUES
# HOUSEPLANTS

*Since the plants that we choose to containerize and grow indoors come from an extremely diverse range of native habitats, there are a similarly large number of techniques by which they can be propagated. Many of these methods, such as raising begonias from a cut leaf, are extremely simple, while others require a little more attention.*

Representatives of a huge diversity of plant groups are grown as houseplants, many of which have tropical or subtropical origins. Some are propagated by methods common to other groups, some use modifications of these techniques, and a few, like orchids and succulents, use techniques that are unique to their group.

## SEED

In most cases, when houseplants can be raised from seed, they are sown in the normal way. The main difference is that higher temperatures are needed to germinate and grow on tropical and subtropical species. As a rule of thumb, seed will germinate at 9°F (5°C) higher than the minimum needed for normal growth.

In nature, most cacti and succulents grow in hot conditions in sharply drained soils, restricting their growing cycle to periods when sufficient moisture is available for growth. Similar conditions must be provided for successful seed germination.

Use a proprietary cactus seed medium or make your own by mixing two parts of soilless seed medium with one part of washed ¼in(5mm) grit.

Water in the seeds by misting with a hand-held sprayer to reduce the risk of overwatering. Sow in late winter so that seedlings make maximum growth to enable them to survive the following winter. Germination will occur at 66°F (19°C) but will be more rapid at 70–81°F (21–27°C).

### Sowing seed of cacti and succulents ...........
The best time to sow seeds of cacti and other succulents is from late winter to late spring. Some fine seed is slow to germinate. Some large seeds may need stratifying for 2 days.

► 1 Prepare a 5in (12.5cm) plastic halfpot by placing a layer of broken clay crocks at the bottom, topped off with a ½in (1cm) layer of mixed gravel and charcoal. Fill the pot to within ½in (1cm) of the pot rim with seed medium. Sow the seed thinly and evenly on the surface of the medium. Water in by misting over with a hand-held sprayer.

◄ 2 Barely cover the medium surface with a layer of fine grit. Enclose the pot in a clear plastic bag, inflate, and seal, and germinate at a temperature of 66–81°F (19–27°C).

### Growing *Tillandsia* from seed ..................
Most bromeliads are sown in seed medium in a pot as for cacti (*see left*), but *Tillandsia* seed germinates better if sown on a bundle of moistened moss and twigs.

▲ Mix moistened, live sphagnum moss throughout a bundle of conifer twigs, and tie with wire or raffia. Sprinkle seeds thinly onto the surface of the bundle. Each tiny seed has a coma, or tuft, of hair and will cling onto the fine strands of damp moss. Water the bundle evenly by misting with a hand-held sprayer. Suspend the bundle in a light, airy place at 81°F (27°C).

## Dividing orchids . . . . . . . . . . . . . . . . . . . . . . . . . . . . .

Some orchids can be propagated by dividing them into one or more pieces. Division should also be used to re-invigorate a mature plant, especially one that has become potbound.

▲ **I** Lift the entire plant (such as the *Cymbidium* shown here) from its pot and separate it into sections by hand, each with at least three healthy pseudobulbs. Remove the leafless back-bulbs, discarding any shriveled ones and retaining plump, green ones for repotting. Trim up each section by removing any dead, shriveled, dark brown roots with a pair of sharp secateurs.

◄ **2** Re-plant each section, setting the oldest pseudobulbs against the back edge of the pot, so that the plump, young ones, which have short, fresh green leaves, lie towards the front center of the pot. The new growth will then have space to spread forwards in the pot. Ensure that the pseudobulbs are planted at the same level as they were before dividing. Fill with orchid medium, tapping the pot gently so that the medium, is shaken down around and between the roots.

### Propagating *Tillandsia* by seed

Tillandsias are bromeliads that, for the most part, grow epiphytically in nature; they do not grow in soil as other plants do, but attach themselves to tree trunks by a reduced root system that serves as anchorage. The leaves absorb food and moisture from mist and rainfall. To germinate successfully, *Tillandsia* seed needs treatment that mimics conditions found in the wild. One way of doing this is to provide a substrate that is similar to the mossy bark upon which they would germinate naturally. Bundle evergreen conifer twigs and living sphagnum moss together and bind it with raffia or twine. Sprinkle the seeds onto the surface and water them thoroughly by misting with a hand-held sprayer. Suspend the bundle in a light, airy but draft-free place at a temperature of 81°F (27°C) and mist over regularly. Seedlings germinate in about 3–4 weeks. When they are large enough to handle, they can be transferred to a "bromeliad tree" or slab of bark.

## DIVISION

Many houseplants, especially clump-forming or herbaceous sorts like *Chlorophytum* or *Aspidistra*, can be divided in early spring as described for herbaceous perennials (*see p. 29*). Water them well, slip the rootball from the pot, and separate it into sections by hand. Re-plant the divisions into clean pots containing the same growing medium as that for the parent plant.

### Dividing orchids

Some orchids, like *Cymbidium* and *Cattleya,* produce new growth from the ends of rhizomes to form stems that swell to form pseudobulbs, which act as water storage organs. This type of orchid can often be increased by division. The process differs slightly from normal division in that, as well as a growing point, each piece needs at least three pseudobulbs if the new plant is to be large enough to establish and survive.

A few orchids will fall apart into sections when removed from the pot. If they do not, cut through the rhizome with a sharp knife to separate the sections.

## STEM CUTTINGS

Some houseplants can be propagated by stem cuttings, although there are variations on the technique. With orchids like *Dendrobium* the process is simply a matter of cutting leafless stems into sections, as shown below,

### Increasing orchids by stem cuttings . . . . . . . . . . . .

A number of orchids, such as *Dendrobium* or *Epidendrum*, produce long, cane-like stems, which make excellent material for propagation by stem cuttings.

◄ **I** Divide some of the parent plant into sections, 10in (25cm) long, by cutting its cane-like stems with a straight cut either at the base or just above a leaf node. Further subdivide each section into 3in (7.5cm) pieces, cutting midway between nodes, so that each piece has at least one node.

▶ **2** Set the pieces neatly on the surface of a seed tray of moistened sphagnum moss and place in a humid environment in good light but out of direct sunlight. The new plantlets should appear after a few weeks.

at the beginning of the growing season. Each piece must have at least one node from which a dormant bud will break. They may take several months to root and can then be potted up individually.

### Cacti and succulents from stem cuttings

With succulent plants like *Kalanchoe* and *Crassula* (*see right*) the technique for stem cuttings is very simple although it differs slightly from other stem cuttings. To reduce the risk of rotting, the wound made in the stem base must be allowed to dry out and callus over before insertion and the cuttings must not be too long. Cuttings are best made in spring at the end of the dormancy to allow a season's growth before the next winter's dormancy.

Select a sturdy stem and remove it from the parent plant with a straight cut just above a bud. Trim it down to 2–3in (5–7.5cm) in length, again with a straight cut at the base, and remove the lowest leaves to leave about 1in (2.5cm) of bare stem. If the wounded stem bleeds, dip it in lukewarm water and staunch the flow from the parent plant's stem by pressing gently with a pad of damp cloth. Avoid all skin contact with the sap; it is highly irritant in many succulent species. Place the cutting in a warm, dry place for about 48 hours.

### Propagating by offsets ........................

Many bromeliads and succulents can be increased by offsets – young rosettes which form at the base of the parent plant that may be detached and rooted.

▶ **I** Bromeliads such as *Neoregelia carolinae* produce daughter offsets after the main rosette has flowered and begun to die back. Leave daughter rosettes in place until they reach one-third of the size of the parent rosette before removing them at their base with a sharp knife.

◀ **2** For succulents, remove the offset and place the cut end just below the surface of the compost. Top-dress with grit to keep it in place and to help avoid rotting at the neck.

### Succulents from stem cuttings .................

Some cacti and other succulents such as *Crassula* can be increased by stem cuttings taken at the end of the dormant season, in spring (or fall for winter-growing plants).

▲ **I** With a straight cut across the base, sever a stem tip bearing 2–3 pairs of leaves, 2–3in (5–7.5cm) long. Remove the lowest leaves, by pulling them off, to retain 1in (2.5cm) of bare stem. Set the cutting in a warm, dry place for about 48 hours to allow the cut to callus over.

▲ **2** Fill a pot with a mixture of equal parts peat and grit or sharp sand. Insert the prepared cutting vertically and centrally in the pot, almost to the depth of the bottom leaves. Top-dress with a ½in (1cm) layer of grit. Then leave in an airy place in bright shade, at 64–75°F (18–24°C).

Insert the cutting just deep enough so that it remains upright in the potting medium and top-dress with a layer of grit. Rooting takes place within two to six weeks and the young plant should then be potted on when the roots fill the pot.

## OFFSETS

Many plants that are grown as houseplants, including succulents and bromeliads, are rosette-formers that produce daughter rosettes around the base of the main parent plant. To increase succulents from daughter rosettes, remove the top-dressing of grit from the base of the parent plant. Cut off the daughter offset with a straight cut, using a sharp knife, just where it arises from the base of the parent. Allow to callus over for about 48 hours in a warm dry place. Insert the rosette into a prepared pot containing a mix of equal parts peat and grit, with the base just below the surface. Top-dress with grit so that the lowest leaves just rest on the surface. As new growth appears, pot on into standard cactus medium, and top-dress with grit.

Many bromeliads produce rosettes that die after they have flowered. As the parent rosette begins to die, a new daughter rosette emerges from its base. When the daughter rosette reaches about one-third of the size of

## Succulents propagation by leaf cuttings . . . . . . . .

Many rosette-forming succulents can be increased easily from cuttings of entire leaves which will produce new plants from dormant buds at their base.

▲ Remove a healthy, young but fully formed leaf from the parent rosette either by pulling off or cutting off at the base with a sharp knife. Allow the leaf base to callus over. Insert the leaf in a prepared pot of cuttings medium at a slight angle, just deep enough so that the cutting remains in position. Top-dress with a layer of grit. Place in a propagator at 64–70°F (18–21°C). The new plantlet appears at the base of the leaf just above the soil surface.

the parent rosette, sever the daughter rosette at the base with a sharp knife. If it has already begun to develop roots, take care to retain them. Set the new rosette into a pot of bromeliad medium, just deeply enough so that it stays in place.

## LEAF CUTTINGS

A number of species and cultivars – usually belonging to the families Begoniaceae, Crassulaceae, and Gesneriaceae – grown as houseplants are capable of forming new plantlets from whole leaves or from sections of the leaves. The phenomenon is seldom seen in hardy woody plants because the cell tissues of their leaves cannot differentiate to form roots and buds.

### Leaf cuttings of succulents

Many succulents, including rosette-formers like *Echeveria*, *Crassula*, *Gasteria*, *Haworthia*, *Sedum*, and *Pachyphytum*, will produce new plantlets from the base of whole leaves. For most, the best time for taking leaf cuttings is spring, just as the plant resumes active growth. As with stem cuttings of succulents, once the leaf has been removed from the parent plant it should be set aside to callus over for about 48 hours.

Select plump, healthy, young but mature leaves and pull them from the parent plant at their base, or remove them with a clean cut made with a sharp blade. Almost fill a 2–3in (5–7.5cm) pot with a mix of equal parts

fine peat moss and sand or grit. Set the callused leaf base on the surface of the cutting mix, and top-dress with a layer of grit.

Place the pot in a partially shaded propagator at 64–70°F (18–21°C). Mist with tepid water to keep the cuttings barely moist (too much water and high humidity will cause rotting).

Rooting takes between three and twelve weeks to occur, although some roots will form on fast-rooting species within a few days. When plantlets appear at the base of the leaf, allow them to grow on for a further two weeks before potting them up in cactus medium.

### Leaf sections

A number of cacti and succulents can be increased by leaf sections. These include succulents like *Sansevieria*, cacti like *Epiphyllum*, with flat, leaf-like stems, and *Opuntia*, with flattened pad-like segments. In the last, each pad is used as an individual cutting.

Select a plump, healthy, fully expanded leaf or stem, and cut it into sections, making a clean straight cut with a sharp knife. For easy handling, make sections between 2–4in (5–10cm) long. Allow the wound to callus over. Use the smallest pot that will contain the cutting comfortably and fill it almost to the rim with proprietary cactus medium or a mix of equal parts peat and grit or sand. Gently push the cutting into the surface of the medium, just deep enough so that the cutting does not fall over. Be sure to insert the cutting with the correct orientation. Top-dress with grit to hold the cutting in place and, if necessary, stake it with a split cane to keep it upright.

## Leaf sections . . . . . . . . . . . . . . . . . . . . . . . . . . . . . . . . . .

Cacti such as *Epiphyllum* and *Schlumbergera*, and succulents like *Sansevieria*, which have fleshy, flattened, stem-like leaves can be propagated by leaf sections.

► Cut a stem-like leaf into sections 2–4in (5–10cm) long, using a straight cut across the leaf. Prepare a 1½–2in (3.5–5cm) pot by filling it with crocks, gravel and charcoal to one-third of the pot's depth and then topping up with gritty medium to within ½in (1cm) of the brim. In each pot, insert a cutting deep enough for the stem to remain vertical; only the base should touch the medium. Top-dress with grit.

## Saintpaulia from whole leaves ················

Saintpaulias, or African violets, can be increased successfully from a complete leaf and its stalk. This can be done at any time of year provided a suitable leaf is available.

◄ **1** From the parent plant, take a young but fully developed leaf with its leaf stalk. Trim the stalk 1–1¼in (2.5–3cm) below the leaf blade. Dibble a hole into a pot filled with a mix of equal parts peat and sharp sand. Insert the stalk so that the leaf blade touches the medium surface.

► **2** Cover the pot with an inverted cloche (a cut-off plastic mineral water bottle is ideal). Set in a warm place in good light but away from direct sunlight. New plantlets will appear at the base of the leaf. Remove the cloche cover when they appear. When they are large enough to handle, separate and pot up individually.

Place the pot in a partially shaded propagator at a temperature of 64–70°F (18–21°C). Mist over with tepid water to keep the cuttings just moist.

Each leaf section takes between three and ten weeks to root and begin to grow away. You may find that slow-growing plants do not root properly before winter. If this happens, keep the cuttings completely dry during the dormant period; they should begin to grow on the following season.

## Taking cuttings from half leaves ···············

Disease can be a major problem when propagating from half leaves so ensure that all tools and other equipment are totally clean and that the parent plant (here *Streptocarpus*) is healthy.

Take an unblemished, mature leaf and cut it longitudinally into two halves by stripping out the midrib. This exposes the leaf veins. Place each leaf half, cut edge downward, in a shallow trench in a prepared seed tray containing a moistened mix of equal parts peat and sand.

Seal in an inflated plastic bag or place in a propagator at 64–75°F (18–24°C). The new plants will appear at intervals along the cut edge. When they are large enough to handle, lift, separate, and pot up.

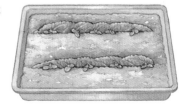

### Propagating by cuttings of entire leaves

A number of rosette-formers with fleshy leaves, like saintpaulias (African violets) and other members of its group (gesneriads) such as *Sinningia* (gloxinias), are easily increased from cuttings of entire leaves. Select young, healthy, fully formed and undamaged leaves; they can be rooted at 68–77°F (20–25°C) at any time of year, but are best taken during periods of active growth. They take three to four weeks to produce roots and a further eight to ten weeks to form plantlets that are sufficiently large to be potted on.

Saintpaulia leaves will also root if the petiole is placed in water, in a dark glass bottle. This is slightly slower than rooting in medium, and the leaf must be potted on as soon as the roots reach 15mm (½in) in length; long roots are too easily damaged when potting up.

### Cuttings of half leaves

Large-leaved gesneriads such as *Streptocarpus* (Cape primrose) are also capable of producing plantlets from their leaves, although in this case slightly different techniques are more successful. The most usual method is from half leaves as shown below left, although they can also be induced to produce new plantlets by scoring through the leaf veins, or sectioning the leaves, as described below for begonias.

### Propagating from leaf squares

Any plant with the capacity to produce plantlets from its leaves can be increased rapidly by leaf squares, but in general the technique is restricted to large-leaved plants

## Propagation from leaf squares ··················

This technique is particularly useful for plants with big leaves, especially begonias, although any leaf that can be propagated vegetatively can be increased in this way.

Take a healthy, young but mature leaf (here *Begonia rex*) and cut out several ½–¾in (1–2cm) squares, each with a strong leaf vein at its center. Place the cut squares on the surface of moistened compost with the veined side in contact with the medium surface. Carefully pin down each leaf square with wire. Enclose the seed tray in an inflated plastic bag and seal tightly. Set in a warm place, at 64–75°F (18–24°C), in good light but away from direct sunlight. Remove the cover once the plantlets appear.

## Increasing plants from cut leaves . . . . . . . . . . . . . .

Plantlet development can be induced in certain plants such as begonias with netted veining on their leaves by making regular slashes across these veins but not cutting up the entire leaf.

◄ 1 Place a healthy, young but mature leaf of a *Begonia rex* flat on a work surface with its underside facing upwards. Make ½in (1cm) cuts at intervals across the main veins.

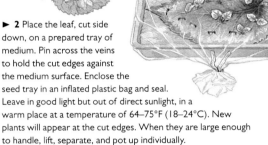

► 2 Place the leaf, cut side down, on a prepared tray of medium. Pin across the veins to hold the cut edges against the medium surface. Enclose the seed tray in an inflated plastic bag and seal. Leave in good light but out of direct sunlight, in a warm place at a temperature of 64–75°F (18–24°C). New plants will appear at the cut edges. When they are large enough to handle, lift, separate, and pot up individually.

with a network of prominent leaf veins, as seen in most large-leaved begonias. The leaves are cut into ¾in (2cm) squares with a sharp blade such as a razor blade; each square must contain a portion of a principal vein. Any torn tissue should be discarded as it will rot and may infect the healthy squares.

The leaf squares are then pinned down on the surface of a moist propagating mix of equal parts peat and sand or perlite with the veins facing downwards. The cut surfaces must make good contact with the propagating mix. Given a temperature of 64–75°F (18–24°C) in a sealed, inflated plastic bag, they will root and form plantlets in about six weeks. Make sure that the plastic bag is kept inflated so that it does not touch the leaf squares.

### Begonias from scored leaves

Propagating from scored leaves is another technique for large-leaved plants with prominent veins. In this case, the undersides of the leaves are scored across the main veins to a length of ½in (1cm) at intervals about 1in (2.5cm) apart. The whole leaf is pinned down, veins downwards, onto a moist propagating medium.

Buds will form at the site of the wounds which then produce roots and, after about six weeks, small clumps

of plantlets. The plantlets can be separated carefully and potted up as soon as they are large enough to handle.

## AIR LAYERING

Air layering is ideal for propagating stout-stemmed species such as *Ficus elastica*, the rubber plant, or *Monstera deliciosa*, both of which will eventually outgrow their planting situation. The parent plant may be either discarded or cut back hard after propagating its replacement.

Air layers are best made towards the top of the plant, since stems less than two years old root most easily. The illustration below shows a sleeve of clear plastic for clarity, but better results may be obtained by using black-and-white plastic with the white side facing outwards since this reduces undesirable temperature fluctuations within the sleeve. New roots may take several months to form and it is essential that the moss wadding remains moist during the rooting period. If necessary, unwrap the sleeve and re-moisten by misting.

### Air-layering a *Ficus elastica* . . . . . . . . . . . . . . . . . . . . . .

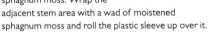

Air-layering is an ideal way of replacing leggy plants that have outgrown their allotted space and has the advantage that the new plant forms roots while still attached to its parent.

► 1 Trim the leaves from a straight piece of stem. Attach a plastic sleeve to the stem, fixing its lower end with adhesive tape. Wound the stem by cutting a tongue into it, with an upward, slanting cut, ¼in (5mm) deep and 1in (2.5cm) long. Apply rooting hormone and wedge the tongue open by padding it with moistened sphagnum moss. Wrap the adjacent stem area with a wad of moistened sphagnum moss and roll the plastic sleeve up over it.

◄ 2 Seal the top of the sleeve with adhesive tape. New roots will grow into the wad of moss. When they fill the plastic sleeve, sever the rooted stem section from the parent plant. Gently loosen the moss ball to separate the roots. Pot up into a prepared pot of medium, firming the medium gently around the roots. Place in a semi-shaded place and water sparingly until the new plant is well established.

# DIRECTORY OF
# HOUSEPLANTS

*Below are instructions and special propagating tips on individual conservatory and houseplants. Unless otherwise specified, follow the detailed instructions given under the propagation section, on pages 132–137.*

## Key

H  Height
S  Spread
🌱  By seed
⚘  Germination
🔧  By division

## Achimenes (Hot water plant)

Gesneriaceae

Some 25 species of perennials, and a number of named hybrids and cultivars, with tuber-like rhizomes, that are valued as houseplants for their long-tubed flowers. The flowers, in a range of colors, are borne above glossy green leaves for long periods in late summer and fall. Plants become dormant in winter, after flowering. They need bright light with shade from direct sun, and a minimum temperature of 50–59°F (10–15°C).

**A. grandiflora.** Upright in habit, this has dark green leaves and bears red-purple flowers with white eyes and purple-dotted throats. H and S to 24in (60cm).

**A. longiflora.** Trailing habit and soft red, lavender, or deep purple flowers. H 10in (25cm), S to 16in (40cm).

### PROPAGATION

🔧 early spring.
Time to flowering – 6 months.
Stem cuttings – spring.
Time to rooting – 2–3 weeks.
Time to flowering – 6 months.

## Agave (Century plant)

Agavaceae

Over 200 species of succulent perennials with rosettes of fleshy, spine-tipped leaves. A number are valued as houseplants for their architectural form. At maturity, they bear long-stemmed, terminal clusters of funnel-shaped flowers, from the center of the rosette. Some species die after flowering, but these can take 40 years to bloom and then usually leave daughter rosettes to grow on. All need good light and a very gritty, sharply drained potting medium. They should be kept almost completely dry in winter.

**A. americana 'Marginata'.** Slow-growing, with evergreen rosettes of lance-shaped, gray-green leaves with creamy-yellow margins. Min. temp. 41°F (5°C). H to 6ft (1.8m), S to 10ft (3m), much less as a houseplant.

**A. victoriae-reginae.** Produces exceptionally neat and symmetrical rosettes of plump, fleshy, oblong-triangular, dark green leaves, outlined with white. Min. temp. 50°F (10°C) H and S 20in (50cm).

### PROPAGATION

🌱 early spring, at 70°F (21°C). Species only.
⚘ up to 3 months.
Offsets – spring or fall.
Time to rooting – 2–4 weeks.

## Aloe

Liliaceae

About 300 species of evergreen, succulent, rosette-forming perennials grown as houseplants for their sculptural form and racemes of tubular flowers. All species need full light and a sharply drained, loam-based potting medium with added grit. They should be kept almost dry in winter. Those featured need a minimum temperature of 10°C (50°F).

**A. aristata.** Clumps of stemless rosettes of lance-shaped, dark green leaves, dotted with tiny soft, white spines, and producing orange-red flowers in fall. H 5in (12.5cm), S 12in (30cm) often much more.

**A. variegata** (Partridge-breasted aloe). Has rosettes of firm, green leaves with bands of greenish white. Pink or red flowers are borne on fleshy stems in summer. H 8in (20cm), S 12in (30cm) often much more.

### PROPAGATION

🌱 early spring, at 70°F (21°C). Species only.
⚘ up to 3 months.
Offsets – spring to early summer. Offsets can often be separated from the parent with roots already formed.
Time to rooting – 2–4 weeks.

## Aspidistra

Liliaceae

Three species of evergreen, rhizomatous perennials of which A. elatior is the most commonly grown. Noted for its tolerance of dry atmospheres, a wide temperature range,

*Achimenes 'Primadonna'*

*Agave victoriae-reginae*

*Aloe aristata*

*Aspidistra elatior 'Variegata'*

*Begonia masoniana*

poor soils, and low light, it has long been used as a houseplant, and can make an elegant specimen when well grown.

*A. elatior.* Produces long-stalked, lance-shaped, glossy, dark green leaves that arise from clustered rhizomes. The insignificant flowers appear at the leaf-stalk bases in late winter or early summer. **'Variegata'** has dark green leaves striped or margined with creamy white. H and S to 24in (60cm).

**PROPAGATION**

✄ early spring.

## Begonia    Begoniaceae

An enormous genus of some 900 species, many of which are grown as houseplants for their foliage or flowers. Most grow best at temperatures of 66–75°F (19–24°C), but, if kept on the dry side in winter, will tolerate 50°F (10°C). All thrive in partial shade and need protection from direct sun. They are divided into several groups.

**Rex-cultorum begonias.** Evergreen, usually rhizomatous perennials, with asymmetrically ovate to lance-shaped leaves in a wide range of colors, many marked, striped, or margined with dark velvety red, bright pink, silvery green, and other metallic shades. Examples: *B.* **'Duartei'**, *B.* **'Helen Lewis'**, *B.* **'Merry Christmas'**.
H and S 6–30in (12–75cm).

**Tuberous begonias.** Deciduous perennials, becoming dormant in winter, producing top-growth annually from a tuber. Some cultivars are used in summer bedding, but many, including the Tuberhybrida, Multiflora, and Pendula types, are also grown as summer-flowering houseplants for their large, brilliantly coloured, usually double, satin-

textured flowers. Examples include *B.* **'Billie Langdon'**, *B.* **'Can-Can'**, *B.* **'Flamboyant'**.
H 24in (60cm), S 18in (45cm).

**Rhizomatous begonias.** Evergreen perennials with creeping rhizomes, grown mainly for their leaves, which often have interesting textures and colours and may be puckered, crested, and/or marked with metallic silver. Examples: *B.* **'Bethlehem Star'**, *B. imperialis*, *B. masoniana* (Iron cross begonia), *B.* **'Norah Bedson'**. H 5–20in (12.5–50cm), S 9–18in (22–45cm).

**Semperflorens begonias.** Fibrous-rooted perennials with mounds of evergreen, glossy, green or bronze, usually rounded leaves, bearing a succession of small, single or double flowers throughout summer. Often used as summer bedding, they also make long-flowering pot plants for the home or conservatory. Examples: *B.* **'Ambassador'**, *B.* **'Organdy'**. H and S 8–16in (20–40cm).

**Cane-stemmed begonias.** Shrubby, fibrous-rooted, evergreen perennials with long, firm, slender stems bearing asymmetrical, often beautifully colored and patterned leaves at the stem nodes. The showy flowers are produced from early spring to summer. Examples: *B. aconitifolia*, *B. albopicta* (Guinea-wing begonia), *B.* **'Orpha C. Fox'**.
H to 4ft (1.2m) or more, S to 14in (35cm).

**Shrub-like begonias.** Shrubby, often succulent-stemmed, evergreen perennials valued for their foliage, which is often beautifully colored and has interesting surface textures. They produce flowers in spring and summer. Examples include: *B.* x *argenteoguttata* (Trout-leaved begonia) and *B. metallica* (Metallic-leaf begonia). H to 6ft (1.8m) many smaller, S 9–24in (22–60cm).

**Winter-flowering begonias.** Compact, bushy, evergreen perennials bearing single, semi-double, or double flowers from fall to spring. They include the **Christmas begonias** (also known as **Cheimantha** or **Lorraine begonias**) and the **Reiger hybrids**. Examples *B.* **'Azotus'**, *B.* **'Kleo'**.
H 8–18in (20–45cm), S 6–24in (15–60cm).

**PROPAGATION**

Leaf cuttings – spring or summer. For small-leaved plants: whole leaf cuttings with a leaf stalk. For larger-leaved plants, leaf squares, or scored-leaf cuttings.
Time to rooting – 2–6 weeks.
✄ spring. Semperflorens begonias.
Rhizome sections – summer. Suitable for Rex-cultorum and rhizomatous begonias.
Time to rooting – 4– 6 weeks.
Tuber sections – late winter to early spring, as new growth begins. Start tubers into growth at 61°F (16°C) on the surface of a tray of sandy medium. As buds swell, section the tubers with a clean sharp knife so that each section has at least one growing bud and some roots. Dust with fungicide, allow to callus over for a few hours, then pot up into moist medium. Do not water again until growth begins; then water sparingly until plants are growing well. Suitable for tuberous begonias.
Basal cuttings – spring, or for winter-flowering begonias, early summer. Start tubers into growth as for sectioning tubers. When the shoots reach 2in (5cm) long, take basal cuttings, each with a small wedge of tuber at the base. Place in a propagator until well rooted, then harden off. Suitable for tuberous and Semperflorens begonias.
Time to rooting – 3–4 weeks.

*Cattleya labiata*

*Chlorophytum comosum* 'Vittatum'

*Cymbidium* Strathdon 'Cooksbridge Noel'

Stem-tip or stem cuttings – spring or summer. For cane-stemmed and shrub-like begonias. Time to rooting – 2–6 weeks.

🌱 late winter to early spring, at 70°F (21°C). Suitable for all species, some tuberous begonias, including Tuberhybrida, Pendula, and Multiflora cultivars, and for Semperflorens begonias. 🌱 15–60 days. Bulbils – set on the surface of a tray of moss peat in spring. Suitable for some species and cultivars of tuberous begonias. (*See also Bulbs, Corms, and Tubers, pp. 64–71.*)

## Cattleya          Orchidaceae

Some 40 species and hundreds of cultivars of epiphytic, evergreen orchids with pseudobulbs and short rhizomes. Valued for their showy flowers, they are suitable for a conservatory. They need filtered light, high humidity, and copious watering in summer. After flowering, give a semi-dry rest and water sparingly. Most need a minimum temperature of 41°F (5°C).

*C. labiata.* Produces racemes of flowers in shades of pink, with ruffled petals and purple lower lips, in fall.
H and S 12in (30cm).

### PROPAGATION

🔧 fall- and winter-flowering species in early spring, spring-flowering species in early summer. Divide summer-flowering species immediately after flowering, or in early spring. If divided too late, they will not develop a good root system before growth slows in fall.
Rhizome sections – each section with at least two or three pseudobulbs, when new shoots and roots appear.
Time to rooting – 6–10 weeks.

## Ceropegia          Asclepiadaceae

Over 200 species of perennials including evergreen and semi-evergreen climbers, trailing plants, and many succulents. A number make attractive houseplants, being tolerant of dry atmospheres, and trailing species, like *C. linearis* var. *woodii* are ideal for hanging baskets. They need a well-drained, leafy potting medium, in bright light with shade from hot sun. Provide a minimum temperature of 50°F (10°C).

*C. linearis* var. *woodii*, syn. *C. woodii*, (Hearts on a string, Rosary vine). Produces long, trailing, slender stems bearing small heart-shaped leaves that are green with pale silvery gray-green or purple markings above and purple beneath. It often produces bulbils in the leaf axils.
H 4in (10cm), S indefinite.

### PROPAGATION

Stem bulbils – spring or early summer. Detach and set on the surface of potting medium. Time to rooting – 2–4 weeks.
Stem-tip cuttings – early summer, 4–6in (10–15cm) long. Root in a mix of equal parts peat and sand in a propagating case at 68–77°F (20–25°C).
Time to rooting 2–4 weeks.

## Chlorophytum          Liliaceae

Among some 200 species of evergreen perennials, the variegated cultivars of *C. comosum* are the most widely grown. An almost indestructible plant, it tolerates neglect, poor soils, hot, cool, damp, and dry conditions, and low light levels.

*C. comosum* (Spider plant, Ribbon plant). Forms clumps of narrowly lance-shaped, bright green leaves and bears tiny 6-petaled

white flowers on cascading stalks that later produce plantlets. 'Variegatum' has white-margined green leaves; 'Vittatum' has green leaves with a cream central stripe. H to 8in (20cm), S 12in (30cm).

### PROPAGATION

🔧 any time between spring and fall.
Plantlets – separate at any time between spring and fall. Encourage them to root while still attached to the parent plant by pinning them onto the surface of a small pot of potting medium.
Time to rooting – 2–4 weeks.

## Cymbidium          Orchidaceae

About 50 species of terrestrial or epiphytic orchids. The most commonly grown are the winter- or spring-flowering hybrids which number in the hundreds. They produce long arching racemes of flowers in a wide range of colours, and have narrow oval to linear leaves. H in a range between about 12–36in (30–90cm), S to 36in (90cm). The dwarf cultivars that grow to 12in (30cm) high are more suitable for growing as houseplants; avoid the larger sturdy plants that can produce long rush-like leaves to about 6ft (1.8m) tall. Cymbidiums are the easiest of the orchids to grow as houseplants and are able to survive temperatures of 46°F (8°C). They benefit from being stood in the garden in a shady spot during summer; this treatment encourages the formation of flower buds.

### PROPAGATION

🔧 large plants in spring.
Back-bulbs – remove when dividing, pot individually, and keep in a propagator with gentle bottom heat.

Dendrobium

Epidendrum radicaris 'Crucifix orchid'

Epiphyllum

## Dendrobium                    Orchidaceae

Over 900 species of deciduous, semi-evergreen, or evergreen, epiphytic and terrestrial orchids, some tiny and moss-like, others with climbing stems to 10ft (3m) in height. The hybrids based on *D. bigibbum* var. *phalaenopsis* and *D. nobile* are most attractive. Hybrids of the former need a great deal of warmth to flower, whereas those of *D. nobile* are more suitable as houseplants. Those listed need a minimum temperature of 50°F (10°C). Water freely and apply a half-strength balanced fertilizer at every third watering. Shield from direct sun and, in early fall, gradually reduce the amount of water until early winter; then cease watering completely until late winter.

*D. infundibulum.* Flowers pure white, with yellow spot on the lip, in spring and summer. H 24in (60cm), S 6in (15cm).

*D. nobile* Rose-pink flowers with a central deep maroon spot, in spring.
H 18in (45cm), S 6in (15cm).

*D. wardiianum.* White flowers with purple petal tips and lips marked maroon and yellow. Spring to fall.
H 12in (30cm), S 6in (15cm).

### PROPAGATION
Stem cuttings – spring, as growth resumes.
Time to rooting 6–10 weeks.
Adventitious growths, or keikeis, with their own roots, appear from the nodes. Remove and pot up during the growing season.

## Dracaena                    Dracaenaceae

About 40 species of evergreen shrubs to tree-like plants with spirally arranged, leathery, glossy, long, linear, and pointed leaves. Some are suitable for cultivation as houseplants if they are kept pruned and are propagated when they become too large. Give full light and provide a minimum temperature of 55°F (13°C). Restrict water during cold periods and in winter.

*D. marginata.* Has red-margined, dark green leaves. 'Tricolor' has green leaves which are margined and veined red and have a cream stripe. H 15ft (5m), S to 10ft (3m).

### PROPAGATION
Air layering – when plant has become leggy, then cut back to soil level.
Time to rooting – 10–16 weeks.
Stem cuttings – cut up main stem into pieces 1–2in (2.5–5cm) long and make into cuttings, using a warm propagator at any time. Alternatively, take stem-tip cuttings.
Time to rooting – 3–6 weeks.

## Epidendrum                    Orchidaceae

Large genus of over 750 species of terrestrial and epiphytic, evergreen orchids, many of which produce cascades of small flowers. They need some summer shade, high humidity, and a minimum of 55°F (13°C). Water copiously in the growing season and very sparingly in winter, when they need as much light as possible.

*E. pseudepidendrum.* Has green flowers with orange lips, in summer and fall.
H to 3ft (1m), S 2ft (60cm).

*E. secundum.* Has long racemes of pink flowers with white- and yellow- marked lips.
H 24in (60cm), S 12in (30cm).

### PROPAGATION
✄ mature plants when pot-bound, as growth resumes in spring.
Stem cuttings – spring, as growth resumes.
Time to rooting – 6–10 weeks.

## Epiphyllum (Orchid cactus)      Cactaceae

About 20 species of mostly epiphytic, rainforest cacti. Most of the plants in cultivation are thought to be hybrids; all are very similar, with flattened stems notched at the edges, at first growing upright then becoming pendent. Frost-tender, they need a minimum temperature of 46°F (8°C). The large flowers are produced in spring and early summer. They are easily grown if they are given richer soil and more water than the desert-type cactus. *E.* 'Gloria' has pinkish-red flowers 4in (10cm) across. *E.* 'Jennifer Ann' has yellow flowers 6in (15cm) across. *E.* 'M.A. Jeans' has smaller blooms 3in (8cm) across. H about 24in (60cm), S 20in (50cm).

### PROPAGATION
Stem cuttings – sections of the stems or single pads root easily in pots of cuttings mediums in a greenhouse.
Time to rooting – 3–6 weeks.

## Epipremnum                    Araceae

About eight species of evergreen climbing plants with large heart-shaped or long-pointed leaves. They need temperatures of 64°F (18°C), moisture-retentive, leafy medium, and the support of a moss pole.
*E. aureum* 'Marble Queen', syn. *Scindapsus aureus* 'Marble Queen'. Evergreen, woody-stemmed climber with leaves streaked with white. H 10–30ft (3–10m) eventually.

### PROPAGATION
Stem cuttings – with 2–3 nodes, with a leaf or without, laid on the medium horizontally or inserted upright. Keep warm and humid in a propagator until roots are formed.
Time to rooting – 3–8 weeks.

141

*Euphorbia pulcherrima*

*Ficus benjamina*

*Kalanchoe daigremontiana*

## Euphorbia  Euphorbiaceae

Very large genus of over 1000 species, including annuals, herbaceous perennials, shrubs, trees, and succulents.

*E. pulcherrima* (Poinsettia). Upright, evergreen shrub. The poinsettias with colored bracts in red, pink, and white, sold by florists are manipulated with chemical retardants, control of day length, or both. To prevent leaf drop, give as much light as possible and a temperature no lower than 46°F (8°C). H and S to 10ft (3m); as a houseplant usually H 8–12in (20–30cm), S 14–16in (35–40cm).

### PROPAGATION

Cuttings – possible, although the resulting plants will not be compact like the parent. Stock plants, which have been dried off after flowering in midwinter, will make small shoots when watered in spring. These can be cut off, the flow of milky sap stopped with hormone rooting powder, and the cuttings then rooted in a warm propagator. Time to rooting 3–8 weeks.

## Ficus (Fig)  Moraceae

About 600 species of mainly evergreen trees, shrubs, and climbers. One of the most well known is the edible fig, *F. carica* (see p. 185). Some species make good foliage houseplants, growing in quite cool conditions in low light. A minimum temperature of 50°F (10°C) is required, possibly more.

*F. benjamina* (Weeping fig). Slender drooping branches bear small, ovate leaves. H to 6ft (1.8m), S 4ft (1.2m) as a houseplant.

*F. elastica* (India rubber tree, Rubber plant). Strong-growing tree with large, shiny, oval leaves, to 10in (25cm) long. The various

forms grown include 'Variegata' which has cream-edged leaves mottled gray-green. H to 6ft (1.8m), S to 20in (50cm) or more.

### PROPAGATION

Cuttings – can be rooted fairly easily given a humid environment at 70°F (21°C). With *F. benjamina* take stem-tip cuttings 4in (10cm) long, with the lower leaves removed, and place around the edge of a pot of medium; they will root in about 6–12 weeks. With *F. elastica*, take stem cuttings with one leaf and a bud (it helps if the leaf is rolled and tied) and put into small pots of medium. Time to rooting – 6–12 weeks. Air layering – the top section of a leggy *F. elastica* can be rooted using this technique. When adequate roots have formed, cut off and pot up. Time to rooting – 10–16 weeks.

## Kalanchoe  Crassulaceae

About 130 species of succulent annuals, perennials, shrubs, climbers, and trees. A number make attractive and unusual houseplants, particularly the bushy perennial succulents with fleshy, rounded leaves and bell-shaped to tubular flowers. Provide a minimum temperature of 50°F (10°C).

*K. blossfeldiana*. Bushy perennial with glossy toothed leaves and clusters of tubular scarlet flowers. The hybrids include red, salmon, and pink. Prefer partial shade. H and S to 16in (40cm).

*K. fedtschenkoi*. Perennial succulent with upright growth bearing fleshy, oval, notched, blue-gray leaves with little plantlets in each notch. Bell-shaped, brownish-pink flowers appear in late winter. Prefers full sun. H to 3ft (1m), S 10in (25cm).

*K. 'Wendy'*. Perennial succulent with glossy

leaves and bell-shaped, deep pink flowers, with yellow rims, in midwinter. A good hanging basket plant in a cool greenhouse. H and S to 12in (30cm).

### PROPAGATION

Stem-tip cuttings – for the majority, put in a gritty medium in a greenhouse at a temperature of 63°F (17°C). Keep relatively dry. Time to rooting – 2–6 weeks. With *K. blossfeldiana*, plantlets drop off the leaves naturally and root on the medium surface or can be removed and planted. Time to rooting – 2–3 weeks.

## Mammillaria  Cactaceae

Some 150 species of mainly spherical cacti, some are cylindrical. Flowering freely, they produce rings of funnel-shaped blooms on their tops. They need a minimum of 41°F (5°C), full sun, and sharply drained medium. Keep dry in the winter to prevent rot.

*M. bocasana* (Snowball cactus). Clump-forming cactus covered with long white hairs. The flowers, borne in summer, can be cream, or pink, often followed by red seedpods. H 2in (5cm), S indefinite.

*M. plumosa*. Spherical cactus that forms clumps of individuals, covered with white hair-like spines; in winter it has creamy-white flowers. H 5in (12cm), S 16in (40cm).

*M. sempervivi*. Dark green spherical cactus, with white spines that have white wool between them. Bears cerise-pink flowers in spring. H and S 3in (7.5cm).

### PROPAGATION

🌱 late winter on gritty, loam-based medium. ⚘ 2–3 weeks.

Offsets – mature offsets with their own roots; pot individually into cactus medium.

*Neoregelia carolinae*

*Peperomia obtusifolia 'Variegata'*

*Philodendron bipinnatifidum*

## Monstera — Araceae

Some 30 species of evergreen climbers.
*M. deliciosa* is a good houseplant for large
spaces, and thrives in low light. Support with
a moss pole and cut back occasionally.
Minimum temperature 59°F (15°C).
*M. deliciosa* (Swiss-cheese plant). Large, shiny
lobed and holed leaves are borne on long
leaf stalks from a stem 1½in (4cm) thick,
from which long brown aerial roots also
grow. **'Variegata'** has splashes of white on
the leaves. H to 20ft (6m).

**PROPAGATION**
Leaf-bud or stem-tip cuttings – pot singly
and place in a plastic bag in a warm place.
Time to rooting – 4–8 weeks.
Air layering – during the growing season.
Time to rooting – 10–16 weeks.

## Neoregelia — Bromeliaceae

About 70 species of evergreen, epiphytic
rosette-forming perennials which hold
closely packed flowers at the rosette's
center. Needs warm, humid conditions with
a minimum of 55°F (13° C) and shade from
full sun. Grow in a mix of granulated bark
and sphagnum moss. Water with care during
the growing season. Keep the center of the
rosette filled with clean water.
*N. carolinae*. Has rosettes of strap-shaped
leaves in the center of which are tubular,
blue-purple flowers, with red bracts.
**'Tricolor'** has green leaves striped with
white that flush pink with age.
H 8–12in (20–30cm), S 16–24in (40–60cm).

**PROPAGATION**
Offsets – pot into small containers of the
growing medium. Root in a propagator with
bottom heat. Time to rooting – 3–6 weeks.

## Peperomia — Piperaceae

Around 400 species of annuals or evergreen
perennials. Those grown as houseplants are
small plants with ornamental foliage. Many
are suitable for growing in small pots and
there are some with a trailing habit that are
good for hanging baskets. The fleshy or
succulent species tend to rot when over-
watered; keep them just moist. Grow in
bright light but protect from full sun. Those
described need a minimum of 50°F (10°C).
*P. marmorata* (Silver heart). Evergreen, bushy
perennial with oval, pointed, fleshy, dull green
leaves, marked with grayish white and
quilted above, reddish below.
H and S to 10in (25cm).
*P. obtusifolia* **'Variegata'**. Evergreen perennial
with thick, fleshy green leaves, irregularly
marked yellow to creamy white at the
margins. Plants occasionally flower.
H and S 10in (25cm).

**PROPAGATION**
✂ large plants, in spring.
Stem cuttings –from large plants, in spring or
early summer. Time to rooting – 2–6 weeks.
Leaf cuttings – entire leaves with or without
a leaf stalk attached.
Time to rooting 2–6 weeks.

## Philodendron — Araceae

About 120 species of evergreen, often
epiphytic, climbing shrubs or trees, grown for
their attractive leaves. They make ideal
houseplants and grow in fairly low light.
Some support, such as a moss pole, into
which the aerial roots can grow should be
provided. Grow in humus-rich soil and water
moderately, less in cooler situations; provide
a minimum temperature of 59°F (15°C).

*P. bipinnatifidum*. Erect, robust, tree-like shrub
with large, deeply lobed, glossy green leaves,
24in (60cm) long, on stalks 24in (60cm) in
length. H to 15ft (5m).
*P. scandens* (Sweetheart plant). Fast-growing
climber with slender stems covered with
attractive small light green, heart-shaped
leaves. Regular pruning may be necessary.
H 12ft (4m).

**PROPAGATION**
Stem cuttings – in a heated propagator.
Time to rooting – 2–6 weeks.
Air layering – for climbers, during the
growing season.
Time to rooting – 10–16 weeks.

## Pilea — Urticaceae

Over 200 species of annuals or evergreen
perennials, some of which are good foliage
houseplants. Grow in well-drained medium,
kept just moist in cool periods; provide a
minimum temperature of 50°F (10°C).
*P. cadieri* (Aluminium plant). Evergreen bushy
perennial with pointed-oval leaves that have
raised silvery patches. H and S 12in (30cm).
*P. involucrata*. Similar to *P. cadieri*, with oval to
round leaves, that have a bronze, corrugated
upper surface and red on the underside.
H 6in (15cm), S 12in (30cm).

**PROPAGATION**
Stem cuttings – spring, in a heated
propagator. Time to rooting – 2–6 weeks.

## Pittosporum — Pittosporaceae

About 70 to 80 species of evergreen, mostly
frost-tender, trees and shrubs.
*P. tobira*. This makes a small rounded tree,
with oval, gray-green leaves, and produces
clusters of fragrant white flowers in late

143

*Rebutia heliosa*

*Saintpaulia*

*Schlumbergera truncata*

spring. It can be confined to a pot if pruned each year after flowering. Min. temp. 32°F (0°C). H to 12–15ft (4–5m), S to 10ft (3m).

**PROPAGATION**

spring. 3 weeks.
Stem-tip cuttings – midsummer, in a propagator. Time to rooting – 6–8 weeks.

### Rebutia                                Cactaceae

About 40 species of spherical or columnar cacti that have green stems with short spines. Seed-grown plants as young as three years old will produce flowers in profusion from their base. They require full sun and well-drained medium. Provide a minimum temperature 40°F (5°C).
*R. aureiflora.* This has dark green stems, often tinged violet-red, with short sharp spines and a longer soft central spine. Late-spring flowers, 1½in (4cm) across, can be yellow, violet, or red. H 4in (10cm), S 8in (20cm).
*R. fiebrigii,* syn. *R. muscula.* Dark green stems are densely covered with soft white hairs. Flowers of bright orange appear in late spring and are 1in (2.5cm) across. H 4in (10cm), S 6in (15cm).

**PROPAGATION**

spring. 2–4 weeks.
Offsets – spring or summer.
Time to rooting – 2–4 weeks.

### Saintpaulia (African violet)    Gesneriaceae

Only six species of evergreen, tender perennials. *S. ionantha* has been used to create many hybrids, which are most commonly grown as houseplants. All form rosettes of rounded, long-stalked, hairy leaves. Flowers range through white, pink,

red, purple, and blue, plain or bicolored some with frilled petals, others double. They need warm growing conditions, with a minimum temperature of 59°F (15°C). Keep out of full sun and water carefully – over-wet compost can cause crown rot and careless watering will mark the foliage. H to 6–8in (15–20cm), S 3–16in (7.5–40cm).

**PROPAGATION**

sow on the surface of peaty medium at a temperature of 70°F (21°C).
3–8 weeks.
Leaf cuttings – at any time; select a healthy undamaged mature leaf, shorten the leaf stalk, insert into cuttings medium with the leaf blade clear of the medium. Then place in a propagator with bottom heat. Small plantlets, which can be separated and grown individually or left to form a multi-crown specimen, will grow from the base.
Time to rooting – 2–6 weeks.

### Sansevieria                          Dracaenaceae

Over 50 species of evergreen perennials, usually stemless with stiff, fleshy, often variegated or mottled leaves growing from short thick rhizomes. They need a minimum temperature of 50°F (10°C) and grow well in sandy loam medium in sun or shade.
*S. trifasciata* (Mother-in-law's tongue). Lance-shaped, pointed, erect leaves are banded horizontally with paler green. 'Laurentii' has yellow margins to the leaves.
H 15–24in (38–60cm), S 20in (50cm).

**PROPAGATION**

or offsets – for variegated plants (leaf cuttings will revert to green).
Time to rooting – 2–4 weeks.
Leaf cuttings – 3in (7.5cm) lengths inserted

into sandy medium and put in a propagator will root freely. Time to rooting – 3–8 weeks.

### Saxifraga                            Saxifragaceae

Over 300 species, most of which are alpines. *S. stolonifera* makes a good houseplant, surviving in poor medium and low light.
*S. stolonifera* (Mother-of-thousands). Leaves are round, hairy, silver-veined, and dark green, reddish purple beneath. It produces long runners with small plantlets on them, making it a good hanging basket plant. Has sprays of small white flowers in midsummer. Frost hardy. H and S to 12in (30cm).

**PROPAGATION**

Easily grown from potted-up plantlets.

### Schlumbergera                       Cactaceae

About six species of epiphytic, bushy cacti with erect, then pendent stems made up of flattened, oblong segments with notched margins. Flowers with petals of varying lengths and prominent stamens are produced on the ends of the stems. The medium needs to be humus-rich and well drained. Grows well in shade, needs less water in cooler periods.
*S. truncata* (Christmas cactus). Purple-pink flowers in fall and winter. Min. temp. 50°F (10°C). H and S to 12in (30cm).

**PROPAGATION**

Stem cuttings – summer, in a shaded greenhouse. Time to rooting – 2–6 weeks.

### Sinningia                            Gesneriaceae

About 20 species of tuberous perennials. They need warm growing conditions and good light but not direct sun. Use a humus-rich medium, and provide a humid

*Stephanotis floribunda*

*Streptocarpus 'Johannes'*

*Tillandsia lindenii*

atmosphere. Minimum temperature 59°F (15°C). When the plants die down after flowering allow the tubers to dry out, then store them frost-free, until the next spring. **S. speciosa** (Gloxinia). This species has many cultivars. They all have large nodding bell-shaped flowers of white, purple, or red, including bicolors and spotted. The leaves are large and oval, velvety on top, red-flushed beneath. H and S to 12in (30cm).

**PROPAGATION**

 on the surface of peaty medium in a propagator at 70°F (21°C) in spring.  3–8 weeks. Tubers – cut into pieces each with a shoot, or remove shoots from tubers in spring and treat as cuttings.
Time to rooting – 2–6 weeks.

## Stephanotis
Asclepiadaceae

Some 15 species of often tall, twining, evergreen shrubs. S. *floribunda* is commonly grown as a houseplant at a minimum temperature of 54°F (12°C). Use loam-based medium and keep it fairly dry in winter. Scale insects can be a problem; clean the leaves regularly to keep them at bay.
**S. floribunda.** Large oval, thick, shiny, opposite leaves and wax-like, fragrant white flowers in clusters in early summer. Support is required for the vigorous shoots; weak shoots should be cut out in spring. H 20ft (6m) or more.

**PROPAGATION**

 early spring. Seed is sometimes produced in cultivation, borne in large egg-shaped pods. Sow in a heated propagator at 68–75°F (20–24°C).  30–90 days. Stem cuttings – of the previous year's growth, root at 64°F (18°C) in a propagator with high humidity. Time to rooting 2–6 weeks.

## Streptocarpus
Gesneriaceae

Some 80 species of annuals and perennials most of which have hairy, veined, wrinkled leaves and tubular flowers with five spreading lobes. S. *rexii* is the parent of many hybrids which are grouped under the name S. × *hybridus*. Those grown as houseplants require humus-rich medium and a humid atmosphere with some shade. Provide a minimum temperature 59°F (15°C). The leaves can be damaged if splashed when watering, especially at low temperatures.
**S. × hybridus.** These have clusters of funnel-shaped flowers of various colours, including magenta, pink, salmon, and white, borne above rosettes of long strap-like leaves, velvety on top and smooth on the underside. H 8–14in (20–35cm), S 18–24in (45–60cm) or more.

**PROPAGATION**

 on the surface of peaty medium at 70°F (21°C).  21–60 days.
Leaf cuttings – selected cultivars. Take from mature leaves either cut across or in half down the main vein, inserting the cut surface into cuttings medium.
Time to rooting – 3–10 weeks.

## Tillandsia
Bromeliaceae

Nearly 400 species of strange plants that almost plastic. Most are epiphytic under natural conditions and should be fastened to bark or cork with nylon fishing line, and a little moss. Most need bright light, high humidity, and high temperatures in the summer; those with few or no roots need spraying with water twice a day during hot weather and keeping completely dry in winter. Minimum temperature 50°F (10°C).

**PROPAGATION**

 or offsets in spring.
 spring, on conifer twig bundles at 81°F (27°C).  21–28 days.

## Tolmeia
(Pig-a-back plant)    Saxifragaceae

A single species of hardy perennial that spreads rapidly. As a houseplant, it tolerates low light levels.
**T. menziesii 'Taff's Gold'** (Pick-a-back-plant). Has trailing, yellow-splashed foliage. Plantlets grow from buds on the top of the leaf stalks.
H to 24in (60cm), S to 6ft (2m).

**PROPAGATION**

Leaf cuttings – insert by the stalk or pin down on the propagating medium.
Time to rooting – 2–3 weeks.

## Tradescantia
Commelinaceae

About 100 species of evergreen perennials. Some make good, easy-to-grow houseplants, particularly the trailing species. Provide a minimum temperature of 41°F (5°C) and bright light, but not direct sun.
**T. fluminensis.** Trailing perennial with leaves that clasp the stem. They are glossy green on top, and purple beneath. '**Aurea**' has longitudinal yellow stripes.
H 6in (15cm), S 8in (20cm).
**T. zebrina.** Trailing perennial with bluish-green leaves banded silver and purple beneath. '**Quadricolor**' has leaves striped with green, pink, red, and white.
H 6in (15cm), S 8in (20cm).

**PROPAGATION**

Stem-tip cuttings – any time. Root in water or medium. Time to rooting – 2–3 weeks.

145

# WATER PLANTS

Water plants are both intriguing and beautiful, and can be appreciated in even the smallest garden if grown in tubs or half-barrels. Their diverse leaf shapes and sizes, and sometimes exotic flowers, are exploited to the fullest extent when reflected in a mirror of clear, still water. Visual appeal, however, is not their only virtue: aquatics of all types – from deep-water plants to floaters – are essential in ensuring clear, clean water and in providing a habitat for amphibians, fish, and other submerged life in the garden pool. Propagation is vital to maintain health and performance, but the techniques are simple and require little in the way of tools and materials.

◄ *Regular propagation of all types of water plants is an essential part of the routine maintenance of a healthy aquatic environment.*

# WHY PROPAGATE
# WATER PLANTS?

*With their feet in the water and their faces in the sun, the majority of water plants grow at a tremendous rate and are, consequently, generally easy to propagate. As in so many other plant groups, younger specimens tend to put on the best show and therefore it makes sense to restock your pond regularly with fresh plants.*

**Many aquatics need frequent propagation for control and to maintain oxygenating efficiency, or to keep them floriferous.**

Most aquatics grow so rapidly that their propagation is more a matter of good husbandry than of increasing quantity; space in the water garden is often limited, so there is seldom a need for more of the same thing. However, constant renewal of young plants is necessary to maintain the efficiency of submerged oxygenators, to maximize the flowering of water lilies, and to sustain the lush foliage of marginals. The techniques are simple: the expertise comes in recognizing when to propagate and which method to use.

## AQUATICS FROM SEED

Since the majority of aquatics are far more easily increased vegetatively, seed is used mainly for propagating annuals such as *Cotula barbata* (bachelor's buttons), and a few marginal and deep-water plants with roots that are difficult to divide, such as *Lysichiton* (the skunk cabbage) and *Orontium* (golden club). Although most water lilies are propagated vegetatively, the dwarf *Nymphaea tetragona* 'Alba' does not produce buds on its root and must therefore be grown from seed.

## DIVIDING YOUR STOCK

Although creeping species with long, sappy stems, such as *Veronica beccabunga*, and those with too few buds on the rootstock, like *Lysichiton*, are difficult to divide successfully, for most marginal and deep-water plants division consists simply of splitting the root system into smaller clumps. For some deep-water plants, such as the hardy water lilies, it involves sectioning the root so that each piece has a growth bud. *Butomus*, the flowering rush, produces tiny bulbils in leaf axils on the woody rootstock which can be detached, or left in place on a small section of root.

## TAKING CUTTINGS

With the exception of rhizomatous, deep-water plants, like water lilies, the majority of submerged plants can be propagated easily from stem cuttings. No great skill

*Lysichiton americana*
The flowering spathes produce a gelatinous mass of seed that germinates freely if surface sown on wet medium. Removing offsets is possible, but not so easy.

*Myosotis scorpioides*
Water forget-me-nots form mats of foliage at the pond margin but to maintain good healthy cover, divide regularly in spring. Seed is also easy, sown *in situ* or in wet medium.

*Menyanthes trifoliata*
Bog bean looks most effective when grown in extensive colonies. Increase your stock by division in spring, or by seed sown as soon as ripe.

*Nymphaea* **species
and cultivars**
Water lilies flower best if propagated and re-planted in fresh soil every third year or so, or when their containers become overcrowded. Increase cultivars by division or bud eyes, the species also by seed.

*Iris sibirica*
To maintain flowering potential, propagate every 3–4 years. This and other water irises (*I. laevigata* and *I. ensata*) divide easily after flowering. Alternatively, sow seed in spring.

or special equipment is required – simply nipping off the stems between finger and thumb and refloating them will often ensure a supply of new plants.

## RUNNERS, OFFSETS, AND PLANTLETS

Floating plants often produce runners and offset plantlets freely and these provide a very easy way of renewing your stock.

## GROWTH FROM BUDS

Many submerged plants develop special overwintering buds, or turions, at the shoot tips which, if detached and rooted in protected conditions, provide a simple means of propagation.

### Containers for Water Plants

*Many aquatics spread so rapidly that they need containment in a small pool. For those needing regular division, growing in a container has the advantage that it is easily lifted to be re-planted when plants become overcrowded. Containers for aquatics are squatter than conventional ones, with flat bottoms for extra stability. The sides are constructed in rigid plastic mesh to permit the exchange of gases between the medium and water that is essential for healthy root function.*

Growing the water lily 'Mrs George C. Hitchcock' in a container makes it easy to lift and propagate regularly.

# PROPAGATING TECHNIQUES
# WATER PLANTS

*The techniques used for propagating aquatic plants need little in the way of specialist equipment. A watertight container – large enough to hold a seed tray – is the main requirement; an unused aquarium makes an excellent propagating case. And for seedlings and very young plants, a greenhouse or coldframe will protect them from frost.*

## SEED

The seed of most aquatics is sown as soon as ripe in summer or early autumn; seed loses viability quickly if it is allowed to dry out. If immediate sowing is inconvenient, store seed in moist peat (or substitute) in a plastic bag, and keep frost-free. *Myosotis*, *Mentha*, and *Alisma* are exceptions; their seed may be stored cool and dry, but should be sown within sixmonths.

## DIVISION

Several techniques are used for the division of aquatics. Most are basically the same as for herbaceous perennials, but can be done from early summer to autumn. It is usually impossible to disentangle roots from the mesh sides of their container, so be prepared to sacrifice it. All new divisions are best kept in partial shade in a frost-free frame or greenhouse for their first winter.

### Marginal plants

Most marginals are divided simply by splitting clumps by hand into two or more portions, as for herbaceous perennials (*see p.29*). Section tough, thick-rooted clumps with a spade. Discard older, less productive, inner pieces and select young, vigorous, outer portions. Pot these up at their original depth into a basket of aquatic growing medium and immerse in a watertight container with water just covering the crown. Cut back aerial growth to within 2in (5cm) of the water level.

### Deep-water plants

*Division of rhizomes:* Fill a small basket with aquatic growing medium and firm thoroughly. Lift the rootstock and hose it clean of soil. Remove old leaves and cut plump rhizomes into sections 2–3in (5–7.5cm) long, each with new growth at their tips. Trim back

---

### Sowing seed . . . . . . . . . . . . . . . . . . . . . . . . . . . . . . . . . . . . . . . . . . . . . . . . . . . . . . . . . . . . . . . .

Seed of deep-water and marginal plants must be sown in conditions similar to those experienced by mature plants in the garden pool. Water levels in the seed container must therefore be controlled, and care taken to ensure that the seed and subsequent seedlings are *never* allowed to dry out.

**Deep-water plants ►**
Seed of deep-water plants must be submerged completely under water. Cover the base of a watertight container with 3in (7.5cm) of aquatic growing medium, level off, and firm lightly. Sow seed thinly and evenly on the surface and cover with ⅛in (3mm) of finely sieved aquatic growing medium. Water in using a fine spray nozzle until the final water level is 1in (2.5cm) above the medium surface.

**Marginal plants ►**
For marginal plants, greater control of the water depth around the seed and seedlings is necessary. Fill a seed pan with aquatic growing medium to within ¼in (5mm) of the rim. Level off and firm the medium gently, then surface-sow the seed thinly and evenly. Cover with ⅛in (3mm) of fine sand, and water in using a fine spray nozzle. Place the seed pan in a watertight container and top up with water until it is level with the medium surface.

**Aftercare**
Place the container in a frost-free greenhouse or frame with part-day sun, to warm the soil, but protect the seedlings from the hottest midday sun. Replenish the water regularly to maintain the same level as at sowing. When the seedlings are large enough to handle, prick out into 3in (7.5cm) containers of aquatic growing medium and return to the watertight container. Keep deep-water plants under 1in (2.5cm) of water and marginals in water that is level with the medium surface. Grow on until the plants are large enough to transfer to a small aquatic planting basket.

thin roots. Press the rhizome on to the medium surface, with the growing tip above the surface at the same angle as it was in the water; do not bury it. Top-dress with 1in (2.5cm) of pea gravel and water in thoroughly. Immerse the basket in a watertight container with the growing point covered by 2–3in (5–7.5cm) of water.

*Division of tubers:* For tuberous water lilies, cut off individual side shoots or "eyes" flush to the tuber. *Nymphaea tuberosa* forms small protuberances on the tuber which can be snapped off. Dust the cut surfaces with charcoal and press onto the firmed surface of aquatic growing medium in a 4in (10cm) container. Treat as for rhizomes.

## STEM CUTTINGS

Stem cuttings are taken from plants in full growth, between early summer and early fall, and usually root within 2–3 weeks. After insertion, immerse in a watertight container with sufficient water to keep the stems below the water surface, and place in a warm, partially shaded site until rooted. Once rooted, transfer to larger containers and introduce to the pond.

## RUNNERS, OFFSETS, AND PLANTLETS

Some tropical water lilies produce small plantlets on the central leaf vein. Remove the leaf from the parent and pin it onto the surface of aquatic growing medium in a container so that the leaf is just submerged. Alternatively, support a container of aquatic growing medium beneath the leaf and pin it down while still attached. Roots develop within 3–5 weeks, and the new plant should be potted on into an individual container and overwintered at its recommended minimum temperature.

*Cyperus papyrus* 'Nanus' forms young plantlets in the flower head. Bend over the flower head and immerse it in a shallow container of aquatic growing medium and water. Within about 3–5 weeks, the plantlet will have rooted. Cut the plantlet away from the parent plant and pot it up separately.

## TURIONS AND WINTER BUDS

Cut off the specialized shoot tips as they develop in mid-fall. Using 3in (7.5cm) containers or seed trays, dibble true turions shallowly into the surface of some aquatic growing medium, or cover hard winter buds

### Stem cuttings

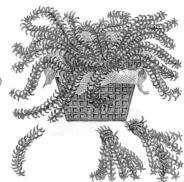

▶ **1** For submerged oxygenators, remove young shoots, 6–9in (15–22cm) long, by hand or with a knife. Bind 8–9 shoots together with florist's wire near the base. Insert each bunch into a 4in (10cm) container of aquatic growing medium, with the tie positioned below the surface of the soil.

▶ **2** Set the pots in a container of water. When rooted, transfer to the pond. For marginals with long soft stems, remove healthy stem tips 3–4in (7.5–10cm) long. Trim just below a node, and remove the leaves from the lowest third. Insert to the depth of the lowest leaves in aquatic growing medium.

### Runners and offsets

Detaching runners or offsets from the parent plant is a very easy method of propagating floating plants, as the new plants already have their own roots.

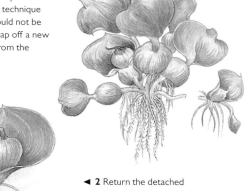

▶ **1** For the majority of floating plants the technique of propagation could not be simpler: simply snap off a new runner or offset from the parent plant.

◀ **2** Return the detached runner or offset immediately to the garden pool, where it will grow rapidly.

with 3–6in (7.5–15cm) of medium. Cover the medium surface with water and overwinter in a frost-free place. As spring growth begins, pot up the plants individually and submerge them to the required growing depth. Once they are growing strongly, set them out into planting baskets or into the pool bottom as appropriate. The turions of floaters, like *Hydrocharis* (frogbit), are simply refloated onto warm shallow water.

# DIRECTORY OF
# WATER PLANTS

*Below are key tips on a selection of water plants that are particularly suitable for propagation. Unless otherwise specified, follow the detailed instuctions given under the propagation section on pages 150–151.*

## Key

H   Height
S   Spread
PD  Depth
🌰  By seed
🌱  Germination
☀  Time to maturity
🔧  By division

## Acorus                                    Acoraceae

Only two species of hardy, rhizomatous, marginal and submerged perennials, grown for their sometimes aromatic foliage. For moist soil or shallow water in sun or light shade.

*A. calamus* (Sweet flag). Deciduous, with fragrant, sword-shaped, wrinkly-margined, olive-green leaves. **'Variegatus'** has olive-green leaves striped cream and yellow. H to 3ft (90cm), S 2ft (60cm), PD to 9in (22cm).

*A. gramineus.* Compact tufts of semi-evergreen, glossy green leaves. The leaves of **'Oborozuki'** are striped yellow. **'Variegatus'**, striped cream and yellow. **'Yodo-no-yuki'**, variegated pale-green. H 3–4in (7.5–10cm), S 10–15cm (4–6in), PD to 9in (22cm).

### PROPAGATION

🔧 spring. Divide overgrown clumps every 3–4 years.   ☀ 4 months.
Section rhizomes of *A. gramineus* cultivars in to 1–2in (2–5cm) lengths.   ☀ 15 months.

## Alisma                                  Alismataceae

Nine species of hardy, rhizomatous, marginal perennials, grown for their loose panicles of flowers and basal rosettes of leaves. Needs full sun or light shade.

*A. plantago-aquatica* (Water plantain). Ribbed, broadly ovate gray to gray-green leaves and tiny pink-white flowers in summer. H to 30in (75cm), S 1½ft (45cm), PD 6in (15cm).

### PROPAGATION

🌰 as soon as ripe or in spring, at 59°F (15°C). As the seed will float, cover with a light dressing of medium. Seed may be stored for up to a year.
🌱 2–5 weeks.   ☀ 15 months.

## Aponogeton                          Aponogetonaceae

Some 44 species of rhizomatous, deep-water perennials, one of which is hardy and is grown for its flowers and foliage. Needs sun or light shade.

*A. distachyos* (Cape pondweed, Water

hawthorn). Floating, strap-like, green leaves above which are borne forked spikes of fragrant white flowers with almost black anthers, in spring and fall. Almost evergreen in mild winters. H 1–3ft (30–90cm), S to 4ft (1.2m), PD 1–3ft (30–90cm).

### PROPAGATION

🔧 spring. Section rhizomes into 2in (5cm) lengths, each with 2–3 buds.   ☀ 24 months.
🌰 as soon as ripe at 55–61°F (13–16°C). Prick out when new tubers are pea-sized.
🌱 3 weeks.   ☀ 12 months.

## Butomus                               Butomaceae

A single species of hardy, rhizomatous, marginal perennial, grown for its flowers and foliage. Prefers full sun.

*B. umbellatus* (Flowering rush). This has rush-like leaves, which are bronze-purple becoming dark green as they mature, and scented, cupped, rose-pink flowers in late summer. H to 5ft (1.5m), S 18in (45cm), PD to 10in (25cm).

### PROPAGATION

🔧 spring.   ☀ 4 months.
Bulbils – spring. Remove root bulbils and pot individually in small containers of aquatic growing medium. Keep under ¾in (2cm) of water.   ☀ 15 months.
🌰 as soon as ripe at 59°F (15°C).
🌱 3–8 weeks.   ☀ 18 months.

## Calla                                      Araceae

One species of hardy, rhizomatous, marginal perennial, grown for its foliage and flowers. Prefers full sun.

*C. palustris* (Bog arum). Dark green leaves and small, white flowers in spring, followed by rounded clusters of red berries in fall H 10in (25cm), S 12in (30cm), PD to 10in (25cm).

*Acorus calamus 'Variegatus'*

*Butomus umbellatus*

*Caltha palustris*

*Eichhornia crassipes*

*Elodea canadensis*

## PROPAGATION

✂ spring. Section rhizomes into 2–4in (5–10cm) lengths, each with a bud, and plant in trays of muddy soil.  ☀ 12 months.
☙ as soon as ripe in late summer at 50°F (10°C). Surface-sow berries in a seed tray and submerge to three-quarters of their depth. Keep wet and frost-free in winter
❦ 6 months.  ☀ 12 months.

## Caltha                          Ranunculaceae

Some 10 species of perennials, including hardy, marginal aquatics valued for their dense clumps of glossy foliage and buttercup-like flowers, borne before the leaves appear in early spring and often producing a second flush in late summer. Prefers full sun.
*C. leptosepala.* This has dark green leaves and silvery-white flowers on leafless stems. It is best at or just above water level in moist soil. H 12in (30cm), S 12in (30cm).
*C. palustris* (Kingcup, Marsh marigold). Produces dark green leaves and masses of cupped, rich golden-yellow flowers on branching stems. H 2ft (60cm), S 18in (45cm), PD to 6in (15cm).

## PROPAGATION

✂ immediately after flowering.  ☀ 9 months.
☙ as soon as ripe at 59°F (15°C). Seed germinates well if the seed pan is placed under a mist propagator.
❦ 2 weeks.  ☀ 12 months.

## Ceratophyllum            Ceratophyllaceae

Some 30 species of half-hardy and hardy, submerged, aquatic perennials, used as oxygenators for water gardens. Prefers sun but tolerates shade.

*C. demersum* (Hornwort). Hardy perennial, often semi-floating, with dense whorls of forked, bristly, dark green leaves produced on slender, brittle stems. H and S indefinite, PD 8–60in (20cm–1.5m).

## PROPAGATION

Stem cuttings – during the growing season, 6–9in (15–23cm) long.  ☀ 2 weeks.
Turions – fall. Break off and weight to the pond bottom to overwinter.
☀ 3 months.

## Egeria                          Hydrocharitaceae

Only two species of semi-evergreen or evergreen, hardy, submerged or floating, aquatic perennials used as oxygenators. They need cutting back regularly to restrict their spread. Requires full sun.
*E. densa,* syn. *Elodea densa* (Argentinian water weed). This species has long, branching stems with whorls of narrow, dark green leaves with pointed tips that usually curl backwards. May produce insignificant white flowers on the water surface in summer. H and S indefinite.

## PROPAGATION

Stem cuttings – summer, 6–9in (15–23cm) long.  ☀ 3 weeks.

## Eichhornia                      Pontederiaceae

Seven species of frost-tender, mainly floating, stoloniferous perennials which spread rapidly in warm climates. Suitable for outdoor pools in temperate climates only during the frost-free summer months. Need full sun and warm water, preferably at 64°F (18°C).
*E. crassipes* (Water hyacinth, Water orchid). Floating, aquatic perennial with spikes of pale lilac flowers in summer above rosettes of

shiny, rounded to ovate, pale green leaves with swollen, air-filled bases. H 6–9in (15–22cm), S 6–12in (15–30cm).

## PROPAGATION

Plantlets – during the growing season. Snap away from the parent plant and re-float immediately. Overwinter young plantlets on wet soil under glass at a minimum air temperature of 45°F (7°C).  ☀ 2 months.

## Elodea                          Hydrocharitaceae

Some 12 species of hardy, submerged, aquatic perennials. *E. canadensis* is widely used as an oxygenator and is invasive if not regularly thinned. Tolerates sun or shade.
*E. canadensis* (Canadian pondweed). Has many-branched stems bearing oblong, pointed, mid-green leaves. H and S indefinite.

## PROPAGATION

Stem cuttings – throughout growing season.
☀ 3 weeks.

## Glyceria                          Gramineae

About 16 species of rhizomatous, aquatic or marginal, perennial grasses grown for their foliage and flowers in summer. Because the most widely grown variety quickly colonizes moist soils or water margins, it is best confined in a tub or basket. Prefers full sun.
*G. maxima* var. *variegata* (Variegated water grass). Has arching, flat linear leaves, strikingly striped with green, cream, and white, usually pink-flushed in spring. The grass-like flowers are green to purple-green. H 32in (80cm), S indefinite, PD to 30in (75cm).

## PROPAGATION

✂ during the growing season. Any division will root easily.  ☀ 4–6 weeks.

153

*Hottonia palustris*

*Houttuynia cordata* 'Chameleon'

*Lysichiton americanus*

## Hottonia
Primulaceae

Just two species of hardy, submerged, aquatic perennials, valued both for their attractive primrose-like flowers, held above the water, and for their feathery light green leaves which act as oxygenators.

*H. palustris* (Water violet). Bears spikes of violet flowers well above the surface of the water in summer; the submerged leaves are finely divided, and bright green. Needs sun and clear, clean water. H to 3ft (90cm), S indefinite, PD 8–24in (20–60cm).

**PROPAGATION**

✎ spring. ☀ 6 weeks.

Stem cuttings – summer. ☀ 8 weeks.

Turions – fruits form turions after pollination. Collect in fall and overwinter in frost-free conditions, on trays of soil covered with 1in (2.5cm) of water. Pot individually in spring. ☀ 6 months.

## Houttuynia
Saururaceae

A single species of rhizomatous, marginal perennial flourishing in the wet margins of pools and streams. The species can be invasive, although its variegated cultivars are less rampant. Prefers partial shade.

*H. cordata.* This has red-stemmed, bluish-green, ovate leaves, with an unusual fragrance when crushed, and bears spikes of small white flowers with white bracts in spring. The green-centered leaves of **'Chameleon'** have heavy markings in vivid crimson and cream. H 6–24in (15–60cm), S indefinite, PD to 4in (10cm).

**PROPAGATION**

✎ spring. Tease the extensive mats of rhizomes and re-plant immediately. ☀ 8–10 weeks.

## Lagarosiphon
Hydrocharitaceae

Some nine species of hardy to half-hardy submerged aquatics grown as oxygenators in indoor and outdoor pools and in aquariums. They prefer full sun.

*L. major,* syn. *Elodea crispa* of gardens (Curly water thyme). Hardy, semi-evergreen perennial with narrow, strongly recurved, fresh green leaves borne in spirals along the stem. Thin regularly to restrict growth. H and S indefinite.

**PROPAGATION**

Stem cuttings – throughout the growing season, 6–9in (15–22cm) long. ☀ 3 weeks.

## Lysichiton
Araceae

Just two species of hardy, moisture-loving or marginal, rhizomatous perennials grown for their striking, arum-like spathes in early spring, which are followed by architectural leaves. They grow best in deep, saturated soil and full sun.

*L. americanus* (Skunk cabbage). Has yellow spathes in early spring, followed by huge, almost stemless, paddle-shaped, glossy green leaves, up to 1.2m (4ft) high. H 3ft (1m), S 30in (75cm), PD 4in (10cm).

*L. camtschatcensis.* More compact than *L. americanus* with white spathes; also blooms slightly later in spring. H 30in (75cm), S 2ft (60cm), PD 4in (10cm).

**PROPAGATION**

🌰 as soon as ripe at 59°F (15°C). As the spadix rots it forms a gelatinous mess containing ripe seeds. Do not extract the seed, but break up the spadix and surface-sow in a large seed tray of aquatic medium. Submerge the tray to half its depth, and replenish the water regularly.

🌱 4–6 weeks. ☀ 18 months.

Offsets – spring. Separation of young plants that form at the base of older rhizomes is possible, but not easy. Pot up into pots of permanently moist medium and keep shaded until established. ☀ 18 months.

## Mentha
Labiatae

Some 25 species of hardy, aromatic, rhizomatous, often invasive perennials, one of which is grown as a water marginal. Prefers a site in full sun.

*M. aquatica* (Water mint). Has strongly fragrant, oval, serrated, dark green leaves, and dense, terminal whorls of tiny lilac flowers in summer. H to 3ft (90cm), S 3ft (90cm) or more, PD 6in (15cm) or more.

**PROPAGATION**

✎ during the growing season. ☀ 8 weeks.

Root cuttings – during the growing season. Place 3in (7.5cm) long root sections horizontally, just below the surface of aquatic medium, and cover with 1in (2.5cm) of water. ☀ 4 months.

## Menyanthes
Menyanthaceae

A single species of hardy, rhizomatous, marginal perennial, grown for its spikes of flowers and spreading foliage.

*M. trifoliata* (Bogbean, Buckbean). Has creeping rhizomes that often float at the water margins, and bears shiny, clover-like, green leaves with long stalks that clasp the rhizome. Small, white flowers open from pink buds in early summer, sometimes again in fall. Needs full sun. H 10in (25cm), S indefinite, PD 0–12in (0–30cm).

*Menyanthes trifoliata*

*Myosotis scorpioides*

*Nymphaea 'Laydeckeri fulgens'*

**PROPAGATION**

🌿 spring. ☀ 6–8 weeks.

🌰 as soon as ripe, at 61°F (16°C). Cover with 1in (2.5cm) of water.

🌱 2–3 weeks. ☀ 6 months.

## Myosotis                    Boraginaceae

Some 50 species of hardy annuals, biennials, and herbaceous perennials, including a number of attractive marginals. They have hairy leaves and salverform flowers. Need sun or light shade.

*M. scorpioides* (Water forget-me-not). Creeping, rhizomatous, marginal perennial with narrowly oval, mid-green leaves, and cymes of bright blue, yellow-eyed flowers in early summer. **'Mermaid'** is more compact and free-flowering. H 9–12in (22–30cm), S 12in (30cm), PD 4in (10cm).

**PROPAGATION**

🌿 spring. Established clumps are easily divided. ☀ 3 months.

Cuttings – nip off young tips about 2–3in (5–7.5cm) long. Treat as softwood cuttings.

🌰 early fall at 59°F (15°C).

🌱 2 weeks. ☀ 6 months.

## Myriophyllum              Haloragidaceae

About 40 species of frost-tender and hardy, submerged aquatics with attractive feathery foliage. They are grown as oxygenators and to provide shelter for fish fry. They need a position in full sun.

*M. aquaticum*, syn. *M. brasiliense*, *M. proserpinacoides* (Parrot feather). Has slender stems clothed in whorls of blue-green leaves that emerge above the water surface. An excellent scrambler for disguising pool edges, it will root in moist soil at the pond margin or may be grown in deeper water in baskets. Half-hardy grown as a marginal plant, frost-hardy as a deep-water aquatic. H to 5ft (1.5m), S indefinite, PD to 5ft (1.5m).

**PROPAGATION**

Stem cuttings – spring. Take cuttings of young growth 5in (12cm) long. ☀ 3 weeks.

## Nelumbo                   Nelumbonaceae

Only two species of half-hardy and frost-tender, rhizomatous, marginal perennials with circular leaves and flamboyant, water lily-like flowers borne well clear of the water surface. Lift the rhizomes in fall and store in damp sand in frost-free conditions. The plants need a position in full sun.

*N. lutea* (American lotus). Has rounded, blue-green waxy leaves, 16–32in (40–80cm) across and, in summer, large, fragrant yellow flowers. H to 6ft (1.8m), S indefinite, PD to 24in (60cm).

*N. nucifera* (Sacred or Oriental lotus). Has bluish-green, waxy leaves, to 32in (80cm) across, and fragrant, pale pink flowers, in summer. There are several superb cultivars, some of which are suitable for half barrels or tubs. H to 5ft (1.5m), S 4ft (1.2m), PD to 24in (60cm).

**PROPAGATION**

🌿 spring. Ensure that there are at least two buds on each section of rhizome and set horizontally just below the medium surface. Keep at a minimum temperature of 41°F (5°C) or they will not produce shoots. ☀ 6 months.

🌰 spring, at 77–86°F (25–30°C). Cover with 2in (5cm) of water.

🌱 4 weeks. ☀ 15 months.

## Nymphaea (Water lily)      Nymphaeaceae

About 50 species of deep-water, usually rhizomatous, sometimes stoloniferous or tuberous, perennials, ranging from tropical species, some of which are night-blooming, to hardy species that will survive long periods beneath ice. They have rounded leaves that float on the water surface and may be red-tinged beneath, and leafless stalks bearing showy flowers usually floating on the water. Many different cultivars are available to suit various depths of water and in a wide range of colors. They need full sun and still water.

**PROPAGATION**

🌿 throughout summer. ☀ 12 months.

Bud eyes – during the growing season. Section small pieces of root with a single bud and plant in a pot or a seed tray. Cover with 2in (5cm) of water. ☀ 24 months.

Plantlets can be removed from certain tender species and cultivars which are viviparous. ☀ 12 months.

## Orontium                        Araceae

One species of hardy, marginal or submerged, deep-rooting, rhizomatous perennial. Leaf color is best in full sun.

*O. aquaticum* (Golden club). Has waxy, blue-green leaves, silvery beneath, which in deep water float on the surface. In spring, bears slender white spadices with yellow tips. H to 18in (45cm), S to 30in (75cm), PD 15–18in (38–45cm).

**PROPAGATION**

🌰 as soon as ripe. Sow in midsummer at 64°F (18°C) and cover with ½–1in (1–2.5cm) of water.

🌱 3–4 week. ☀ 12 months.

155

*Peltandra undulata*

*Pontederia cordata*

*Ranunculus lingua 'Grandiflorus'*

## Peltandra          Araceae

Only two species of hardy, marginal perennials, with prominent surface rhizomes, suitable for muddy margins or very shallow water, in full sun. They scramble over the soil surface, rooting at the nodes.

*P. sagittifolia*, syn. *P. alba* (White arrow arum). Has long-stalked, bright green leaves and arum-like white spathes in early summer, followed by fleshy red berries. H 18in (45cm), S 24in (60cm), PD 8in (20cm).

*P. undulata*, syn. *P. virginica* (Green arrow arum). This is similar to *P. sagittifolia*, but its leaves are strongly veined and it produces greener and narrower flowers that are followed by green berries. H 36in (90cm), S 24in (60cm), PD 8in (20cm).

### PROPAGATION
✎ spring. Section rhizomes into 4–8in (10–20cm) lengths as the buds may be well dispersed.    ☀ 12 months.

## Persicaria, syn. Polygonum Polygonaceae

Some 50–80 hardy and tender species of annuals and usually rhizomatous or stoloniferous perennials, many of which are moisture-lovers. They need full sun.

*P. amphibia*, syn. *Polygonum amphibium* (Amphibious bistort, Willow grass). Hardy, semi-aquatic, stem-rooting perennial, useful for pools with fluctuating water levels. Lance-shaped leaves and pink flowers. H 1–2ft (30–60cm), S indefinite, PD 1–3ft (30cm–1m).

### PROPAGATION
Stem cuttings – spring and summer.
☀ 3 months.
✎ in the growing season. Separate stems that root at the nodes on contact with soil.
☀ 2 months.

## Pontederia          Pontederiaceae

Five species of hardy and frost-tender, rhizomatous, marginal and aquatic perennials with spikes of colorful flowers and attractive leaves. They need full sun.

*P. cordata* (Pickerel weed). Hardy, with a thick, creeping rhizome and neat habit. It has upright, lance-shaped, glossy olive-green leaves, and bears spikes of soft blue flowers at the top of the stem in late summer. H 18–24in (45–60cm) or more, S to 30in (75cm), PD 5in (12.5cm).

### PROPAGATION
✎ late spring. Divide when the plant begins active growth. Use a very sharp knife because the rather spongy fibrous rootstock is inclined to rot if pulled apart carelessly.
☀ 3 months.
🌰 as soon as ripe, at 54–59°F (12–15°C). Seed is best harvested green; it loses viability very quickly.
🌱 1–2 weeks.    ☀ 9 months.

## Potamogeton          Potamogetonaceae

About 80–100 species of hardy and tender, marginal and deep-water, often vigorous, rhizomatous perennials, with submerged and floating leaves. Some are suitable for growing as oxygenators or aquarium plants. They need full sun.

*P. crispus* (Curled pondweed). Hardy and efficient oxygenator with wavy-edged, almost translucent leaves, varying from green to reddish brown. It may be used in cold-water aquariums and outdoor pools. S indefinite, PD 2ft (60cm) or more.

### PROPAGATION
Stem cuttings – spring and early summer.
☀ 4 weeks.

## Ranunculus          Ranunculaceae

About 400 hardy and tender annuals, biennials, and perennials, including moisture-loving and aquatic species that prefer full sun.

*R. aquatilis* (Water crowfoot, Water buttercup). Hardy, submerged, clump-forming perennial with segmented submerged leaves and flatter, kidney-shaped, three-lobed floating leaves. White flowers with yellow throats appear on the surface in midsummer. S 2–3ft (60–90cm) or more, PD 6–24in (15–60cm).

### PROPAGATION
Stem cuttings – midsummer after flowering.
☀ 12 weeks.
*R. lingua* (Greater spearwort). Hardy, marginal perennial with upright red stems and long-stalked, arrow-shaped leaves. The flowering shoots bear linear or lance-shaped leaves, and cupped, yellow flowers in early summer. H 5ft (1.5m), S 6ft (1.8m), PD 6–9in (15–22cm).

### PROPAGATION
✎ spring. Use a sharp knife to divide the thick, rather sappy rootstock.    ☀ 3 months.

## Sagittaria          Alismataceae

Some 20 hardy and frost-tender, tuberous-rooted, mainly marginal perennials. The leaves may be floating or submerged and are often attractive. Decorative species are useful in wildlife pools; waterfowl eat the tuberous roots. They prefer full sun.

*S. latifolia* (Duck potato). Hardy, vigorous, marginal perennial with short, branching tubers. It has arrow-shaped to ovate leaves and bears whorls of 3-petaled white flowers in summer. H 16–24in (40–60cm), S 36in (90cm), PD 3–5in (8–12cm).

*Sagittaria sagittifolia*

*Saururus cernuus*

*Sparganium erectum*

**S. sagittifolia** (Japanese arrowhead). Hardy, tuberous perennial with arrow-shaped leaves and handsome, 3-petalled white flowers with a purple blotch at the base. H 36in (90cm), S indefinite, PD 3–5in (7.5–12cm).

**PROPAGATION**

✂ spring and summer.  ☀ 3 months.
☙ as soon as ripe, at 59°F (15°C), in a partially immersed seed tray or container. ♈ 2 weeks.  ☀ 9 months.

## Saururus                    Saururaceae

Just two hardy species of shallow-water, marginal, clump-forming perennials grown for their distinctive small, fragrant flowers. Prefers a sunny position but tolerates light shade.

**S. cernuus** (American swamp lily, Lizard's tail). Heart-shaped, bright green leaves and tiny, fragrant white flowers in long spikes in late summer. H 12–24in (30–60cm), S 12in (30cm), PD 4in (10cm).

**PROPAGATION**

✂ spring. Divide every 3–4 years to maintain vigor.  ☀ 3 months.
Offsets – summer. Separate offsets from the margins of the clump.  ☀ 6 months.

## Sparganium                  Sparganiaceae

Some 21 species of deciduous or semi-evergreen, mostly hardy, rhizomatous perennials producing slender rush-like brown-green leaves and inconspicuous flowers. They are capable of forming large spreading clumps at the water's edge, which make ideal nesting sites and provide winter food for waterfowl.

**S. erectum**, syn. *S. ramosum* (Bur reed). Hardy, semi-evergreen species with extensive rhizomatous roots bearing rosettes of long sword-shaped leaves. The branched inflorescence has rounded female flowers beneath several smaller male flowers in the same flower spike. H 3–4ft (90cm–1.2m), S indefinite, PD to 18in (45cm).

**PROPAGATION**

✂ spring. Divide the tough rootstock with a spade.  ☀ 3 months.
☙ as soon as ripe. Extract seed from the prickly round fruit in summer, and germinate at 59°F (15°C) in partially submerged seed pans.
♈ 3 weeks.  ☀ 12 months.

## Stratiotes              Hydrocharitaceae

One species of hardy, semi-evergreen, submerged and floating, stoloniferous perennial suitable for a sunny pond.

**S. aloides** (Water aloe, Water soldier). Rosettes of prickly, sword-shaped, olive-green leaves, which rise to the water surface to flower. The cup-shaped, white male and female flowers grow on separate plants. The rosettes sink after flowering, then produce auxillary shoots which resurface in late summer to grow into new plants with winter buds or turions. H to 20in (50cm), S indefinite, PD 12in (30cm)or more.

**PROPAGATION**

Runners – spring. Separate runners and re-float immediately.  ☀ 3 months.
Turions – fall. Collect in fall and overwinter on mud, covered with 6in (15cm) of water under glass at 40°F (4°C). Scatter on the pond surface in spring.  ☀ 3 months.

## Utricularia            Lentibulariaceae

About 180 species of hardy and frost-tender, epiphytic, terrestrial, and floating, rootless perennials. They are insectivorous. The aquatic species thrive in stagnant water in the wild and need soft (acid) water in cultivation and full sun.

**U. vulgaris** (Greater bladderwort). Hardy, stoloniferous perennial with narrowly bell-shaped, rich golden-yellow flowers carried well above the water surface on strong stems. Small insectivorous bladders are held in a network of feathery, bronzed leaves. H 12in (30cm), S indefinite, PD 12–24in (30–60cm).

**PROPAGATION**

✂ early summer. Tease the mats of floating foliage and re-float immediately.
☀ 6 weeks.

## Veronica               Scrophulariaceae

Some 250 species of annuals, herbaceous perennials, and mainly deciduous subshrubs including several hardy marginals.

**V. beccabunga** (Brooklime). Hardy, with creeping fleshy stems bearing spikes of white-centered, dark blue flowers and shining, rounded, fleshy leaves. It is excellent for scrambling over muddy banks, rooting as it goes. H 4–12in (10–30cm), S indefinite, PD to 5in (12cm).

**PROPAGATION**

Stem cuttings – midsummer.  ☀ 3 months.
✂ summer.  ☀ 2 months.
☙ as soon as ripe. Collect the rounded seed capsules and sow immediately at 59°F (15°C) in partially submerged seed trays.
♈ 2 weeks.  ☀ 6 months.

157

# ERBS

Ornamental and useful, herbs combine the delights of the flower garden with the satisfying activity of producing fresh ingredients for the kitchen that are far superior in flavor to their dried or frozen equivalents.

As a bonus, the delightful fragrances of herbs can be enjoyed in passing, making the garden a more relaxing place. Step into the garden and pick a few sprigs of herbs and you enter a different world, alive with invigorating scents and age-old associations.

Above all, herbs are among the easiest of plants to grow and propagate. This chapter describes how to increase and replenish your collection of herbs of all kinds, for culinary, medicinal, and cosmetic applications.

◄ *A mixed herb garden can include annuals, perennials, and shrubs that have been propagated by the full range of propagation techniques.*

# WHY PROPAGATE
# HERBS?

*Fresh herbs have a strength of aroma that gives intense flavor for kitchen use and the highest levels of active ingredients for teas and tonics. Propagating your own herbs in the garden will give you a wide choice of types, and will ensure that you can harvest them at exactly the right time for optimum freshness.*

**Most culinary herbs yield better flavors when young and succulent, so propagate regularly to maintain your supply.**

Many herbs are short-lived plants that need to be replaced annually or after only a few years. Some are annuals, or are treated as such, or biennials. Others, like rosemary, produce their best growth when young and become woody with age, so are best renewed regularly.

Whatever their ornamental value, herbs can also be treated as a crop. They may be grown like vegetables and sown at intervals, discarded after cropping, or forced for out-of-season use, regardless of their type.

Propagating herbs enables you to grow them in quantity for storing – either dried or frozen – for use in cooking, pot-pourri, or herbal remedies throughout the period when they are not available fresh. Herbs include a range of plant types, and the key to propagation is to know what kind of plant – annual, biennial, perennial, or shrub – you are dealing with.

## ENSURING SUPPLY

Propagation by seed is used for annual and biennial herbs, but can also be used to raise perennials and shrubby types. The advantage is that, for herbs such as parsley that are treated as a crop, you can sow them at intervals throughout the growing season to ensure a continuous supply of good-quality material. This also guarantees disease-free plants.

## REJUVENATING YOUR PLANTS

Division is the best method for clump-forming perennial herbs and rhizomatous types that can become woody or congested and less productive with age. All plants propagated by this and other vegetative methods will be identical to the parent. Besides increasing your stock, division benefits the plant by stopping it from becoming congested and unproductive. Most plants need to be routinely divided every 3–5 years.

(1)

***Melissa officinalis* 'Aurea'**
Variegated lemon balm will rapidly outgrow a container. It does not come true from seed, so divide in autumn or spring, or try rooting cuttings of non-flowering shoots in water.

(2)

***Oreganum vulgare* 'Aureum'**
Golden marjoram needs renewing every third growing season. Divide in spring or autumn, take basal cuttings in spring, or stem tip cuttings in early summer.

# TAKING CUTTINGS

Cuttings can be taken of some perennials and the majority of shrubby herbs. The treatment of those cuttings depends largely on the time of year they are taken, since cuttings struck early in the season are generally in need of more aftercare than those taken a few months later. Softwood cuttings are usually taken in spring, semi-ripe cuttings in summer, and stem-tip cuttings whenever suitable material is available. In addition, some herbs root more readily from cuttings taken at specific times of the year. A number of the perennial herbs can also be increased by root cuttings.

# MOUND LAYERING

This is a useful method for creating replacement stock of shrubby herbs such as lavender and rosemary. Its primary use is to propagate from plants that have become bare and woody at the base with age.

## *Herbs in Containers*

*All of the common culinary herbs – such as rosemary, bay, tarragon, parsley, chives, and mint – can be grown as short-term container plants. Strawberry pots are only suitable for growing small herbs that tolerate dry conditions, as competition for space and moisture is fierce when eight or nine plants are grown in the same container. Herbs in pots are in their prime in their first year. Small, shrubby and perennial herbs may look good for a second season, if pruned and well fed. But if you want good-quality herbs in subsequent seasons, be prepared to propagate and re-plant in fresh medium.*

A pot of mixed herbs can provide all your culinary needs for most of the year. Potted herbs can be harvested for much of the year.

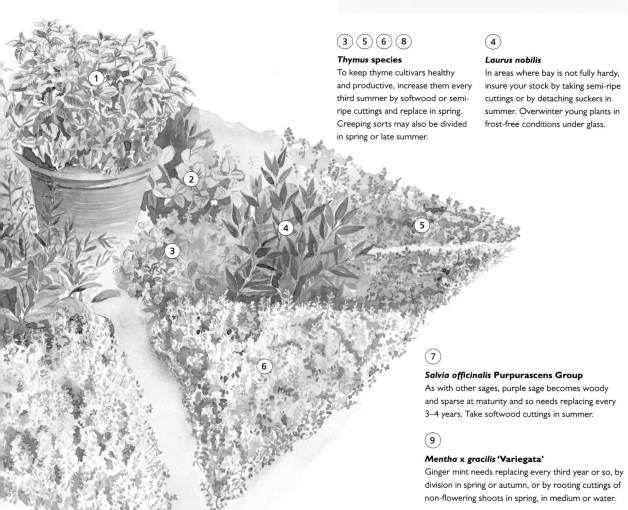

**③ ⑤ ⑥ ⑧**

### *Thymus species*

To keep thyme cultivars healthy and productive, increase them every third summer by softwood or semi-ripe cuttings and replace in spring. Creeping sorts may also be divided in spring or late summer.

**④**

### *Laurus nobilis*

In areas where bay is not fully hardy, insure your stock by taking semi-ripe cuttings or by detaching suckers in summer. Overwinter young plants in frost-free conditions under glass.

**⑦**

### *Salvia officinalis* **Purpurascens Group**

As with other sages, purple sage becomes woody and sparse at maturity and so needs replacing every 3–4 years. Take softwood cuttings in summer.

**⑨**

### *Mentha* x *gracilis* **'Variegata'**

Ginger mint needs replacing every third year or so, by division in spring or autumn, or by rooting cuttings of non-flowering shoots in spring, in medium or water.

# PROPAGATING TECHNIQUES
# HERBS

*The term herb covers such a wide range of plant types, and the key to successful propagation is to know what kind of plant – annual, biennial, perennial, shrub – you are dealing with. In some cases, a choice of techniques may be available to you, and it is often worth trying more than one as insurance against failure.*

Herbs are among the easiest of garden plants to grow and propagate and, by propagating regularly, you can be sure of a continuous supply.

## SEED

Since, unlike vegetables, many of the herbs grown in gardens are true species that have not been subjected to breeding programs. As a result, you can harvest your

### Drying seed heads ...........................

The technique for harvesting and drying seeds is essentially the same whether they are to be saved and sown the following year, or put to culinary use in the kitchen.

▶ **1** Harvest seeds promptly just as they begin to change color from green to brown. If you wait until they are brown and fully ripe the seeds are likely to fall and be lost. On a fine dry day when the dew has dried, cut the seed head with a good length of stalk to provide enough energy for ripening to continue.

◀ **2** Bundle the stalks and seed heads into small bunches and tie with a length of string or soft twine. Hang the bunches upside down in a dry airy place with the seed heads in paper bags secured with string. The seeds will then be caught in the bag as they ripen and fall.

own seed from year to year. For plants whose leaves are normally harvested regularly (and thus do not flower or set seed), allow a couple of plants to grow unharvested, so that seed is produced from summer onward.

Gather the seed when ripe and keep it dry in a paper bag until you are ready to sow it in spring or fall. Many hardy herbs can be sown *in situ*. Rake over the soil to make a fine tilth and make seed furrows 1in (2.5cm) deep. Sow the seed either individually or (in with very fine seed) in pinches. Cover the seed and water in well.

Some seed needs warmth to germinate. (*For the requirements of individual herbs, see the plant directory, pp. 166–171.*) You can also sow the seed of some hardy herbs in warmth early in the season for early crops. Fill seed trays or 3in (7.5cm) containers with a seed germination medium and firm it gently. Water the medium and allow it to drain. Sow the seed on the surface and cover to its own depth with medium. Very fine seed need not be covered. Water with a fine mister and allow to drain. Put the trays or pots into a closed, heated propagator. Once the seedlings are large enough to handle, pot them on and gradually harden them off by opening the propagator for progressively longer spells.

The seed of herbs that resent root disturbance and do not transplant easily can be sown individually in plug trays. Keep the seedlings in the trays until they are sufficiently developed for planting out.

## DIVISION

In spring or fall, lift clumps and shake them free of soil. Either tease them apart with your hands or with a hand fork. On tough plants, you may need to use a knife. To divide large clumps, the best method is to insert two forks back-to-back into the center of the clump and prise it apart into smaller sections. Discard any woody, overgrown sections (usually those from the middle of the clump) and replant the healthiest pieces. Water them in well.

## CUTTINGS

Some perennial and most shrubby herbs can be increased by various types of cutting, and many can be increased by more than one type of cutting as the season progresses by various types of cutting.

*Stem-tip cuttings:* These can be taken from perennials throughout the growing season. Select a non-flowering side shoot and cut it just above a node to a length of 3–5in (7.5–12.5cm). Trim the cutting just below a node and strip away the leaves from the lower third. Immerse the cutting in a fungicidal solution. Dip the base in hormone rooting powder and insert it to one-third of its length in a 3in (7.5cm) pot containing a mixture of equal parts coarse sand, perlite or vermiculite, and peat. Tent the cuttings with a plastic bag supported by stakes.

*Softwood cuttings:* Softwood cuttings of shrubby herbs are taken from the current season's growth in late spring and early summer and are treated in the same way as stem-tip cuttings. Inspect the cuttings regularly for any signs of fungal disease. Discard any affected cuttings and spray the remainder with a fungicide.

*Semi-ripe cuttings:* Take semi-ripe cuttings also from the current season's growth from mid- to late summer, just as the base of the current year's growth is beginning to firm and change from green to brown. If the base of the cutting is woody, remove a sliver of wood at one side about ½–1in (1–2.5cm) long before treating with a fungicide. Continue as for stem-tip cuttings.

Alternatively, where the side shoots are of a suitable length, take heeled cuttings. Grasp the stem at its point of origin on the main stem and pull it away to tear off a piece of bark. Trim this with a knife, then remove the lower leaves and continue as for stem-tip cuttings.

In warm weather, keep the cuttings shaded. In cold weather, you can speed up rooting by placing them in a heated propagator. Semi-ripe cuttings need not necessarily be tented in polyethylene but can be placed in a coldframe outdoors, though they will be slower to root. Ensure that they are watered regularly.

## Forcing herbs for winter use ....................

It is possible to extend the season for some herbs with protection. Chives, mint, and tarragon naturally die down in winter but can be lifted, divided and forced for winter use.

▶ I In early autumn, lift the clump and remove any dead or dying stems. Divide the clump into manageable portions by separating by hand or, for tough-rooted plants, by cutting with a sharp knife.

▶ 2 Pot up the divisions into pots containing a loam-based potting medium. Water in well. Keep in a light, frost-free place and harvest new shoots regularly when they reach about 4in (10cm) in height.

Cuttings given bottom heat should root within 4–6 weeks. As soon as they show signs of new growth, pot them individually and gradually harden them off. Semi-ripe cuttings without bottom heat should generally root by fall but are best left undisturbed until spring. Store them over winter in a coldframe (above freezing) and pot them up the following spring.

*Root cuttings:* Some perennials can be increased by root cuttings. When the plant is fully dormant, in mid- to late winter, lift it and remove a healthy section of root of pencil thickness (you will find that the best material is generally at the center of the crown). Cut the root into sections about 2–4in (5–10cm) long. To ensure that the cuttings are inserted correctly, cut the top of the root (the end nearest the crown) straight across and the base with a sloping cut. Each root should yield several cuttings. Dust the cut edges with fungicide. Insert them into moist rooting medium with their tops level with the medium surface and cover them with coarse sand. Then, place them in coldframe until new growth emerges, which is usually in early spring. Topgrowth is often produced before the roots, so check for rooting before potting up. When the cuttings are well rooted, pot them up individually and grow them on.

## MOUND LAYERING

This is a useful method for shrubby herbs that have become woody at the base and that do not respond well to hard pruning to rejuvenate them.

In spring, prepare a friable mixture of soil, potting medium or peat moss, and coarse sand. Mound the mixture around the plant, working between the stems so that only 2–4in (2.5–5cm) of the tips are visible.

### Planting up a herb wheel .................................................................................

Dig over the whole ground thoroughly, removing all weeds and large stones. Cover the soil with a weed-suppressing, inorganic mulch, such as black plastic sheeting, and on this add a thick layer of ornamental mulch such as coarse bark, pebbles, or sand. Define the edge of the wheel and each "spoke" with bricks or similar edging material, then fill each sector with a different herb.

Planting plan for a culinary herb wheel

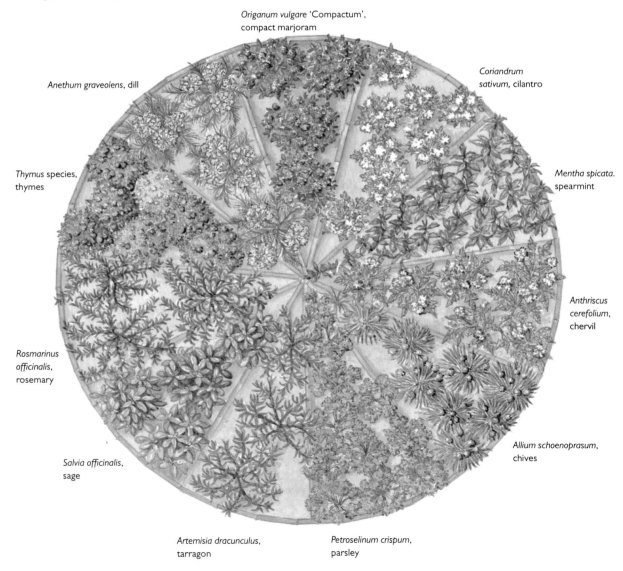

Origanum vulgare 'Compactum', compact marjoram

Anethum graveolens, dill

Coriandrum sativum, cilantro

Thymus species, thymes

Mentha spicata, spearmint

Anthriscus cerefolium, chervil

Rosmarinus officinalis, rosemary

Allium schoenoprasum, chives

Salvia officinalis, sage

Artemisia dracunculus, tarragon

Petroselinum crispum, parsley

Keep the mound well watered during dry spells in summer. Replenish the mound if heavy rains wash the top surface away. The stems will normally have rooted by fall. Scrape the surrounding soil away gently with your fingers and sever the new plants from the parent, cutting just below each new root system.

## MAINTAINING A HERB WHEEL

The herb wheel (*shown left*) can be kept in productive condition by planning for renewal and replacement. Clockwise from the top, renew the herbs as follows:
*Origanum vulgare* 'Compactum' (Compact marjoram). Hardy perennial. Divide and then replant after two growing seasons.
*Coriandrum sativum* (Coriander). Annual. Renew from seed in spring and again in summer.
*Mentha spicata (*Spearmint). Hardy perennial. Divide and re-plant each spring.
*Anthriscus cerefolium* (Chervil). Annual. Renew from seed in spring and again in summer.
*Allium schoenoprasum* (Chives). Hardy perennial. Divide and re-plant after two growing seasons.
*Petroselinum crispum* (Parsley). Hardy annual/biennial. Renew from seed each spring and fall.
*Artemisia dracunculus* (Tarragon). Hardy perennial. Divide and re-plant after two growing seasons.
*Salvia officinalis* (Sage). Evergreen shrub. Take cuttings in the third summer to re-plant the following spring.
*Rosmarinus officinalis* (Rosemary). Evergreen shrub. Take cuttings in the third summer to re-plant the following spring.
*Thymus* species and cultivars (Thyme). Take cuttings in the third summer to re-plant the following spring.
*Anethum graveolens* (Dill). Annual. Renew from seed in spring and again in summer.

## HERBS IN CONTAINERS

Growing herbs in containers not only turns them into a decorative feature, but is often the only choice for those without space for a herb garden. Containers are not only ornamental in themselves, but can be positioned for maximum enjoyment and effect wherever is most convenient. If you place them against a wall, or bring them under cover during the winter, herbs are protected against the elements, and enjoy relative freedom from some of their worst enemies –

### Propagation shortcut . . . . . . . . . . . . . . . . . . . . . . .
Rooting stem-tip cuttings in water is a simple shortcut. Nothing roots as quickly as mint, but it is also worth trying marjoram,

Place a few stem tips in water, having removed any leaves that will be below water. When roots are 1–2in (2.5–5cm) long, carefully pot up each cutting separately.

mud-splashing and slug attack. Propagate and re-plant regularly to keep containers looking good all season, or try adding different herbs for a change.

Containerized herbs must be fed and watered regularly, yet never allowed to become waterlogged. Water from the top and stand the pot in a small dish to increase water uptake if the medium becomes very dry. Most potting media have no more than six weeks' supply of nutrients. After this, feed plants every week with a liquid fertilizer. If overcrowded, herbs dry out quickly, exhaust nutrients, and fall prey to pests and diseases. Success lies in planting the right kind and number of herbs to a pot, and knowing when to propagate and replace them.

All young herbs fit nicely into a 3in (7.5cm) pot, but their rates of growth and eventual size will differ greatly. Always treat a containerized herb as a short-term proposition. As a rule, replace annuals at least once a year; perennials after two years; and shrubs after three years. Exceptions are bay, lemon verbena, and myrtle, which can be kept healthy in pots for many years.

• Check final height and spread, and whether the herb is annual, biennial, or perennial so that you have an idea how often it will need replacing.

• Do not overplant containers.

• Ensure that the container has a drainage hole and broken pieces of clay at the bottom.

• Use loam-based potting medium for better moisture retention.

• Add a top layer of coarse sand for additional weight and to prevent mud-splashing.

# DIRECTORY OF
# HERBS

*Below are key tips on a selection of herbs that are particularly suitable for propagation. Unless otherwise specified, follow the detailed instructions given under the propagation section on pages 162–165.*

## Key

| | |
|---|---|
| H | Height |
| S | Spread |
| HA | Hardy annual |
| HHA | Half-hardy annual |
| HB | Hardy biennial |
| HP | Hardy perennial |
| HSh | Hardy shrub |
| HHSh | Half-hardy shrub |
| TH | Time to harvest |
| ✄ | Division |
| ☙ | Seed |

### Allium                                    Alliaceae

About 700 species of rhizomatous and bulbous perennials with onion-scented leaves (*see also Bulbs directory p.64 and Vegetables directory, p.202*). Two are used as herbs; they tolerate semi-shade.

*A. schoenoprasum* (Chives). Has deciduous, grass-like, mildly onion-flavored leaves. May be forced for winter use. **'Forescate'** is larger, with pink flowers. HP. H 18in (45cm), S 12–18in (30–45cm).

*A. tuberosum* (Chinese chives, Garlic chives, Cuchay). Has flat, garlic-flavored leaves, and

*Allium schoenoprasum*

white star-shaped flowers in late summer. HP. H 12–18in (30–45cm), 12in (30cm).

**PROPAGATION**
✄ fall.
☙ *in situ* in spring.
TH throughout growing season.

### Aloysia                                    Verbenaceae

Some 37 shrubs with either deciduous or evergreen, aromatic leaves. Only *Aloysia triphylla,* is commonly grown.

*A. triphylla*, syn. *Lippia citriodora* (Lemon verbena). This has deciduous, lemon-scented leaves and tiny pale lilac flowers in summer. Needs shelter and is hardy to about 23°F (-5°C). Freeze-damaged plants usually survive, re-sprouting from the base in late spring or early summer. HHSh. H and S to 10ft (3m), but seldom exceeding 4ft (1.2m).

**PROPAGATION**
Heel cuttings – spring.
Softwood cuttings – early summer.
TH throughout the growing season.

### Anethum                         Apiaceae/Umbelliferae

A genus of only two species, one a biennial, one an annual. Both have very attractive, feathery, aniseed-scented leaves.

*Angelica archangelica*

*A. graveolens* (Dill). Umbels of tiny yellow flowers above the leaves are followed by flat, oval seeds. Leaves and seeds have a similar flavor but the seeds are stronger. **'Dukat'** has luxuriant foliage. **'Mammoth'** has sparse foliage but large heads of seeds. Dill bolts if overcrowded or too dry, and self-seeds in good conditions, although seedlings do not transplant easily. Dill and fennel can cross-pollinate to produce a hybrid. HA. H 2–3ft (60–90cm), S 6–12in (15–30cm).

**PROPAGATION**
☙ *in situ*, spring and early summer. Sow at monthly intervals for a succession of young leaves. Alternatively, sow in pots or trays at 45–50°F (7–10°C) for cutting like mustard and cress.
TH leaves before flowering. Cut seed heads as they start to ripen and hang upside down in a paper bag.

### Angelica                         Apiaceae/Umbelliferae

About 50 species of herbaceous biennials and perennials. One is grown as a herb.

*A. archangelica* (Angelica). Imposing specimen plant with lush, aromatic foliage and large umbels of green-white flowers. Needs sun or semi-shade in fertile, moist soil. Self-sows prolifically, so remove unwanted seed heads before they ripen. HB. H 5–10ft (1.5–3m), S 3ft (1m).

**PROPAGATION**
☙ *in situ* in fall or spring. Home-grown seed must be sown in the fall – it loses viability within 3 months.

### Anthriscus                       Apiaceae/Umbelliferae

Some 12 species of annuals, biennials, and herbaceous perennials with small flowers in umbels and ferny leaves. Only chervil is usually grown for its anise-flavored leaves.

*Artemisia dracunculus*

*Borago officinalis*

*Chamaemelum nobile*

*A. cerefolium* (Chervil). Chervil has fern-like, anise-flavored leaves. Grow in sun or semi-shade in cool, moist soil; it bolts if overcrowded or too dry. Take off flower buds to prolong leaf production. HA/HB. H 12–18in (30–45cm), S 9–12in (22–30cm).

**PROPAGATION**
🌱 sow *in situ* at monthly intervals, in spring and summer, or at 45–50°F (7–10°C) under glass in early fall for winter leaves.
TH before flowering.

## Armoracia      Brassicaceae/Cruciferae

Just three species of perennials with deep, woody or fleshy taproots, variable leaves and small flowers. Only horseradish is cultivated.
*A. rusticana* (Horseradish). Dock-like leaves grow from a thick, pungent tap root, which can be difficult to eradicate once established. **'Variegata'** is less invasive, with cream-variegated leaves. Tolerates semi-shade. HP. H 24in (60cm), S indefinite.

**PROPAGATION**
🪓 fall or early spring. Take root cuttings when plants are dormant.
🌱 species only, *in situ*, in spring.
TH lift roots from summer to late fall. Store in boxes of sand for winter use.

## Artemisia      Asteraceae/Compositae

Around 300 species of evergreen or deciduous annuals, perennials, and shrubs, with usually aromatic, often gray or silvery leaves *(see also Herbaceous Perennials directory, p. 35)*. Several are classed as herbs.
*A. abrotanum* (Southernwood, Lad's love). Deciduous or semi-evergreen shrub grown for its feathery, gray-green, pungent leaves. It rarely flowers in cold climates. Leaves retain

their scent on drying. HSh. H 2–3ft (60–90cm), S 12–24in (30–60cm).
*A. dracunculus* (French tarragon). Deciduous subshrubby perennial with narrow, green, anise-flavored leaves. **subsp.** *dracunculoides* (Russian tarragon) has an inferior flavor. Remove flower stems to prolong leaf production. May be forced for winter. HP/HHP. The subspecies is hardier than the species. H and S 18–24in (45–60cm).

**PROPAGATION**
🪓 early spring.
Cuttings – *A. abrotanum* semi-ripe, with a heel in late summer. *A. dracunculus* basal in spring.
TH throughout growing season.

## Borago      Boriginaceae

Three species of annuals and perennials, of which only borage is usually grown.
*B. officinalis* (Borage). Borage has cucumber-flavored leaves and edible, star-shaped blue flowers. Needs to be grown in moist but well-drained soil. Self-sows freely. HA. H 18–36in (45–90cm), S 6–12in (15–30cm).

**PROPAGATION**
🌱 *in situ* from spring to midsummer.

## Carum      Apiaceae/Umbelliferae

Some 30 species of tap-rooted biennials and herbaceous perennials. Caraway is the only species generally in cultivation.
*C. carvi* (Caraway). Upright plant with carrot-like foliage and umbles of tiny white flowers, then aromatic seeds. Seedlings do not transplant well. Seeds may not ripen from spring sowings in a cool summer. HB. H 60–90cm (2–3ft), S 30cm (12in).

**PROPAGATION**
🌱 *in situ* in fall or spring.
TH cut seed heads from summer to fall, and hang upside down in paper bags.

## Chamaemelum      Asteraceae/Compositae

Some four species of aromatic annuals and evergreen perennials with attractive feathery leaves and white daisy flowers.
*C. nobile*, syn. *Anthemis nobilis* (Roman chamomile). Mat-forming with bright green, apple-scented foliage and daisy-like flowers. Prefer well-drained soil in full sun. **'Flore Pleno'** has double white flowers. **'Treneague'** is non-flowering and recommended for chamomile lawns. The plants may become rather bare in cold winters but usually recover. Handling chamomile may cause dermatitis. HP. H to 6in (15cm), S 18in (45cm).

**PROPAGATION**
🌱 *in situ* in spring.
🪓 spring or fall.
Cuttings – stem-tip cuttings of side shoots in summer. Cultivars by cuttings only.

## Coriandrum      Apiaceae/Umbelliferae

Two annual species with scented foliage and small flowers. Coriander is used in cooking.
*C. sativum* (Coriander, Cilantro). Has strongly aromatic leaves and white flowers, then round seeds. Leaves and seeds have different aromas and uses. Grow in sun if seeds are required; for leaf crops, grow in semi-shade. HA. H 24in (60cm), S 12–18in (30–45cm).

**PROPAGATION**
🌱 *in situ* in spring for seed crops; at monthly intervals, from spring to summer for leaf crops. Leaves can also be produced

*Foeniculum vulgare*

*Lauris nobilis*

*Lavandula* 'Hidcote'

at 50–55°F (10–13°C) under glass.
TH leaves before flowering. Cut seed heads
as they ripen in summer and fall, and hang
upside down in paper bags.

## Cuminum · Apiaceae/Umbelliferae

Only two species of annuals with very fine
dark green foliage and umbels of white or
pink flowers, followed by oval seeds.
**C. cyminum** (Cumin). Cumin has dark green,
finely divided leaves and umbels of tiny white
flowers, followed by pungent, rather bitter
seeds. Needs a long, warm growing season
for seeds to ripen. HHA.
H 6–12in (15–30cm), S 3–4in (7.5–10cm).

### PROPAGATION
🌰 *in situ* in late spring.
TH cut seed heads as ripe in summer and
fall; hang upside down in paper bags.

## Cymbopogon · Poaceae/Graminae

Some 56 perennials, usually evergreen. They
are aromatic and their narrow leaves grow in
tufted groups; their flowers are like narrow
grassy seed heads. Lemon grass, the only
species cultivated, is used in cooking.
**C. citratus** (Lemon grass). Large, tropical grass
with thick stem bases and long, lemon-
scented leaves. The stem bases are used
whole, or finely chopped. The leaves are used
for herb tea. In cold areas, it is easily grown in
large pots of well-drained potting medium
on a sunny windowsill; it grows only to a
manageable size in pots, but is very much
larger grown outdoors in subtropical or
tropical gardens. Small plants are occasionally
offered for sale in the vegetable departments
of supermarkets. TP. Minimum temperature
45°F (7°C ). H 3–5ft (1–1.5m), S 3ft (1m).

### PROPAGATION
🔪 spring.
TH cut off individual stems throughout the
growing season.

## Foeniculum (Fennel) · Apiaceae/Umbelliferae

A single species of aromatic perennial.
**F. vulgare** (Fennel). An upright plant with
anise-scented foliage and umbels of tiny
flowers, followed by aromatic seeds. Not
reliably hardy in heavy soil and cold,
wet winters. **'Purpureum'**, (Bronze fennel)
is hardier. Comes true from seed. HP.
H 6ft (1.8m), S 18in (45cm).

### PROPAGATION
Seed – *in situ*, spring. Self-sows in light soils.
🔪 spring.
TH Leaves before flowering. Cut seed
heads as they ripen in summer and fall and
hang upside down in paper bags.

## Galium (Bedstraw) · Rubiaceae

Some 400 species of annuals and perennials,
usually with whorls of leaves and tubular
star-shaped, white, pink, or yellow flowers.
*G. odoratum* is the only one grown.
**G. odoratum** (Sweet woodruff). This
has whorls of narrow, fresh bright
green leaves and in spring and early
summer it bears pretty heads of pure
white, star-shaped flowers. Needs shade
or partial shade and moist soil. On drying,
leaves develop a delicious scent of new-
mown hay. HP. H 6–10in (15–25cm),
S indefinite.

### PROPAGATION
🔪 autumn or spring.
TH when in flower.

## Hyssopus (Hyssop) · Labiatae/Lamiaceae

Some five species of aromatic herbaceous
perennials and evergreen or semi-evergreen
shrubs with narrow leaves and spikes of
tubular flowers above the foliage.
**H. officinalis** (Hyssop). This is a semi-
evergreen shrub producing leaves with a
sage-mint flavor, and deep blue flowers in
summer. There are pink- and white-flowered
forms. **subsp. aristatus** (Rock hyssop) has a
dwarf habit, useful for edging and containers.
HSh. H and S 18–24in (45–60cm).

### PROPAGATION
🌰 *in situ*, fall or spring.
Softwood cuttings – summer.
TH throughout growing season.

## Laurus · Lauraceae

Only two species of evergreen plants: one
a tree, the other, bay, a shrub or small tree.
**L. nobilis** (Bay). This has leathery, evergreen,
pointed leaves. An excellent plant for
containers, it thrives in a relatively small pot,
tolerates fluctuations in watering, and
responds well to growing as a standard or
clipping. Needs protection from frost and
wind in areas with hard winters. H/HHSh.
In containers H 10–40ft (3–12m), S to 6ft
(1.8m), S 10ft (3m), 3–4ft (90–120cm).

### PROPAGATION
Semi-ripe cuttings – summer.
Suckers – detach in summer.
TH all year, but best in summer for drying.

## Lavandula (Lavender) · Lamiaceae/Labiatae

About 25 species of evergreen shrubs and
subshrubs with aromatic leaves and spikes of
fragrant tubular flowers often in shades of
blue or pink to purple, usually held on long

*Melissa officinalis*

*Mentha x gracilis* 'Variegata'

*Monarda* 'Cambridge Scarlet'

stalks above the plant. They prefer a site that has well-drained soil and is sunny. *(See also Shrubs, Trees, and Climbers directory, p.121.)*
*L. angustifolia* (Common lavender, English lavender). Has downy narrow leaves and spikes of purple flowers. HSh. H to 3ft (1m), S to 4ft (1.2m). **'Hidcote'** has dark purple flowers. HSh. H 24in (60cm), S 30in (75cm).
*L. x intermedia* (Lavandin). Variable rounded, sometimes lax shrub with gray-green leaves, and long spikes of very fragrant flowers. HSh. H and S 18–36in (45–90cm). **'Grappenhall'** has lavender-blue flowers. HSh. H 3–4ft (1–1.2m), S 5ft (1.5m).

**PROPAGATION**
*(See directions when grown as a shrub, p.121.)*
TH cut as the flowers begin to open.

## *Levisticum*    Apiaceae/Umbelliferae
A single species of herbaceous perennial.
*L. officinale* (Lovage). Triangular to diamond-shaped leaves with a celery or yeast flavor. Provide moist fertile soil in sun or partial shade. HP. H 6ft (1.8m), S 3ft (1m).

**PROPAGATION**
🌱 sow *in situ* in fall.
✂ spring.
TH throughout the growing season, but best before flowering.

## *Matricaria*    Asteracae/Compositae
Some five annual herbs with divided leaves and daisy-like flowers.
*M. recutita* (German chamomile). Has finely divided aromatic leaves. HA. H 24in (60cm), S 6–12in (15–30cm).

**PROPAGATION**
🌱 *in situ*, in fall or spring.
TH when flowers first open.

## *Melissa* (Balm)    Lamiaceae/Labiatae
Three species of herbaceous perennials with oval, serrated leaves smelling strongly of lemon. In cultivation for over 2,000 years.
*M. officinalis* (Lemon balm). Has wrinkled, nettle-like leaves. **'Aurea'** has yellow-variegated leaves. Grow in sun or semi-shade in moist soil. Cut stems back after flowering to encourage new leaves. HP.
H 24–75cm (60–75cm), S 12–18in (30–45cm).

**PROPAGATION**
🌱 species only, or divide in autumn/spring.
TH before flowering.

## *Mentha* (Mint)    Lamiaceae/Labiatae
Some 25 herbaceous perennials, most regarded as herbs, with fleshy creeping roots and aromatic serrated leaves. The flowers are small and tubular to bell-shaped. Grow mint in sun or semi-shade in moist soil. Mints are invasive, so plant in a large pot or sink an old container or heavy-duty plastic bag in the ground to control spread. HP. H 2–3ft (60–90cm), S indefinite.
*M. x gracilis* 'Variegata' (Ginger mint). Mild, spicy spearmint flavor.
*M. x piperita* (Peppermint). Purple-flushed leaves with a strong peppermint flavor. **'Citrata'** (Eau-de-Cologne mint) has a lavender-mint aroma.
*M. spicata* (Spearmint). Bright green, sweetly scented leaves.
*M. suaveolens* (Applemint). Rounded, woolly leaves with a mild spearmint flavor. **'Variegata'** (Pineapple mint) is more ornamental, with cream-splashed leaves.
*M. x villosa var. alopecuroides* (Bowles' mint). Large woolly spearmint-flavored leaves.

**PROPAGATION**
✂ fall or spring.
Stem-tip cuttings – non-flowering shoots in spring or summer.
TH before flowering.

## *Monarda* (Bergamot)    Lamiaceae/Labiatae
About 15 species of annuals and clump-forming rhizomatous perennials with aromatic leaves and whorls of tubular flowers, often with colored bracts.
*M. didyma*. Has bright red flowers and leaves smelling like Earl Grey tea. Most plants in cultivation are hybrids like **'Cambridge Scarlet'** and **'Croftway Pink'**. Grow in sun in fertile, moist soil. HP. H 24–36in (60–90cm), S 12–24in (30–60cm).

**PROPAGATION**
✂ spring.
Softwood cuttings – early summer.
🌱 spring at 61–70°F (16–21°C).
TH cut leaves before flowering, and flowers as they open.

## *Myrrhis*    Apiaceae/Umbelliferae
A single species of herbaceous perennial.
*M. odorata* (Sweet Cicely). Has fern-like, bright green leaves with a sweet, anise flavor, which can be used as a flavoring, and small star-shaped white flowers in summer. Grow in a sunny position or in shade in cool, moist, humus-rich soil. Self-seeds freely in good conditions. HP. H 36in (90cm), S 24–36in (60–90cm).

**PROPAGATION**
🌱 *in situ* in fall or spring (often slow to germinate).
✂ fall or spring.
TH throughout growing season.

**169**

*Ocimum basilicum*

*Origanum vulgare*

*Petroselinum crispum*

## *Myrtus* (Myrtle)    Myrtaceae

Two species of evergreen tree and shrubs.
*M. communis* (Common myrtle). A bushy
shrub with evergreen leaves, which have a
juniper-like aroma, and fragrant white flowers
in summer. **'Variegata'** has cream and gray-
green variegation. **subsp. *tarentina*** is dwarf,
compact, and free-flowering; it is good for
containers. Needs neutral to alkaline soil.
HHSh. It is hardy to about 23°F (-5°C).
'Variegata' is less hardy. H and S to 6ft (2m),
2–3ft (60–90cm) in containers.

### PROPAGATION

🌱 at 61–70°F (16–21°C) in spring.
Semi-ripe cuttings – late summer.
TH throughout growing season.

## *Ocimum*    Lamiaceae/Labiatae

Some 35 species of annuals and evergreen
perennials which have aromatic leaves. Most
have culinary or medicinal uses. Basil is the
most commonly grown.
*O. basilicum* (Basil). This has bright green,
clove-scented leaves and small white flowers.
**'African Blue'** is a large-growing hybrid. H
3–4ft ( 90–120cm). It is propagated by
cuttings from overwintered plants. **'Anise'**
has purple-flushed foliage, pink flowers, and
an aniseed flavor. **var. *citriodorum*** has citrus-
scented leaves. **var. *crispum*** (Curly or
Lettuce-leaf basil) has large, wrinkled leaves.
**'Dark Opal'**, syn. 'Purpureum', has purple-
black leaves. **var. *minimum*** (Bush basil) is
hardier, with small leaves and a compact
habit. H 6–12in (15–30cm). Grow in sun in
fertile, well-drained to dry soil. TA/TP.
Minimum temperature 55°F (13°C). Prone
to botrytis when cool and damp. H 24–36in
(60–90cm), S 12–18in (15–30cm).

### PROPAGATION

🌱 late spring, at 55–65°F (13–18°C).
TH throughout growing season.

## *Origanum*    Lamiaceae/Labiatae

About 20 species of often rhizomatous,
herbaceous perennials and evergreen or
deciduous subshrubs, mainly flowering in
summer and with aromatic leaves. Grow
*Origanum* in well-drained to dry, neutral to
alkaline soil. Use leaves and flowering sprigs
in Italian, Greek, and Mexican dishes. The
flavor is best when dried.
*O. majorana* (Sweet marjoram). This has gray-
green, strongly aromatic leaves and tiny white
flowers in bobble-like heads. Good for
growing in containers. HHSh, usually grown
as an annual. H 32in (80cm), S 18in (45cm).
*O. onites* (Pot marjoram). This has mildly
aromatic leaves and white or purple-flushed
flowers in summer and early fall. HP.
H and S 24in (60cm).
*O. vulgare* (Oregano). Variable in aroma and
appearance, the flavor is more pungent in
hot, dry conditions. It bears purple-pink
flowers in summer. **'Aureum'** has bright
yellow-green leaves and mauve flowers and
is best in semi-shade to avoid scorching.
**'Compactum'** has a dense, rounded habit, to
6in (15cm) tall, and is good for edging and
containers. HP. H and S 18in (45cm).

### PROPAGATION

🌱 species only, at 50–55°F (10–13°C) in
early spring, or *in situ* in late spring.
✂ fall or spring.
Basal cuttings – spring.
Tip cuttings – early summer.
TH leaves throughout growing season;
flowering sprigs when flowers first open.

## *Petroselinum*    Apiaceae/Umbelliferae

Three species of biennials with toothed
divided leaves and tap-roots. Cultivars of
*P. crispum* are commonly grown. *P. crispum*
(Parsley). The true species has been
superseded by curly-leaved cultivars such as
**'Moss Curled'**, **'Afro'**, and **'Clivi'**. Forms with
flat leaves are Italian or French parsley. They
are larger and more strongly flavored. Grow
in sun or semi-shade in fertile, well-drained,
neutral to alkaline soil. Protect winter crops
under plastic covers or in a coldframe.
HB/HA. Italian and French parsley are
hardier. H 12–24in (30–60cm),
S 12in (30cm).

### PROPAGATION

🌱 *in situ*, at intervals from early spring to
late summer for a year-round supply. For
early crops, sow at 50–55°F (10–13°C)
under glass. Soak seed overnight in
hot water to speed germination, which
otherwise takes 3–6 weeks.
TH throughout growing season in the first
year; before flowering in the second year.

## *Pimpinella*    Apiaceae/Umbelliferae

Some 150 species of annuals or herbaceous
perennials with hairy stems and umbels of
tiny star-shaped flowers, usually white or
yellow, followed by fruits.
*P. anisum* (Anise). Aromatic, divided leaves
and flat heads of tiny white flowers followed
by aniseed-flavored seeds. HHA.
H 12–20in (30–50cm), S 10–18in (25–45cm).

### PROPAGATION

🌱 sow *in situ* in spring.
TH leaves throughout growing season. Cut
seed heads as they ripen and hang upside
down in paper bags.

*Rosmarinus officinalis*

*Salvia officinalis* 'Icterina'

*Thymus x citriodorus* 'Silver Queen'

## Rosmarinus    Lamiaceae/Labiatae

Two species of aromatic evergreen shrubs with narrow leaves and whorls of flowers.
*R. officinalis* (Rosemary). This shrub has spiky, resinous, highly aromatic foliage and pale blue flowers in spring. **'Miss Jessopp's Upright'** has a vigorous, erect habit. **'Sissinghurst Blue'** is free-flowering, upright, and relatively hardy. Grow in well-drained to dry, neutral to alkaline soil. HSh. Hardy to about 14°F (-10°C), perhaps more with sharp drainage. H and S 3–6ft (1–2m).

### PROPAGATION
 spring, at 50°F (10°C), species only.
Semi-ripe cuttings – summer.
TH all year, but best in growing season.

## Rumex    Polygonaceae

Some 200 annuals, biennials, or herbaceous perennials, often with tap roots or rhizomes. The cultivated forms usually have basal leaves and flowers in clusters on erect stems. Needs sun or semi-shade, in moist soil.
*R. acetosa* (Sorrel). Has acidic, leaves. HP. H 12–18in (30–45cm), S 10–18in (25–45cm).
*R. scutatus* (French sorrel, Buckler-leaf sorrel). Smaller, less acidic leaves than *R. acetosa* and a mat-forming, invasive habit. HP. H 6–9in (15–23cm), S 4ft (1.2m).

### PROPAGATION
 *in situ* in spring.
 fall or spring.

## Salvia (Sage)    Lamiaceae/Labiatae

About 900 species of annuals, biennials, herbaceous and evergreen perennials, and shrubs. (*See also Herbaceous Perennials directory, p. 44*.) *S. officinalis* is the main herbal species.

*S. officinalis*. Velvety, evergreen leaves with a scent of camphor, and purple-blue to pink or white flowers. **'Icterina'** has yellow-variegated leaves. **Purpurascens Group** (Red sage, Purple sage) has purple-gray leaves. Grow in well-drained to dry, neutral to alkaline soil. Replace plants every 3–4 years when they become woody. HSh. Less hardy in heavy soils and in cold, wet winters. H 24–32in (60–80cm), S 24–36in (60–90cm).

### PROPAGATION
 spring, at 55–65°F (13–18°C). Species only.
Softwood cuttings – summer.
TH all year but best before flowering.

## Satureja (Savory)    Lamiaceae/Labiatae

Some 30 species of annuals, herbaceous perennials, and subshrubs with aromatic leaves and spikes of flowers. The two listed are most commonly grown. Needs sun in well-drained to dry, neutral to alkaline soil.
*S. hortensis* (Summer savory). Small, narrow, leaves taste peppery. Tiny lilac to white flowers. HA. H 10in (25cm), S 12in (30cm).
*S. montana* (Winter savory). Evergreen, with a less pleasant flavor than *S. hortensis*. HSh. H 16in (40cm), S 8in (20cm).

### PROPAGATION
 *in situ* in spring.
Softwood cutting – spring, for **S. montana**.
TH summer.

## Thymus (Thyme)    Lamiaceae/Labiatae

Some 350 species of evergreen perennials, shrubs, and subshrubs, with aromatic leaves and clusters of tubular flowers. Many species are cultivated as herbs or as ornamentals.

Grow in well-drained to dry, neutral to alkaline soil. HSh. Less hardy in heavy soils or cold wet winters.
*T. x citriodorus* (Lemon thyme). Lilac flowers and a lemon flavor. **'Silver Queen'** has variegated leaves. H 10–12in (4–5cm), S 24in (60cm).
*T. herba-barona* (Caraway thyme). Creeping stems and minute, spicy, dark green leaves. H 2–4in (5–10cm), S 24in (60cm).
*T. serpyllum* (Wild thyme). Prostrate with pink-mauve flowers. Aroma like *T. vulgaris*. H 2–3in (5–7.5cm), S 12–36in (30–90cm).
*T. vulgaris* (Common thyme). Variable, with dark green to gray-green, strongly aromatic leaves and mauve to white flowers. H 12–18in (30–45cm), S 24in (60cm).

### PROPAGATION
 spring, at 50°F (10°C), or in a coldframe. Species only.
Softwood cuttings – early summer.
Semi-ripe cuttings – midsummer.
 creeping thymes. Spring or late summer.
TH summer.

## Zingiber (Ginger)    Zingiberaceae

Some 100 herbaceous perennials with aromatic rhizomes and reed-like leaves.
*Z. officinale* (Common ginger). The ginger root used in cooking. Needs fertile, well-drained, neutral to alkaline soil, in sun or semi-shade, with warmth and high humidity. It grows well in pots although it will not crop heavily in the greenhouse in cool areas. TP. H 3–4ft (1–1.2m), S indefinite.

### PROPAGATION
 divide and plant rhizomes at 75°F (24°C) in early spring.
TH fall.

# RUIT

The fruit garden can be an idyllic place, alive with bees and other pollinating insects in spring, and providing a feast for the eye as well as the table as the fruits develop and ripen from summer to fall.

Most of the fruit sold in grocery stores and supermarkets has been bred for disease-resistance, heavy cropping, and long shelf life; flavor is a lower priority for commercial growers. However, in your own fruit garden, where you are growing for pleasure and not profit, you can concentrate on the taste and genuine quality of the produce.

Such is the range of fruit plants that a wide variety of propagation techniques must be used to ensure the continued life of the plants.

◄ *Keeping a well-stocked fruit garden involves a range of propagating techniques; some are needed annually, and others, like grafting, are perhaps needed only once in the life of the garden.*

173

# WHY PROPAGATE
# FRUIT?

*The flavor of home-grown fruit, fully ripened on the plant, is infinitely superior to anything that can be bought in a store, yet it is probably the easiest of the kitchen garden crops to produce. All you need is well-prepared ground and young, vigorous plants, which are simple to propagate for yourself using a variety of straightforward techniques.*

Propagation is an important and integral part of fruit growing. Young plants often produce better-quality fruits. Many soft fruits, particularly strawberries, are short-lived and need replacing regularly, and most fruit-bearing trees, though long-lived, need replacing as they age or if they are damaged and become less productive.

Nearly all fruit plants are hybrid cultivars selected especially for their fruit quality and heavy cropping. Because of this, seed is seldom used except to produce new hybrids, and clonal methods are most common.

## BETTER STRAWBERRY CROPS
Propagation by runners is used for strawberries, which are short-lived herbaceous perennials best replaced after three years or less. Regular propagation helps maintain virus-free stock and, if fresh plants are established in late summer, good crops will be assured the next year.

## TIP LAYERING
This method can be used for cane fruits that produce a quantity of flexible stems from their base. This is the way that blackberries spread naturally in the wild, and the technique is virtually fool-proof.

## A VERSATILE TECHNIQUE
Propagation by cuttings is a straightforward method for a number of fruit plants. Hardwood cuttings taken in autumn are the easiest and can be used for a number of subjects including currants, gooseberries, figs, and vines. Softwood cuttings can be used for blueberries and vines among others, and are quicker to root but require bottom heat and high humidity for best results.

The methods of leaf-bud and three-bud cuttings maximize the use of available cuttings material, but are slower to develop than conventional cuttings.

Apples

Raspberries

Blackberries

Red currants

**The techniques for propagating fruit range from the simplest pegging down of strawberry runners to sophisticated forms of grafting that need skill and practice to perfect.**

Strawberries

## MAKING USE OF SUCKERS
Naturally suckering plants, such as raspberries and hazelnuts, can be increased by severing the suckers. This also benefits the parent, which could otherwise become overgrown, less productive, and more prone to disease. This method should only be used on disease-free stock.

# THE ATTRACTION OF GRAFTING

Tree fruits are usually propagated by grafting onto a rootstock. Cuttings are seldom successful. The rootstock that is chosen can also promote improved disease resistance and affect the rate of growth. Using a dwarfing rootstock, for instance, will enable you to grow in a small space trees that would otherwise be too big. It is also quite possible to graft material from several different cultivars onto one rootstock to make a "companion" tree (*see box*).

The main grafting methods are chip- or T-budding, in late summer, and whip-and-tongue grafting, which is done from late winter to spring. Using both methods on any tree to be propagated increases your chance of success without losing growing time, as if the former is not successful, the latter can be carried out in the following spring.

*Companion Trees*

*Very few apples are self-fertile and so do not pollinate reliably or consistently, and the standard advice is that more than one apple has to be planted to ensure cross-pollination and fruiting. In small gardens with limited space, this may not be possible. One answer to this is to plant a "companion" or "several-in-one" tree. A companion tree is one in which branches of an established popular cultivar have been grafted with one or two other cultivars that are compatible pollinators. This also gives a choice of fruits in a limited area, and ensures that there is seldom a glut of one variety.*

A companion tree grafted with two different mutually-compatible cultivars of apple.

**① Apples**
Almost invariably grown as grafted plants, apples are exceptionally long-lived and seldom need re-propagating. They do not come true and take many years to fruit if they are grown from seed.

**② Raspberries**
These fruit can be increased by suckers, but only take suckers if the parent plant is healthy and virus-free. Diseased stock should be replaced by certified virus-free stock from the garden center.

**③ Blackberries**
These and other related cane fruits increase naturally by tip layers and this is by far the easiest means of propagation. Propagation by leaf-bud cuttings is useful where greater numbers are required.

**④ Red currants**
Along with black and white currants, red currants are among the fruiting shrubs that can be increased simply by hardwood cuttings, but are only likely to need replacement after 10–20 years.

**⑤ Strawberries**
Strawberries are best propagated every second or third year as the best crops are obtained from young plants. Regular propagation also reduces the risks of virus infection.

# PROPAGATING TECHNIQUES
# FRUIT

*The methods that one can employ for the propagation of fruit are as numerous and diverse as the species to which they apply. However, very few techniques will present problems to the dedicated gardener or require much in the way of special equipment. Whichever of the many following methods you opt for, the result should always be the same: better-quality fruit.*

## SOFT FRUIT

Soft fruits are mainly those that are grown on bushes or canes, like currants, raspberries and blackberries, as opposed to tree-borne fruits.

### Propagation of strawberries by runners

Unique among the soft fruits, strawberries, *Fragaria × ananassa*, are herbaceous perennials. They are prone to soil-borne viruses and to fungi like strawberry red core, and should be grown in rotation: that is, new plantings should be made in clean ground that has not grown strawberries before. Strawberries should not be grown in the same ground for more than three years, and a supply of fresh plants should be maintained by regular propagation. Ideally, where space in the garden permits, one-, two-, and three-year-old strawberry plants should be grown and harvested successively, so that the oldest three-year-old plants can be discarded after fruiting and burned.

### Strawberry runners . . . . . . . . . . . . . . . . . . . . . . . . . . .

The crown buds of strawberry plants develop stems that spread horizontally over the ground. These runners readily root at the leaf nodes to form new plants.

► Runners will be produced from mid- to late summer. Peg them down using U-shaped lengths of wire, either onto the open ground or (preferably, to keep the new plants free from soil-borne diseases) into a 3in (7.5cm) pot filled with pasteurized, loam-based medium plunged into the ground, with the surface of the medium level with the soil. When the runners are well rooted, usually by late summer, they can be severed from the parent plant.

When first planting a strawberry bed, it is essential to obtain certified virus-free stocks. When propagating from established plants, use only healthy, disease-free material. Strawberries can be affected by several viruses, usually spread by aphids and other sap-sucking insects, which cause symptoms including stunting, mottling, spotting, yellowing, or puckering of the leaves.

Although runners can be taken from fruiting plants, it is best to set aside a few plants specifically for runner production and to grow them in a separate bed well away from fruits grown for cropping. A propagating bed can be expected to be productive for two years. Plant them 3ft (1m) apart in a nursery bed and remove any flowers so that all their energy goes into runner production. Runners will be ready to peg down between mid- and late summer. Keep the soil moist to encourage the runners to root. Once well rooted, usually by late summer, young plants can be set out in their cropping site to crop in the following season. In areas with very cold winters, it is advisable to delay planting out young plants until spring, but, in this case, remove the flowers in the first season to allow them to build up into strong plants before cropping.

### Tip layering

Blackberries, *Rubus fruticosus*, and a number of other hybrid berries that have arching canes can be propagated by this simple method. A limited number of new plants can be made, since only well-placed stems can be used, each producing one new plant. Where larger numbers of plants are required, leaf-bud cuttings or three-bud cuttings can be used (*see p. 178*).

From late summer to early fall, select strong, healthy canes, preferably toward the edge of the plant. Bring the tips down to soil level and, at the point where they

## Tip layering . . . . . . . . . . . . . . . . . . . . . . . . . . . . . .

This method of propagation is invaluable with various members of the *Rubus* family because their stem tips when buried will swell, develop roots, and form new plants.

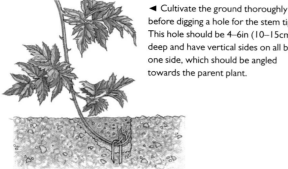

◄ Cultivate the ground thoroughly before digging a hole for the stem tip. This hole should be 4–6in (10–15cm) deep and have vertical sides on all but one side, which should be angled towards the parent plant.

meet the ground, dig a hole 4–6in (10–15cm) deep. On heavy soils, work in a little sand to improve drainage. Bury the stem tips, securing them with wires bent into U-shapes if necessary.

Keep the soil moist. The tip should root within 3–4 weeks and a new shoot will develop from it. The following spring, sever the rooted tip from the parent plant, retaining about 9in (22cm) of the stem, and plant it out in a new position.

## Hardwood cuttings

Soft fruits such as currants (black, red, and white) and gooseberries, figs, and grapes can be increased by hardwood cuttings. Where the root pest *Phylloxera vastratrix* is present, grapes must be grafted onto a *Phylloxera*-resistant rootstock.

Most cuttings root best if taken in early fall, at or just before leaf fall, before they are fully dormant. This enables some root initials to form before the onset of winter, allowing rapid root growth the following spring. In cold climates, figs and grapes need to be rooted under glass or in a closed coldframe.

*Currants (red and white) and gooseberries:* Red and white currants and gooseberries are usually grown as open-centered bushes, cordons, or fans.

Take cuttings 12–15in (30–38cm) long, cutting with an angled cut just above a bud. Trim the base of the cuttings with a straight cut just below a bud. Remove the tips of the cuttings if they are very soft. On gooseberries, trim off large thorns for ease of handling.

On red and white currants, remove the lower buds to prevent suckers arising from below ground level, leaving four good buds at the top. Gooseberries do not root as easily, so, to promote root initiation retain all of the buds on the stem and the top two leaves, and dip the base of the cutting in hormone rooting powder.

Prepare a slit or straight-backed trench 6–9in (15–22cm) deep. On heavy soils, run sharp sand into the bottom to aid drainage. Insert the cuttings to half their length, 6in (15cm) apart, and firm them in. In cold climates, lay a 3ft (1m) wide strip of black plastic along the nursery bed, burying about 6in (15cm) on either side. This prevents the soil surface from freezing and also retains moisture and discourages weed germination.

## Hardwood cuttings . . . . . . . . . . . . . . . . . . . . . . . . . . .

The length of a hardwood cutting and how it is prepared depends on how readily a particular type of fruit develops roots – a longer cutting is needed for more reluctant types.

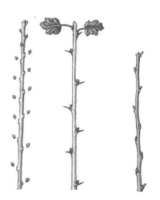

► The gooseberry cutting (*center*) is similar in length to a red currant one (*left*), but also has two leaves at the tip to assist in root production. All but four buds are removed on red and white currants to prevent suckers forming below ground, whereas the buds on the black currant cutting (*right*) are left intact.

Make holes of the appropriate depth and insert the cuttings through the plastic into the soil.

Transplant the rooted cuttings to their final positions the following fall if they have made good growth. After lifting, rub off any buds below ground level to discourage suckering. If the cuttings are undeveloped, prune them lightly and transplant them to a nursery bed, or pot them up and grow on for another season.

*Black currants:* Select well-ripened shoots of pencil thickness from the current season's growth. Take cuttings 8–10in (20–25cm) in length, cutting just above a bud at the base. Trim the bases just below a bud and remove any soft wood at the tip, cutting just above a bud. Strip off any leaves, but retain all the buds.

**Raspberry suckers** . . . . . . . . . . . . . . . . . . . . . . . . . . . . .

Any rasberry plant that shows signs of disease should not be used as propagation material, even if its suckers are apparently healthy. If at all uncertain, dig up and discard the plant.

◀ In spring and summer, allow suckers to grow up away from the main row. In the fall, lift them carefully, taking care not to damage the root system as you do so. Shorten the stakes to about 9in (22cm) and plant them out in their final positions.

Prepare a trench as for red and white currants. Insert the cuttings 6in (15cm) apart, with the second bud from the top of the cutting level with the soil surface. Firm them in. Transplant the rooted cuttings to their final positions the following fall.

*Grapes:* Grapes root easily from hardwood cuttings taken in midwinter when the plant is dormant. The sap bleeds if stems are cut while in active growth. Select stout stems of ripe wood, just over pencil thickness, from the current season's growth and cut into 2–3ft (60–90cm) lengths. Remove soft material from the tips. If you cannot make cuttings immediately, heel them in to two-thirds of their length in a coldframe in a sheltered site ; they can be stored for up to six weeks.

To root directly in a coldframe, dig in sand at the base to improve drainage. Cut sections from the stems, each with three buds (i.e. three-bud cuttings). Trim the cuttings with a straight cut just below the lowest bud and a sloping cut just above the topmost one. Insert them to two-thirds of their length, 6in (15cm) apart. For a greater quantity of new plants, root two-bud cuttings inserted to half their length in 3in (7.5cm) pots of loam-based medium and place these in a cold greenhouse.

Where there is less material available or if more plants are required, single-bud cuttings, known as grape eyes, can be used. Make a cut above a bud and another 2in (5cm) below, and insert these into a mix of equal parts peat and sand with the bud just resting on the surface. For the best results, place the cuttings in a closed propagator with bottom heat of 64°F (18°C).

For optimum growth, pot the cuttings on in stages once rooted and grow them on under glass for a year. Keep them under cover for the second winter and plant them out the following spring, if sufficiently developed.

### Leaf-bud cuttings

These are suitable for producing large numbers of plants from blackberries and hybrid berries.

Prepare a rooting medium of equal parts peat and sand in 3in (7.5cm) pots, or fill a coldframe to a depth of 3in (7.5cm) with the same mixture.

From mid- to late summer, take cuttings 12–18in (30–45cm) long from canes with healthy buds and leaves. Shorten the tips to remove any soft, immature buds. Cut the stems into sections about 1in (2.5cm) long, each with a bud halfway along. Insert the cuttings into the rooting medium so that the bud is just touching the soil surface. Keep them humid until rooted, usually within 6–8 weeks, either by tenting the pots in polyethylene or by keeping the coldframe closed. Harden them off and overwinter in a coldframe, ventilating during mild spells. The following spring, plant them out 12in (30cm) apart in rows, or pot them on and grow them on for a year before setting out.

### Softwood cuttings

Softwood cuttings can be used for blueberries, grapes, and blackberries, and for hybrid berries.

Take cuttings 4–6in (10–15cm) long of the current season's growth in midsummer, cutting just above a bud. Trim the cutting just below a bud and remove the lower leaves, leaving the top three or four. Immerse them in a fungicidal solution. Dip the base in hormone rooting powder and insert into a mixture of equal parts sand and peat in 3in (7.5cm) pots. Root in a closed propagator at 64–68°F (18–20°C), out of direct sunlight. They should root in 4–6 weeks. Check for fungal infection regularly, remove affected cuttings, and spray the remainder with fungicidal solution.

Once new topgrowth emerges, pot up the cuttings individually (blueberries must be potted up into lime-free growing medium). Gradually harden off the cuttings in a coldframe, keeping them shaded from hot sun. Apply a high-potash liquid fertilizer every ten days until late summer. Grow on for a year before planting them out into their final positions.

## Suckers

Fruit plants most commonly increased in this way are raspberries, hazelnuts and filberts. Unless required for propagation, remove suckers as they appear, since they divert the parent plant's energies from fruit production.

*Raspberries:* Raspberries are very susceptible to virus infection and any plants that show signs of disease should not be used as propagation material, even if the suckers are apparently healthy. Symptoms to watch out for include mottled or crinkled leaves, stunted growth, small crumbly fruit, and reduced yield.

*Hazelnuts and filberts:* Remove most of the suckers that appear, leaving the strongest to develop naturally, preferably those farthest from the tree. In fall, after leaf fall, carefully dig around the suckers, check that it has a good root system, and sever it from the parent below ground. Plant out in a nursery bed 18in (45cm) apart, or pot them up and grow on for a year. The following fall, transplant the plants to their final positions.

*Pineapples:* These can also be increased by suckers or slips: that is, suckers that appear in the leaf axils, or at the base of the fruit or stem. The crown shoot that appears at the top of the fruit can also be used as propagating material. Pineapples are tropical fruits that need high humidity and a temperature of 64–80°F (18–20°C) to produce fruit. They are grown outdoors only in tropical regions; elsewhere they need a warm greenhouse and full light.

## TREE FRUIT

Most tree fruits are propagated by grafting onto rootstocks of a closely related species or genus, a technique that has been used for thousands of years. There are good reasons why they are propagated in this way.

They do not come true from seed – every seedling is distinct and many are of inferior quality. They are slow to crop when grown from seed – apples may take 7–10 years; peaches take 4–5 years. Nor, with some exceptions, do they strike easily from cuttings.

Apple rootstocks are derived from cultivated apples or other *Malus* species. Pears are commonly grafted onto quince (*Cydonia oblonga*), which produces dwarf trees that crop early; pear rootstocks themselves are

## Pineapples

A pineapple plant can be increased from cuttings, "slips" (which arise from immediately beneath the fruit), or suckers (which occur on the stem and at the base of the stem).

▲ **Propagation by cuttings**
I Carefully cut out the complete crown shoot. Dip in fungicide, then allow to dry for a few days.

▲ 2 Fill a pot with cuttings medium and insert the pineapple cutting. Set in a temperature of 64°F (18°C) while roots develop.

▲ **Propagation by "slips"**
Sever the "slip" from immediately under the fruit. remove the lower leaves, then treat as for a cutting.

▲ **Propagation by suckers**
Using a sharp knife, cut off each sucker close to the stem. Then treat as above, for a cutting.

## Rootstocks

Most fruit trees are grafted onto the rootstock of a different cultivar to control their height and fruiting vigor.

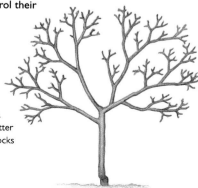

► When buying a fruit tree, always check the type of rootstock on which a tree is grafted and the type of soil that rootstock requires. For a small garden it is generally better to select the dwarfing rootstocks M27 and M9 than one of the larger, more vigorous ones.

## Tree forms ...............................................................................................................................

Trees can be trained from an early age into a variety of shapes to increase their flowering and fruiting potential, to confine their growth to fit a particular area, or to make them more accessible for picking. In a small garden, it is particularly useful to grow a fruit tree against a wall or fence, in a cordon, espalier, or fan, which can look highly attractive.

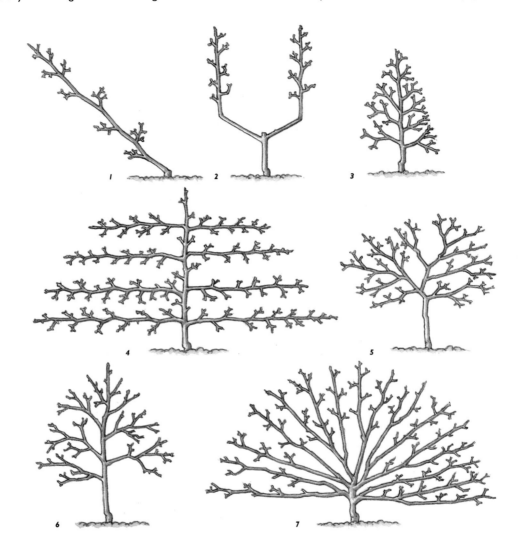

**1** The cordon is an excellent tree form in a small garden, where its main stem can be trained obliquely, thereby saving space and height. The most popular rootstock for an apple cordon in a very fertile soil is M9; elsewhere, use M26 or MM106. For pears, choose Quince C or A.

**2** Cordons may also be trained with two or more stems. Obliquely trained double cordons tend to be more fruitful and less vigorous than those trained vertically.

**3** Dwarf pyramid apple trees are generally grown on M9 and M26 rootstock and pears on Quince C or A. Planted closely together, they make compact trees, 6–7ft (1.8–2.1m) in height, which are easy to crop.

**4** Espaliers have one or more pairs of opposite stems trained at right angles to a vertical main stem. Such a tree form makes very economical

use of space. Espalier apples are generally grafted on MM106 or MM111 rootstock, and pears on Quince A.

**5** This goblet-shaped tree with its open center is best suited to gardens where space is relatively unrestricted. This tree form is grown on M9 to MM111 apple rootstock and Quince C and A pear stocks.

**6** A spindlebush tree is cone-shaped and has its central leader removed at an early age. Thereafter most branches are trained almost horizontally by tying them to pegs in the ground to encourage fruit production. Spindlebush apples are grown on M9, M26, and MM106 rootstocks and pears on Quince C and A.

**7** The fan is best trained against a wall or fence, where it will not only make excellent use of space but also be easy to pick. Apple fans are generally grown on MM106 rootstock and pears on Quince A.

difficult to propagate and not sufficiently dwarfing for most gardens. Cherries are grafted onto rootstocks of related *Prunus* species. Peaches, nectarines, apricots, and almonds are commonly grafted onto plum rootstocks.

Some cultivars are not compatible with all rootstocks. For example, not all pears can be grafted directly onto quince and an interstock of a compatible cultivar must be used such as 'Beurré Hardy' or 'Doyenné du Comice'. Peaches, nectarines, and apricots are not directly compatible with the dwarfing rootstock Pixy, but Pixy can be used with an interstock of the compatible rootstock St Julien A.

The most common rootstocks in order of ascending vigor are as follows:

| **Apples** | M27 | M9 | M26 | | MM106 | MM111 |
|---|---|---|---|---|---|---|
| **Pears** | | | Quince C | | Quince A | |
| **Plums** | | | Pixy | | St Julien A | |
| **Peaches/** | | | (with Pixy | | St Julien A | |
| **Nectarines** | | | interstock) | | | |
| **Apricots** | | | | | | |
| **Cherries** | | Edebritz | | | Colt | F12/1 |
| | | (Tabel) | | | | |

The choice of rootstock will control the size of the tree grafted onto it, and the various rootstocks have been selected to provide good anchorage, disease resistance, and early cropping. Rootstocks are propagated by layering, stooling, or in some cases by cuttings.

## TECHNIQUES OF GRAFTING

Grafting unites two plants so that they form a strong union and grow together. The plant to be propagated is called the scion; the plant onto which it is grafted is the rootstock. For any graft to be successful, there are three criteria that must be met. The timing must be correct and the material for stock and scion must be compatible, of similar size, and in good condition. There must be good contact between the cambial layers of rootstock and scion. In addition, all cuts made for grafting must be made with a sharp knife. Although grafting is a relatively advanced technique, attention to detail and practice will eventually lead to success.

The most common and successful methods used for tree fruit are bud-grafting (by chip-budding and T-budding) and whip-and-tongue grafting.

Your choice of method will be dictated to some extent by the type of plant you are trying to propagate and the time at your disposal. Budding is done from mid- to late summer; whip-and-tongue grafting in late winter or early spring.

### Preparing the rootstocks

Buy good-quality rootstocks and plant them 18in (45cm) apart in rows 3ft (1m) apart in a nursery bed until you are ready to start propagating. For best results, ensure that the rootstocks are well established before budding or grafting. Rootstocks planted in the dormant season will be ready for budding in mid- to late summer and for whip-and-tongue grafting late the next winter.

On planting, remove any side shoots from the bottom 12in (30cm) of trunk. During the growing season, rub out any side shoots that develop below 12in (30cm). Water the rootstocks well during the summer, so that they are growing actively and the bark will lift easily (especially important for T-budding).

It is vital to select good-quality material for the rootstock and for the parent plant to be propagated. With few exceptions, viruses in tree fruit are graft transmitted, so it is always best to propagate from guaranteed virus-free stock for both rootstock and scion to ensure that the resulting plant is virus-free for the whole of its life. However, on older and rare cultivars, guaranteed virus-free stock may not be available. In this case, propagate from the healthiest-looking material available.

### Bud-grafting

Bud-grafting, which includes chip-budding and T-budding, is economical with material, since only a single bud of the scion is need for each new plant. More importantly, it produces high-quality trees with strong graft unions.

Chip-budding is generally more successful than T-budding as it provides better cambial contact between scion and rootstock and there is less risk of infection by the canker *Nectria galligena* and other pathogens that can enter through wounded tissue. With T-budding, the bud is enclosed beneath the bark of the rootstock and so the spores of any pathogens will be trapped in a good position to infect the wound.

It is important to lose as little time as possible between cutting the scion and carrying out grafting, since any loss of moisture will work against a good union taking place.

In late summer, cut well-ripened stems from the sunny side of the tree. These bud sticks must be vigorous, of pencil thickness, and with dark green leaves and well-developed buds. They should just be turning woody at the base. Strip the stems of leaves immediately to prevent water loss by transpiration, but leave the petioles (leaf stalks) intact; remove any stipulary leaves (small leaves near the base of the petiole). Wrap the stems carefully in a damp cloth and bud-graft straight away.

It is always preferable to bud immediately after collecting the scions, but if necessary, you can store the stems for a few days – at the most – in a dry, clean plastic bag in the refrigerator.

## Chip-budding . . . . . . . . . . . . . . . . . . . . . . . . . . . . . . . .

Always take scion material from a vigorous, healthy stem of the current season's growth and with well-developed buds. Clear this bud stick of any leaves and trim off the soft tip.

► **I** Make a downward, angled cut at about 45°, ¾in (2cm) below the bud. Then make a shallow cut, starting ¾in (2cm) above the bud and bringing the blade down to meet the first cut.

► **2** Locate a smooth area of stem on the rootstock (free from any buds), about 9in (22cm) from the base and preferably on the shady side (where the cuts are less likely to dry out). Make a concave cut corresponding to the bud.

► **3** Put the bud in place. Ideally they should fit exactly, but if, as is likely, the rootstock is broader than the bud, place the bud to one side so that there is cambial contact between the rootstock and scion on one side.

► **4** Tie the two firmly with grafting tape, but not so tightly as to constrict the bud. After 4–6 weeks, callus tissue should form around the bud, indicating that the union has "taken." Remove the tape carefully.

## Chip-budding

Budding is carried out at a height of between 6–12in (15–30cm) on the rootstock, with 9in (23cm) being the preferred height.

Remove any hard wood at the bottom of the scion and any immature wood at the top. Working from the bottom of the scion, locate healthy buds, avoiding any buds with multiple leaves, which may be fruit buds. Shorten the petioles to about ¼in (5mm).

To remove a bud, first make a downward-angled cut at 45°, ¼in (5mm) deep, at about ¾in (2cm) below the bud. Make the second cut about 1½in (4cm) above the first cut and about ¾in (2cm) above the bud so that it slopes downward to meet the first cut. Take care not to damage the bud when making this cut. Remove the bud chip, handling it by its bud to avoid contaminating the cambium layer.

On the rootstock, select a smooth, clean piece of stem free of any buds at about 9in (22cm) high. Make a shallow, slightly concave cut corresponding in shape to that of the bud chip and put the bud chip in place. Ideally, the cuts should match, but if the cut on the rootstock is larger than the bud, as is often the case with a thick rootstock, place the bud to one side so that there is cambial contact between rootstock and scion on one side only.

Tie the graft firmly with grafting tape. Trap the end of the tape below the bud and wrap the bud with only a slight overlap, tying off just above the bud with a half-hitch. The tie should be firm but not constricting. It is essential to pay careful attention to detail at this stage, since more grafts fail from poor tying than from poor grafting technique. Keep the grafting tape in place until callus tissue forms around the bud. This usually takes four to six weeks. Remove the grafting tape by cutting the tie behind the bud and unwrapping it carefully.

If the bud has taken successfully, the bud will appear healthy and the petiole will remain plump and healthy until fall, when it will drop off with the other leaves on the rootstock. If budding has failed, the petiole will shrivel but remain attached.

The following spring, head back the rootstock with a backward-sloping cut just above the grafted bud. A strong shoot should develop. For the plant's benefit, remove any flowers that form, but take care not to damage the shoots behind them. If the new leading shoot does not grow straight, encourage it to do so by attaching it to a vertical stake.

Allow any side shoots that emerge on the rootstock below the bud to develop until 2–3in (5–7.5cm) long (they help feed the plant), then remove them. By the following fall, the new tree should be ready to plant shoot does not grow straight, encourage it to do so by attaching it to a vertical cane inserted close to the stem with soft twine.

## T-budding

Preparing the rootstock and collecting scion material is the same as for chip-budding, but the budding technique is slightly different. To ensure success with T-budding, it is vital that the rootstock is growing well so that the bark will lift easily for insertion of the scion bud. If there is prolonged dry weather in the two weeks before budding, keep the rootstocks well irrigated.

Collect the bud sticks and prepare them as described for chip-budding, but trim the petiole to ½in (1cm) instead of ¼in (5mm). Starting from the top of the budstick, locate a healthy bud and make a shallow cut, beginning ¾in (2cm) below the bud. Draw the knife upward under the bud and lift the knife so that the bud is removed with a long tail of bark.

Choose a smooth piece of stem on the rootstock at around 9in (22cm) above soil level, and make a horizontal cut, by pressing firmly with a clean sharp knife. Next make a vertical cut 1in (2.5cm) long, to join up with the center of the horizontal cut. Gently lift the bark using the spatula end of a budding knife, or the flat of a knife blade. Insert the bud into the T-shaped cut and slide it down into position, using the petiole as a "handle". Trim off the excess tail of bark level with the horizontal bar of the "T," then trim off the petiole. Wrap the graft with grafting tape as for chip-budding, but do not enclose the bud. Put a little petroleum jelly on the bud to protect it from the red bud borer (*Resseliella oculiperda*). Aftercare is as for chip-budding.

The rootstock should be headed back the following spring by cutting back with a sloping cut just above the grafted bud. Aftercare is the same as for chip-budding, and by the following fall the young tree will be ready to plant out in position to begin training into the required form.

## T-budding

The secret of success for T-budding is to ensure that the rootstock is kept well watered so that its bark can be lifted without damage before the bud is inserted.

▶ **1** Collect and prepare the material as for chip-budding, but leave a length of ½in (1cm) of the petioles on the scion. To remove the buds, starting from the top of the scion, make a shallow, upward cut starting ¾in (2cm) below the bud, draw the knife under the bud, and then lift the knife, removing the bud with a long "tail" of bark at the top (this is to help you to handle the bud).

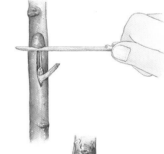

◀ **2** Find a smooth area on the rootstock about 9in (22cm) from ground level, and make a horizontal cut. Next make a 1in (2.5cm) long vertical cut up to and joining the center of the first cut, in a "T" shape.

▶ **3** Gently lift the bark using the end of the budding knife or the back of the blade. Insert the bud and slide it down to fit snugly, using the tail of bark as a handle. Trim off the excess tail level with the horizontal cut on the rootstock.

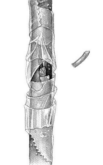

▶ **4** Bind around the budded area with grafting tape but leave the bud exposed. Trim the petiole.

### Aftercare
Aftercare is as for chip-budding (see p.182).

Bud-grafting by chip- or T-budding generally produces better trees than whip-and-tongue grafting, but if budding fails, whip-and-tongue grafting can be carried out the next spring and a tree will be produced at the same time as one budded the previous summer.

## Whip-and-tongue grafting .....................

If the rootstock is thicker than the scion, move it to one side until there is good contact between the cambial layers. This will encourage a quick union between the two parts.

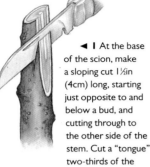

◀ **I** At the base of the scion, make a sloping cut 1½in (4cm) long, starting just opposite to and below a bud, and cutting through to the other side of the stem. Cut a "tongue" two-thirds of the way up the cut and about ½in (1cm) long. Shorten the scion to three or our buds (including the stock bud).

▲ **2** Make corresponding cuts on the rootstock (i.e. with a groove to receive the tongue), and fit the two elements together.

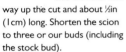

▼ **3** Fit the scion groove into the rootstock tongue, aligning the cambial layers carefully.

► **4** Bind the graft with grafting tape. Seal any exposed cut surfaces with grafting wax. The graft should take within 4–6 weeks, and fresh growth should emerge from the union. Cut the tie behind the stock bud and then carefully unwrap it.

### Whip-and tongue-grafting

This is the most common and reliable method of grafting rootstocks using dormant scions. It is carried out in late winter to early spring, when the sap is beginning to rise in the rootstocks and its buds are just starting to swell.

Scions are collected while fully dormant in midwinter. Choose healthy, well-ripened stems of pencil thickness from the previous year's growth and take cuttings 9in (22cm) long, cutting just above a bud. Keep them cool by burying them in bundles of five or six to three-quarters of their length in a sheltered part of the garden, or (preferably) put them in a sealed, dry plastic bag and store in the bottom of the refrigerator.

As the buds start to break on the rootstocks, cut the stem tip back with a sloping cut just above a clear smooth part of the stem about 9in (22cm) from the base. Trim off any side shoots. At the top of the rootstock, make an upward, sloping cut, 1½in (4cm) long. Make a second downward cut, about one-third of the way down the first cut, to create a slit about ½in (1cm) deep. This forms the "tongue" that will hold the scion. At all stages, avoid touching the cambium to prevent contamination.

Prepare the scion by removing any soft wood at the tip. Make a 1½in (4cm) sloping cut behind the bud (the stock bud) and cut a "tongue" two-thirds of the way up the first cut, about ½in (1cm) long. Trim the scion to three or four buds, including the stock bud.

Unite the rootstock and scion by slotting the scion carefully into the tongue at the top of the rootstock, ensuring that the cut cambial surfaces are matched as nearly as possible. Wrap with grafting tape as described for chip-budding, and seal any exposed surfaces with grafting wax. The graft union should form within 4–6 weeks, after which time the grafting tape should be carefully removed by cutting its tie and unwrapping it gently.

In spring, the buds on the scion will begin to shoot. Select the strongest shoot (usually from the topmost bud) and allow only this one to develop. Pinch the remainder back to about 3in (7.5cm).

Any side shoots on the rootstock should also be removed once they are about 3in (7.5cm) long. If necessary, tie the leading shoot to a stake to encourage upright growth. You can then either transplant the new tree to its final position the following fall, or leave it in place while you prune and train it to the form you wish it to take.

### Grafting over

This is a technique that may be used to introduce a new pollinating cultivar to the orchard or simply to replace old cultivars with new and more productive ones. It is mainly used in commercial orchards to graft new scions onto established apple or pear trees, having the advantage that the new cultivar fruits relatively quickly, because it involves grafting onto a well-established root-and-branch system. For the amateur the easiest way of doing this is by bark grafting.

## Bark grafting · · · · · · · · · · · · · · · · · · · · · · · · · · · · · ·

Before bark grafting, it is important to pollard all but two of the main branches on the rootstock to within 24–30in (60–75 cm) of the main branch framework, in early spring.

◄ **1** Make one or more vertical cuts, 1in ( 2.5cm) long, into the bark at the top of the pollarded tree, depending on how many scions are to be grafted. With the back of the knife, ease the bark away from the cambial layer.

► **2** Prepare each scion by reducing it to three buds, making an angled cut at the top, just above a bud, and a more tapering one, 1in ( 2.5cm) long, at the base. Insert each scion into a bark slit, facing its cut edge inwards.

◄ **3** When all the scions have been inserted, bind them in place with soft twine. Then seal all exposed surfaces with grafting wax.

The receiving tree is prepared by cutting back most of the main framework branches in early spring to within 24in (60cm) of the trunk, leaving one or two central branches to continue to draw the sap upward. This reduces the number of water shoots that would otherwise proliferate from around the wounds. Alternatively, the main branches may be left intact and one or two cut back to receive scions.

Prepare the scions by taking cuttings of the well-ripened wood of the previous season's growth. Trim them down to three buds with a slanting top cut just above a bud and a sloping cut at the base, about 1in (2.5cm) long.

Prepare the rootstock branches by making a vertical slit, 1in (2.5cm) long, through the bark at the top of the branch. Slide the scion into the slit with the sloping cut facing inward to make cambial contact with the rootstock. Insert two or three scions at even spacings

around the branch. Bind with soft twine and seal the exposed surfaces with grafting wax. Once the graft union has formed, remove the binding. Select the strongest of the scions and remove the remainder. If more than one shoot grows from the scion, retain the most vigorous shoot and shorten the remainder to about 3in (7.5cm). The new shoot should begin to fruit within 3–4 years.

### Double-working

Double-working is a modification of grafting used mainly for pears, some of which are not directly compatible with the quince that is commonly used as a pear rootstock. It involves using an "interstock" of a cultivar such as 'Beurré Hardy' that is mutually compatible with the rootstock and the intended fruiting scion. The most frequently used technique is double chip-budding, although double grafting using a whip–and–tongue graft is also used.

In the first year, the interstock is chip-budded onto one side of the quince rootstock and allowed to grow away. In the second year, the interstock receives a bud chip of the fruiting scion. This is inserted on the opposite side to the first bud chip, and about 2in (5cm) above it. This gives a stronger union and ultimately a straighter stem than if the bud chips were inserted on the same side.

Once the scion takes and begins to grow away, the interstock is cut back just above the new bud. Any shoots that arise on the interstock should be rubbed out as soon as seen.

## Double working · · · · · · · · · · · · · · · · · · · · · · · · · · · · · ·

Some pear cultivars are not directly compatible with quince rootstocks, so the rootstock is grafted with a comptable interstock, and the instertock is then grafted with the cultivar.

◄ **Double budding**
This method involves chip-budding an interstock onto a quince rootstock. The following year the selected cultivar is chip-budded onto the opposite side of the interstock.

► **Double grafting**
This is similar to double budding except that the interstock is whip-and-tongue grafted onto the rootstock and scion of the selected cultivar.

# DIRECTORY OF
# FRUIT

*Below are key tips on a selection of fruits that are particularly suitable for propagation. Unless otherwise specified, follow the detailed instructions given under the propagation section on pages 176–185.*

## Key

CR    Chilling requirement

🌰    Seed

## *Actinidia* (Kiwi)     Actinidiaceae

Some 40 mainly deciduous climbers.
*A. deliciosa*, syn. *A. chinensis* (Kiwi fruit). Deciduous, trailing climber grown for its soft, green-fleshed fruit and as an ornamental. It grows best in areas with mild winters and warm springs and summers, and needs fertile, well-drained soil, shelter, and sun. It is hardy but young growth can be damaged by spring frosts. In cool temperate areas, it is best grown in a cool greenhouse if fruit are required. Mostly dioecious, male and female plants are needed for pollination, at a ratio of one male plant to five females. Hermaphrodite cultivars include '**Jenny**'.
CR: 400 hours below 45°F (7°C).
H 30ft (10m).

### PROPAGATION

Hardwood cuttings – late summer.
Softwood cuttings – spring.
T-budding or whip-and-tongue grafting can be used but do not always take well, as *Actinidia* tends to "bleed" and this interferes with the union of the grafted material.

## *Ananas* (Pineapple)     Bromeliaceae

About six tender, evergreen perennials, grown as ornamentals and for their fruits.
*A. comosus* (Pineapple). Has spiky dark leaves and bears edible pineapples. Needs full sun, temperatures of 64–86°F (18–30°C), and humidity at 60–80% to grow and fruit well. In temperate regions, it needs a warm greenhouse or a special frame to produce edible fruit. H 3ft (1m), S 20in (50cm).

### PROPAGATION

Suckers from the leaf axils or slips arising from the fruit stem or fruit crowns. Remove with a sharp knife, treat with fungicide, and allow to dry for a few days, then insert into a sandy, well-drained medium.
Scooping – scoop out the crown shoot with a knife, taking care not to damage the shoot base. Remove any flesh and the lower leaves, treat with fungicide and allow to dry for 2–3 days before potting up as above. Root all in a propagator or under plastic to maintain humidity. Rooting takes about 4 weeks.

## *Citrus*     Rutaceae

Some 16 subtropical, evergreen shrubs and trees with fragrant flowers and edible fruits. Most, including *C. limon* (Lemon) and

*C. sinensis* (Sweet orange), need optimum temperatures of 58–86°F (15–30°C) to grow and fruit well. In temperate areas, grow under glass during the cooler months and stand outdoors in summer. Plant in fertile, well-drained, slightly acid soil. The most usual tree forms are bush and standard (an attractive form for a pot). Citrus are self-fertile but may need hand pollination.

### PROPAGATION

🌰 as soon as ripe. Most citrus are polyembryonic, and offspring are very similar to the parent. They take 7–10 years to fruit.
T-budding, onto a seedling rootstock (use a similar seedling to the cultivar that is being budded).

## *Corylus* (Hazel)     Betulaceae

Up to 15 hardy, deciduous trees and shrubs.
*C. avellana* (Hazel) and *C. maxima* (Filbert) produce edible nuts. They grow and crop best in cool moist summers. Plant in slightly acid, moist but well-drained, not-too-fertile soil, in a sheltered site in dappled shade. They are best grown as open-centered bushes on a short leg of about 24in (60cm). They are wind-pollinated and, although self-fertile, they benefit from pollinators as male and female flowers on the same tree do not always mature at the same time. The cultivar '**Cosford**' is an excellent pollinator.
CR: 800–,200 hours below 45°F (7°C).

### PROPAGATION

Suckers – winter.
Simple layering – fall.

## *Cydonia* (Quince)     Rosaceae

A single species of small deciduous tree, *C. oblonga*, grown for its edible fruit.
*C. oblonga*. Has white to pink flowers followed by pear-shaped fruit. It grows best in areas with mild winters and hot summers.

*Actinidia deliciosa*

*Citrus sinensis*

*Corylus avellana*

*Cydonia oblonga* 'Bereczki'

*Fragaria* 'Pegasus'

Grow in moist but well-drained, slightly acid soil. It can be grown as a bush tree in the open in warm areas; in colder areas, fan-train against a sheltered sunny wall. It is self-fertile, but fruits better with a pollinator.
CR: 100–450 hours below 45°F (7°C).
H and S 15ft (5m).

**PROPAGATION**
Chip- or T-budding, or whip-and-tongue grafting, onto 'Quince C' or 'Quince A' (see **Pyrus**, pear, p. 188).
Hardwood cuttings – late fall. Less satisfactory than grafting; the resulting tree will be vigorous and prone to suckering.

### *Diospyros* (Persimmon)　Ebenaceae
Some 450 evergreen or deciduous trees and shrubs, all of which produce edible fruit.
**D. kaki.** This can be grown as an ornamental for its glossy leaves with good fall colors, but to set and ripen fruit it needs a lot of warmth and sunshine (1400 hours).
It will grow well with a minimum of 50°F (10°C ), but requires 61–72°F (16–22°C) in fall for adequate ripening. Grow in well-drained, fertile soil. It can be grown as a bush or spindlebush tree under glass and is suitable for pots. Persimmon is dioecious, some are self-fertile (producing seedless fruits) and others need pollinators. It may be seed-raised, but can take 5–10 years to produce fruit of dubious quality. It must be propagated vegetatively for good fruit.
CR: 100–200 hours below 45°F (7°C).
H to 30ft (10m), S 20ft (6m).

**PROPAGATION**
Chip- or T-budded, or whip-and-tongue grafted, onto seedling rootstocks of *D. kaki* or *D. virginiana*. It can also be propagated by

softwood cuttings but these need heat and high humidity (preferably under mist) to root.

### *Ficus* (Fig)　Moraceae
Genus of 800 often ornamental, frost-tender and hardy, mainly evergreen shrubs, trees and climbers. *F. carica* is grown for its fruit.
**F. carica.** Deciduous tree or large shrub with lobed leaves and brown or purple fruit near the tips of the previous year's shoots. It is best in climates with mild winters and long hot summers, where it may be grown as a bush tree. In cooler areas, fan-train against a warm sunny wall. Plant in humus-rich, well-drained soil. It needs root restriction to control vigor, and is ideal for a container, although it must be brought into a frost-free place over winter. No pollinator is needed.
CR: 100–300 hours below 45°F (7°C).
H 10ft (3m), S 12ft (4m).

**PROPAGATION**
Hardwood cuttings – early fall.

### *Fragaria* (Strawberry)　Rosaceae
About 12 hardy herbaceous perennials. Most species are grown for their edible fruits.
**F. x ananassa cultivars.** These low-growing, spreading perennials tolerate a range of climates. Most fruit according to day length, but some day-neutral cultivars grow and fruit according to temperature and are suited to warmer regions. In cool climates, day-neutral types behave as perpetual cultivars. They thrive in fertile, well-drained soils, but not on shallow chalk. Most are self-fertile.

**PROPAGATION**
Runners – mid- to late summer. Some perpetual cultivars produce few runners; propagate by division.

Alpine strawberries do not produce runners; sow seed in fall or spring.

### *Juglans* (Walnut)　Juglandaceae
Some 15 species of deciduous trees producing pinnate leaves and edible nuts.
**J. regia** (English walnut, Persian walnut). This grows and fruits best in areas with warm springs and hot summers. In cooler climates the young growth can be damaged by late spring frosts, and fruit may not ripen fully in short, relatively cool summers. Even in Britain, however, the young unripe fruit can be excellent for pickling. Grow in fertile, moist but well-drained, near-neutral soil. Walnuts are monoecious and self-fertile but wind-pollinated, and male flowers often release their pollen before the females are ready to receive it. Two or more cultivars should be grown to ensure pollination; the cultivar **'Franquette'** is a good pollinator.
CR: 500–1000 hours below 45°F (7°C ).
H 40–60ft (12–18m), S 45ft (15m).

**PROPAGATION**
Chip-budding or whip-and-tongue grafting onto seedling rootstocks of *J. regia*.

### *Malus* (Apple, Crabapple)　Rosaceae
Genus of 35 hardy, deciduous trees and shrubs, grown for their flowers, mainly in spring, and some also for their fruit.
**M. sylvestris var. domestica.** This is the basic species from which apples of all kinds have been developed. Apples are hardy over a wide climatic range but cultivars must be selected to suit your climate, and sizes to fit the space available. Apples are tolerant of most well-drained soils, but to crop well need a sheltered site in sun. Apples are trained in

*Malus 'Discovery'*

*Passiflora caerulea*

*Persea americana*

many forms, each using a different rootstock:

**M27**: very dwarfing; for dwarf bush, spindle-bush, dwarf pyramid, and horizontal cordon.

**M9**: dwarfing; for dwarf bush, spindlebush, dwarf pyramid, and cordon.

**M26**: semi-dwarfing; for bush, spindlebush, pyramid, cordon, and pot cultivation.

**MM106**: semi-vigorous; for bush, spindlebush, pyramid, and espalier.

**MM111**: semi-vigorous; for bush, half-standard and espalier.

Most are not self-fertile and need cross-pollination by other cultivars flowering at the same time. Most are compatible, but some closely related cultivars are incompatible, e.g. 'Cox's Orange Pippin' with 'Holstein'. Triploid cultivars (with three sets of chromosomes instead of the more usual two) are ineffective as pollinators and must be grown with two diploid pollinators. The following are arranged in compatible pollination groups; cultivars can be pollinated by those in the same or adjacent groups.

T = Triploid      B = Biennial bearing

Group 1: 'Gravenstein' (T), 'Lord Suffield', 'Stark Earliest', 'Vista Bella' (B).

Group 2: 'Adam's Pearmain', 'Beauty of Bath', 'Bismark', 'Bolero', 'Cheddar Cross', 'Egremont Russet', 'George Cave', 'Idared', 'Irish Peach', 'Keswick Codlin', 'Lord Lambourne', 'McIntosh Red', 'Melba', 'Norfolk Beauty', 'Reverend W. Wilks' (B), 'Ribston Pippin' (T), 'St Edmunds Pippin', 'Warner's King'.

Group 3: 'Allington Pippin', 'Belle de Boskoop' (T), 'Blenheim Orange' (T, B), 'Bountiful', 'Bramley's Seedling' (T), 'Charles Ross', 'Cox's Orange Pippin', 'Crispin', 'Discovery', 'Elstar', 'Emneth Early', syn. 'Early

Victoria', 'Epicure', 'Falstaff', 'Fiesta', 'Fortune', 'Granny Smith', 'Grenadier', 'Holstein', 'James Grieve', 'Jonagold' (T), 'Jonathan', 'Katy', 'Lane's Prince Albert', 'Lord Grosvenor', 'Merton Knave', 'Merton Worcester', 'Peasgood's Nonsuch', 'Spartan', 'Sturmer Pippin', 'Sunset', 'Worcester Pearmain'.

Group 4: 'Autumn Pearmain', 'Cox's Pomona', 'Delicious', 'Duke of Devonshire', 'Dumelow's Seedling', 'Ellison's Orange', 'Gala', 'Gloster 69', 'Golden Delicious', 'Jester', 'Lord Derby', 'Monarch' (B), 'Orleans Reinette', 'Superb' (Laxton's) (B).

Group 5: 'Gascoyne's Scarlet', 'King of the Pippins' (B), 'Merton Beauty', 'Newton Wonder', 'Royal Jubilee', 'Suntan' (T), 'William Crump'.

Group 6: 'Ben Pool', 'Court Pendu Plat', 'Edward VIII'.

Group 7: 'Crawley Beauty'.

CR: 900 hours min. below 45°F (7°C).

**PROPAGATION**

Chip or T-budding, or whip-and-tongue grafting. It is also possible to graft new cultivars onto over-healthy mature trees by top-working or frame-working (*see p. 184*).

## *Mespilus* (Medlar)                    Rosaceae

A single species of hardy, deciduous tree. *M. germanica*. Has white flowers, apple-like fruit, and attractive fall foliage. It grows and fruits best in areas with warm summers. In colder areas, provide a sunny, sheltered site. Thrives in almost any moist but well-drained soil. Can be grown as a standard; bushes and half-standards are also cultivated. Self-fertile. CR: 100–450 hours below 45°F (7°C ).

H 20ft (6m), S 25ft (8m).

**PROPAGATION**

Chip-or T-budding, or whip-and-tongue grafting, onto *Cydonia oblonga* or *Crataegus* rootstocks.

## *Morus* (Mulberry)                    Moraceae

Some 12 hardy, deciduous trees, grown for their edible raspberry-like fruits and their foliage, which is used to feed silkworms. *M. nigra* (Black mulberry). Has dark purple fruit. Best in a sheltered, sunny site, in moist but well-drained, slightly acid soil. Usually grown as bush or standard trees. Self-fertile. CR: 800–1200 hours below 45°F (7°C). H to 40ft (12m), S 50ft (15m).

**PROPAGATION**

Heeled softwood cuttings – early summer. Heeled hardwood cuttings – fall. Mulberry is not as easy to root from hardwood cuttings as is often supposed.

Simple layering – fall or spring.

## *Passiflora* (Passion flower)          Passifloraceae

Genus of 500 evergreen or semi-evergreen climbing plants, grown for their flowers and edible fruits (passion fruit). Many also have attractive lobed leaves. *P. edulis*. Has white flowers and purple fruits; **f.** *flavicarpa* has yellow fruits. Both are frost-tender and need minimum winter temperatures of 45–50°F (7–10°C). Greenhouse cultivation is necessary in temperate regions. Grow in fertile, but not too nitrogen-rich, well-drained soil in full sun; high-nitrogen soils produce foliage growth at the expense of flowers. As climbers, they need to be trained on wires, with two leaders per plant. They are self-fertile but hand-pollination is necessary. H 15ft (5m).

*Prunus armeniaca*

*Prunus avium* 'Bigarreau Napoléon'

*Prunus domestica* 'Victoria'

## PROPAGATION

🌱 named cultivars are seldom available in temperate areas. Seed is extracted from ripe fruit and fermented for 3–4 days. Sow in pots or trays, at 68°F (20°C). Transplant to their permanent pots when they are 10in (25cm) tall.

Softwood cuttings – early summer.

## *Persea* (Avocado)     Lauraceae

Genus of 150 frost-tender evergreen trees. *P. americana* is grown for its edible fruits.

*P. americana* (Avocado). Has pear-shaped fruits with leathery skin and soft, green flesh. It needs high humidity and temperatures of 68–82°F (20–28°C) to flower and fruit. In temperate areas it must be grown under glass in a warm greenhouse and even there may not fruit freely. Needs fertile, well-drained soil and full sun. It is best pot-grown as a bush and requires little pruning or training. Self-fertile but fruits better if more than one cultivar, with overlapping flowering times, is grown. H to 60ft (20m).

## PROPAGATION

🌱 in temperate areas, seed is usually the only practicable means of propagation since clonal material – cultivars that have been bred and selected for good fruiting – is seldom available. (If clonal material can be obtained it can be grafted onto seedling rootstocks using a side-wedge graft.) Select seed from ripe fruits and soak in hot (not boiling) water for about 30 mins. Remove the top ½in (1cm) from the pointed end of the seed and dip the wound in fungicide. Place the seed in a pot of well-drained medium with the cut top just above soil level. It should germinate in 3–4 weeks.

## *Prunus*     Rosaceae

Genus of 430 hardy, flowering trees and shrubs; some are grown for their edible fruits. Mostly deciduous, although a few are evergreen. Height depends on the rootstock.

*P. armeniaca* (Apricot). They flower early in spring, and in cold areas are best fan-trained on a sunny wall or grown in a cold greenhouse to protect the flowers from frost damage. Grow in deep, fertile, slightly alkaline loam. They can be grown in the same tree forms as plums. They are self-fertile. Hand-pollinate the flowers with a soft brush. CR: 350–900 hours below 45°F (7°C ).

## PROPAGATION

Chip- or T-budding, or whip-and-tongue grafting, onto 'St Julien A'. If grown on the dwarfing rootstock 'Pixy' use an interstock of 'St Julien A'.

*P. avium* (Sweet cherry). Bears fruit at the base of the previous year's wood and on spurs on older wood. Grow in deep, fertile, well-drained soil, in shelter. Can be grown as fans or bush trees. Most are self-sterile and only cross-compatible with certain cultivars. The following are compatible groups.

Group 1: **'Early Rivers'**, **'Noir de Guben'**.

Group 2: **'Merton Favourite'**, **'Merton Glory'**.

Group 3: **'Merchant'**, **'Roundel'**, **'Starkrimson'**, **'Van'** (SF).

Group 4: **'Governor Wood'**, **'Lapins'** (SF), **'Bigarreau Napoléon'**, **'Stella'** (SF), **'Sunburst'** (SF).

Group 5: **'Bigarreau Gaucher'**, **'Bradbourne Black'**.

CR: 800–1200 hours below 45°F (7°C).

## PROPAGATION

Chip- or T-budding, or whip-and-tongue grafting. The most reliable rootstock is the semi-vigorous **'Colt'**; there are newer dwarfing rootstocks such as **'Edebritz'**, syn. **'Tabel'**.

*P. cerasus* (Sour cherry). **These** fruit on the previous year's wood and are best grown as fans or pyramids. They thrive on a shady wall, in deep, fertile, well-drained soil. Most are self-fertile; **'Morello'** is very reliable; **'Nabella'** and **'Montmorency'** are also recommended. CR: 800–1200 hours below 45°F (7°C).

## PROPAGATION

Chip- or T- budding, or whip-and-tongue grafting. The most suitable rootstock is 'Colt'.

*P. domestica* (Plum and Gage), *P. institia* (Bullace and Damson) and *P. cerasifera* (Cherry plum). All are grown in the same way. They are hardy but flower early and so fruit best in areas with warm springs. They should be grown as pyramids, which are less prone to branch breakage than bush or half-standard trees. They can also be fan-trained against a warm wall and, using the dwarfing rootstock 'Pixy', as cordons. Some, including most bullaces and damsons and all cherry plums, are self-fertile, but all benefit from cross-pollination. Pollination groups:

Group 1: **'Angelina Burdett'**, **'Blue Rock'**, **'Heron'**, **'Jefferson'**, **'Mallard'**, **'Monarch'** (SF), **'Utility'**.

Group 2: **'Ariel'**, **'Avalon'**, **'Brahy's Greengage'** (SF), **'Coe's Golden Drop'**, **'Denniston's Superb'** (SF), **'President'**, **'Warwickshire Drooper'** (SF).

Group 3: **'Allgrove's Superb'**, **'Bountiful'** (SF), **'Bryanston Gage'**, **'Czar'** (SF), **'Early Laxton'**, **'Early Rivers'**, **'Edwards'** (SF), **'Golden Transparent'** (SF), **'Goldfinch'**, **'Herman'** (SF), **'Late Orange'**, **'Laxton's Cropper'** (SF), **'Laxton's Gage'** (SF), **'Merryweather Damson'** (SF), **'Merton Gem'**, **'Opal'** (SF),

*Prunus persica*

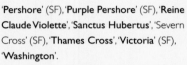

*Pyrus 'Conference'*

*Ribes sativum*

'Pershore' (SF), 'Purple Pershore' (SF), 'Reine Claude Violette', 'Sanctus Hubertus', 'Severn Cross' (SF), 'Thames Cross', 'Victoria' (SF), 'Washington'.

Group 4: 'Blaisdon Red' (SF), 'Bradley's King Damson' (SF), 'Cambridge Gage', 'Count Althann's Gage', 'Early Transparent Gage' (SF), 'Farleigh Damson', 'Giant Prune' (SF), 'Kirke's Ontario' (SF), 'Oullins Golden Gage' (SF), 'Wyedale'.

Group 5: 'Belle de Louvain' (SF), 'Blue Tit' (SF), 'Excalibur', 'Late Transparent', 'Marjorie's Seedling' (SF), 'Old Greengage', 'Pond's Seedling', 'Prune Damson' (SF).

There are three incompatibility groups:

Group 1 'Allgrove's Superb', 'Coe's Golden Drop', 'Coe's Violet Gage', 'Crimson Drop', 'Jefferson'.

Group 2 'Cambridge Gage', 'Late Orange', 'Old Greengage', 'President'.

Group 3 'Blue Rock', 'Early Rivers'.

CR: 700–1000 hours below 45°F (7°C).

### PROPAGATION

Chip or T-budding, or whip-and-tongue grafting, onto 'St. Julien A', or onto the dwarfing rootstock 'Pixy' for cordons.

*P. dulcis* (Almond). This is very ornamental when in flower; the flowers are produced on the bare shoots in spring. It is closely related to the peach, and since it also flowers early in spring, it only crops well in areas that have warm springs and hot summers. Grow in neutral to slightly acid, well-drained fertile soil in a warm, sheltered, sunny site. Usually grown as bush or half-standard trees. Although usually self-fertile, they crop better if two cultivars are grown for pollination.

CR: 300–500 hours below 45°F (7°C).

### PROPAGATION

Chip- or T-budding, or whip-and-tongue grafting. The best rootstock is 'St Julien A', but peach and almond seedlings are still sometimes used.

*P. persica* (Peach) and var. *nectarina* (Nectarine). These are both hardy, but grow and fruit best in areas with warm springs and hot summers. They flower early, needing frost protection when flowering, and often require hand pollination. Grow in a sheltered, sunny site, in deep, fertile, slightly acid soil. In cool temperate areas, they are best fan-trained against a warm sunny wall, or grown in a cold greenhouse. In warmer areas, they can be cultivated as bush trees. With few exceptions, they are self-fertile.

CR: 600–900 hours below 45°F (7°C).

### PROPAGATION

Chip- or T-budding, or whip-and-tongue grafting, onto 'St Julien A'. They may be double-worked onto 'Pixy' using an interstock of 'St Julien A'.

## *Pyrus* (Pear)                    Rosaceae

About 30 species of hardy, deciduous trees, some grown for their fruit. *P. communis* var. *sativa* (Common pear). Basic species from which all kinds of pears have been developed. Best in areas with warm springs and hot summers. In cooler areas, select late-flowering cultivars; the flowers may escape frost-damage. Grow in fertile, well-drained soil in full sun. Two main rootstocks; 'Quince C' is semi-dwarfing and is suitable for cordons, bush and pyramid forms; 'Quince A' is semi-vigorous and is used for bush, half-standard, pyramid, and espalier forms. Most are not self-fertile; need cross-pollination.

With few exceptions, most cultivars flowering at the same time are compatible. Triploid cultivars are ineffective as pollinators and must be grown with at least two diploid cultivars. Pollination groups:

Group 1: 'Brockworth Park', 'Maréchal de la Cour' (T), 'Précoce de Trévoux'.

Group 2: 'Doyenné D'Eté', 'Duchesse d'Angoulême', 'Emile d'Heyst', 'Louise Bonne of Jersey', 'Packham's Triumph', 'Passe Crassanne', 'Seckle'.

Group 3: 'Beurré Hardy', 'Beurré Superfin', 'Black Worcester', 'Concorde', 'Conference', 'Dr Jules Guyot', 'Durondeau', 'Fondante d'Automne', 'Hessle', 'Jargonelle' (T), 'Josephine de Malines', 'Merton Pride' (T), 'Souvenir du Congress', 'Thompson's', 'William's Bon Chrétien'.

Group 4: 'Beth', 'Beurré Bedford' (MS), 'Bristol Cross', 'Catillac' (T), 'Clapp's Favourite', 'Doyenné du Comice', 'Glou Morceau', 'Gorham', 'Nouveau Poiteau', 'Onward', 'Pitmaston Duchess' (T), 'Winter Nelis'.

There are also two incompatibility groups:

Group 1: 'Fondante d'Automne', 'Laxton's Progress', 'Louise Bonne of Jersey', 'Précoce de Trévoux', 'Seckle', 'William's Bon Chrétien'.

Group 2: 'Beurré d'Amanlis', 'Conference'.

CR: 600–900 hours below 7°C (45°F).

### PROPAGATION

Chip- or T-budding, or whip-and-tongue grafting. Some are not compatible with Quince rootstock and are double-worked using a mutually compatible interstock such as 'Beurré Hardy' or 'Doyenné du Comice'. Graft the interstock in spring, and bud the required cultivar in the same or following summer.

*Rubus fruticosus* 'Oregon Thornless'

*Rubus idaeus* 'Autumn Bliss'

*Vitis* 'Richtenstein'

## *Ribes* (Currant and Gooseberry)

Grossulariaceae

Some 150 hardy, deciduous or evergreen shrubs, some grown for fruit. *R. nigrum* (Black currant), *R. sativum* (Red and White currant). Black currants fruit on previous year's wood and are grown as stooled bushes, pruned on a replacement system. Red and white currants fruit on base of previous year's wood and on older wood; they are grown as bushes, standards, cordons, and fans. Hardy and grow best in damp, cool temperates. Grow in sun, in fertile, well-cultivated soil. Self-fertile. H and S 3–5ft (1–1.5m).

### PROPAGATION

Hardwood cuttings – late fall. For black currants, 8–10in (20–25cm) long, inserted with two buds above ground level. For red and white currants, 12–15in (30–38cm) long. Remove the lower buds, leaving the top four, and insert to half their depth.

*R. uva-crispa* (Gooseberry). Fruits are produced at base of previous year's wood and on older wood. Best in cool moist climates. Grow in sun, in fertile, well-cultivated soil. They are grown as bushes, standards, cordons, and fans. Self-fertile. H and S 3–5ft (1–1.5m).

### PROPAGATION

Hardwood cuttings – early to late fall. Leave the top two leaves and all the buds and insert to half their depth. To produce standards, chip-budding or whip-and-tongue grafting are used; graft onto *R. odoratum* or *R. divaricatum* at 3–4ft (90–120cm) high.

## *Rubus* (Blackberry, Bramble)   Rosaceae

Genus of 250 species that include woody-stemmed, scrambling climbers grown for their edible fruits. Blackberries are hardy and crop reliably; many blackberry cultivars are hybrids with other *Rubus* species. Hybrid berries include the raspberry × blackberry hybrids – loganberry and tayberry, as well as boysenberry and king's acre berry. They produce good-quality fruit, but not all are as hardy as the blackberry and they are less suited to cultivation in cold, exposed areas. The boysenberry is best in warmer, drier conditions. All fruit in late summer or early fall, on canes made the previous year, and all are self-fertile. Grow in sun or part-shade in fertile soil. They are trained in various ways, including fan, weave, and rope; all these techniques involve separating fruiting canes from those that will fruit the following year.

### PROPAGATION

Tip-layering – summer.

Leaf-bud or 3-bud cuttings can be used where greater numbers are required.

*R. idaeus* (Raspberry). Hardy and grows best in cool damp climates. Plant in moderately fertile soil in sun. Raspberries fall into two groups: summer-fruiting cultivars which fruit on the previous year's canes, in early to midsummer, e.g. 'Glen Moy' and 'Leo'; and fall-fruiting, or primocane, cultivars, which fruit from late summer to fall, on the current year's growth, e.g. 'Autumn Bliss' and 'Heritage'. Autumn-fruiting raspberries are unsuitable for more northerly locations as the fruit seldom ripens satisfactorily. Raspberries are self-fertile.

### PROPAGATION

Suckers – late fall, when dormant. Ensure that parent plants are healthy and virus-free. Root cuttings – winter.

## *Vaccinium*   Ericaceae

Some 450 species of shrubs, small trees, and vines, some with edible fruit. American blueberries are produced by high-bush blueberry hybrids of *V. corymbosum* and *V. australe*. Hardy, thriving in cool moist climates. The bushes need sun and moist but well-drained acid soil, pH 4–5.5. They can be grown in tubs of ericaceous medium. Blueberries are grown as stooled bushes, and fruit on 2–3-year-old wood. Once established, some of the oldest wood is cut out to the base annually in winter. Blueberries are partially self-fertile but crop best if more than one cultivar is grown. CR: 700–1200 hours below 45°F (7°C).

### PROPAGATION

Softwood cuttings – early to midsummer.

## *Vitis* (Grape)   Vitaceae

About 65 deciduous, mostly climbing species, some producing edible fruit. Grape vines grow best in warm temperate climates with long, hot, dry summers. In cooler climates, provide very sheltered sites, or grow by a sunny wall. In even colder areas, grow in a cool greenhouse. They prefer well-drained, humus-rich soil. Grapes are self-fertile. They can be trained in many ways on permanent frameworks or using the Guyot system, which is widely used in vineyards.

### PROPAGATION

Hardwood cuttings – winter, 8in (20cm) long. Also by vine-eye cuttings, or by 3-bud cuttings.

Softwood cuttings – early summer.

In countries where the pest *Phylloxera* is present, vines are whip-and-tongue grafted onto resistant rootstocks.

191

# VEGETABLES

There is enormous satisfaction to be gained from growing your own produce to eat and share, and vegetables from your own garden will be far superior in quality and taste to any you can buy in the store. Whereas commercial growers aim for yield and uniformity, you can grow for freshness and flavor, and do not need to restrict yourself to cultivars that have a long shelf life.

Regular propagation ensures the best of all types of vegetable, from simple hardy crops such as cabbages and carrots to tender eggplants and the like that need warmth to germinate and a long growing season. Propagating will help you plan sowing, planting, and harvesting times to get the best from your garden all year.

◄ *The succession of vegetable seeds, sets, and tubers that were sown and planted during the spring and early summer begin to yield harvests from midsummer onward.*

193

# WHY PROPAGATE
# VEGETABLES?

*To grow your own vegetables is one of the most satisfying experiences a gardener can have. Not only does produce cooked fresh from the kitchen garden taste wonderful, but it also is packed full of vitamins and minerals. Self-sufficiency may be beyond most people's means these days, but we can all find space for a few choice crops.*

**Almost all vegetables are increased by seed, but success depends on timing and the use of correct sowing techniques.**

Most vegetables are started every year from seed, tubers, or bulbs. Plantlets are usually available from garden centers, but raising them yourself is not difficult and gives a much better choice of cultivars. It also lessens the risk of introducing soil-borne diseases such as clubroot to the garden. Some vegetables are perennial and others are biennials or can be grown as such, sown one year for harvesting the next. Time your sowings so that you have early crops from overwintered seedlings, mid-season crops from seed sown early in the year in warmth, and later crops from seed sown *in situ*.

## THE ADVANTAGE OF SEED

Propagation by seed ensures healthy crops and allows you to experiment with different cultivars. Try several cultivars of the same vegetable, for instance, and assess which gives the heaviest crops, provides the best eating, and is easiest to grow in the conditions that prevail in your garden. Most gardeners quickly find out which is their favorite tomato, for instance, basing their judgment on taste, yield, and ease of cultivation, but you should always be ready to experiment with new cultivars that may turn out to be superior.

Specific cultivars have been developed for harvesting at certain times of year. Winter cabbage cultivars, for example, are hardier than summer ones. Cultivars may also mature at different times. Early cultivars mature more quickly, and thus can be useful both for early crops and for areas with short summers.

F1 hybrid cultivars give more vigorous, uniform plants of reliable quality. However, crops tend to be ready for harvest

**Cabbage**
All cabbages are treated as annuals and grown from seed, but different cultivars are bred to crop at different times of the year and so sowing dates vary according to cropping season.

**Ruby chard**
Chards are hardy and vigorous biennials, that, from spring sowings, will crop until the following spring. To ensure continuity of supply, sow successively in spring, midsummer, and late summer.

**Onions**
Onions need a long season to crop well. Choose between seed or sets; seed is inexpensive but slow; sets are quicker and easier but the available cultivar range is more restricted.

**Carrots**
Although truly biennial, carrots are grown as annuals. Early ones are sown as soon as the soil warms up in spring; maincrop carrots are sown between late spring and early

at the same time, which can be a problem unless they store well. Quick-growing vegetables such as radishes can be sown in succession for continuity of cropping.

## BULBS AND TUBERS

Shallots and onions are usually grown from bulbs, as is garlic. Bulbs, or sets, are robust and easy to handle, and generally give good yields. Onion sets in particular mature earlier and give higher yields than seeds sown *in situ*. However, there is a more limited choice of cultivars available and plants are more likely to flower and seed. Potatoes and Jerusalem artichokes are nearly always grown from seed tubers bought in fresh each year.

## PERENNIAL VEGETABLES

Globe artichokes and rhubarb are usually propagated by divisions or offsets of mature plants. As with all vegetative methods, this guarantees that the resulting

*"Cut-and-come-again" crops*

*Many leafy vegetables re-grow after they are first cut, giving two or more cuts from one sowing. One great advantage of "cut-and-come-again" cropping is that harvesting takes place when plants are immature – sometimes even at the seedling stage – giving very quick crops. Most can also be spaced closely (with seedlings, only 2–2.5cm [¾–1in] apart), so make good use of space. Vegetables that respond well as immature plants include salad-bowl lettuce and curly kale. Good seedling crops include salad rocket, cress, and chicory. In cold*

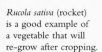

*areas, plastic covers will protect early sowings.*

*Rucola sativa (rocket) is a good example of a vegetable that will re-grow after cropping.*

plants are true to type, although they can also be grown quite successfully from seed. Asparagus, which is a long-lived perennial, can be grown from seed or crowns.

**5**

**Curly kale**
Kale is among the hardiest of brassicas and may be grown as an annual or biennial; sow crops for summer in early spring, sow in late spring for winter and spring harvest.

**6**

**Fava beans**
For very early crops, sow hardy cultivars in a sheltered site, in fall or late winter with plastic cover protection. Sow early and late crops in an open site, between spring and early summer.

**7**

**Runner beans**
Frost-tender perennial climbers grown as annuals. Sow *in situ* when danger of frost has passed and the soil is warm, but for good crops in cold areas, sow earlier under glass.

# PROPAGATING TECHNIQUES
# VEGETABLES

*In the majority of cases, vegetables should be propagated from good-quality annual seed, but seed can deteriorate if kept for any length of time, and some plants do not give their best when raised in this way. In these instances, vegetative methods can be deployed instead. Rhubarb, for example, is commonly propagated by division.*

Most vegetables are easy to propagate and can be produced with the minimum of equipment. Some form of crop protection widens your options and lengthens the growing season. Every crop has a minimum germination temperature, and higher temperatures usually speed both germination and growth. However, for some crops (such as lettuce) germination is inhibited above an upper temperature limit. Minimum germination temperatures, and upper

### Plastic covers for vegetables ...................
When planning a vegetable garden, you need to consider the local climate and soil conditions, as well as whether you wish to extend the natural climate by using plastic covers.

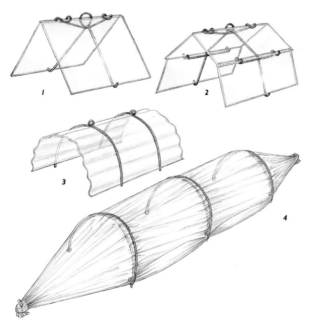

**1** A tent cover made from glass is inexpensive and easy to erect.
**2** The extra height of a barn cover is useful for tall plants.
**3** A corrugated plastic cover is cheap but becomes brittle with age.
**4** A plastic-covered tunnel cover completely covers the plants.

temperature limits, where these are low enough to be encountered, are given in the plant directory (*see pp. 200–209*) .

Some crops (such as beet) should not be sown too early where late frosts are likely, as exposing young plants to cold can cause premature bolting. Others, such as Chinese cabbage, are sensitive to day-length and grow best if sown after midsummer.

### Protected cropping
Various techniques can be used to extend the growing season for vegetable crops. Crop covers can raise soil and air temperatures and protect young plants from wind and frost. A combination of methods can extend the growing season at both ends of the year and may enable tender vegetables to be grown in colder areas.

*Clear plastic sheeting*: A most effective way of raising soil temperatures early in the year. Place it over the seedbed a few days before sowing, and replace it after sowing to maintain temperatures. Remove after germination.

*Black plastic*: Is less efficient in raising soil temperatures significantly, but suppresses weed growth and keeps a seedbed moist and clean, ready for sowing.

*White plastic*: This reflects heat and light, and prevents soil warming – useful where summer soil temperatures are too high for heat-sensitive crops such as lettuce.

*Floating mulches*: These films lie on the crop to give it protection. They are permeable to light, air, and water, and lightweight enough to be supported by the plants. They warm the soil and air around the plants and filter strong winds. There are several types. Perforated polyethylene film warms the soil rapidly, but offers little protection from frost and does not let in much rain. It is useful for speeding germination and protecting young crops early in the year, but, later on, can create undesirable heat and humidity. Horticultural fleece is much softer, more permeable to light, air, and water, and gives several degrees of frost protection. It is useful for both early- and late-sown hardy crops and can, if necessary, be left on until harvest or over winter. It also gives protection from flying pests such as aphids.

*Plastic covers and tunnels*: These give good frost protection and can cover a seedbed, or seedlings in plug trays. They are useful for tender vegetables like tomatoes and green beans, as well as early- and late-season hardy crops. Various designs are available (*see opposite page*).

## SEED

Most vegetables are raised from seed annually. Since nearly all are cultivars, many do not come true from garden-collected seed. Buy fresh seed each year and, if it is not sown at once, keep it cool and dry. All seed deteriorates with age, especially in a warm moist atmosphere, so check the dates on seed packets.

*Pelleted seeds:* These have a protective coating to make them easy to place; this disintegrates if the soil is kept moist. Pelleted seeds are useful for crops with small seed sown *in situ*, and helps avoid wasteful thinning later.

*Pregerminated seed:* This is seed that has just started to germinate, each seed showing a tiny root. It is a useful way of starting off vegetables that are difficult to germinate. You can also pregerminate seed yourself.

*Primed seed:* This is seed brought to the point of germination, then dried and packeted. It is useful for crops such as carrots and parsley that germinate slowly.

*Treated seed:* Seed of certain crops is sometimes coated with chemicals to help combat diseases.

### Sowing *in situ*

This is the main method for hardy, closely spaced crops like carrots that do not transplant easily. Before sowing, a suitable "tilth" must be created – a crumbly structure

### Sowing seeds in a narrow furrow ..............
Seeds should be covered with their own depth of soil, so any furrow must be only twice the seed depth. If they are planted too deep, successful germination will be delayed or reduced.

◄ **1** Mark the position of the row with a line held taught between two stakes, and make a furrow along the line using the corner of a hoe or spade. Alternatively, press the narrow edge of a board into the ground. Either space the seeds thinly and evenly along the row, or put three or four seeds together at each point where a plant is required and thin the seedlings that emerge.

► **2** After sowing, cover the furrow by gently pushing back the soil with the aid of a rake, a hoe, or your fingers.

free from clods and stones. The finer the seed, the finer the tilth required, so that seeds make good contact with the soil to take up moisture.

To improve soil structure, incorporate organic matter the fall before sowing. the following spring, remove the surplus from the surface. On heavy clay, dig the soil over roughly before winter, leaving large clods to be broken down by freeze-and-thaw cycles.

Just before sowing, rake the topmost layer of soil to create a level, fine-crumbed surface. Keep cultivation to the minimum, however, as overworking the soil can damage the structure. Never try to rake out a seedbed when the soil is excessively wet or dry.

Water dry ground the evening before sowing. If the ground is wet and rain is forecast, put plastic covers over the area a few days beforehand to help it to dry out.

### Sowing in narrow furrows

The usual method of sowing *in situ* is in narrow furrows. The depth of the furrow depends on the size of the seed. Small seeds need shallower furrows. About ½–¾in (1–2cm) is ideal for lettuce, for example, the

## Sowing seeds in a broad furrow ...............

Broad furrows are suitable for such crops as early carrots and scallions, which are grown close together in an intensive manner and cropped before they are fully mature.

▶ **I** Use pegs and string to mark out parallel furrows, 9in (22cm) wide. The exact distance needed between each furrow will depend on the crop being sown. Then, with a hoe or narrow spade, create flat-based furrows to the required depth.

◀ **2** Row-sow the seeds in three parallel rows within the furrows by placing large seeds individually or by sowing three or four smaller seeds together at each point and thinning the seedlings later. Then cover the seeds with their own depth of soil, using a hoe.

deeper limit being best for light soils. Larger seeds, such as peas, can be sown deeper (about 1–1½in (2.5–4cm)). (*For the sowing depths of other vegetables, see the plant directory, pp.200–209.*) Make sure the depth is uniform along the furrow, or germination will be uneven.

### Sowing in broad drills

Flat-bottomed furrows, usually 3-6in (7.5–15cm) across, are useful where a wide, dense row is required – for peas, early carrots, scallions, and "cut-and-come-again" crops, for example.

If the soil at the base of the furrow is dry, water carefully just along the bottom. Sow the seeds and cover them with dry soil. If the soil is still wet when you come to sow, make the furrows as described but at the maximum recommended depth, then line them with spent seed or potting medium, and cover the seeds with the same mixture.

*Broadcasting seed:* Broadcasting, where seed is scattered over the surface and raked in, is a method sometimes used for quick, closely spaced crops, but is only advisable on relatively weed-free soil, since it may not be immediately apparent whether the emerging seedlings are of the germinating crop or of alien weeds.

### Sowing in a seedbed

To save space, slow-growing hardy vegetables such as Brussels sprouts and leeks can be sown in a seedbed and transplanted later to their final positions. The ground they will eventually occupy can meanwhile be used for another crop. Only a small area needs to be prepared as a seedbed, and this is easier to protect from the weather and pests. Choose an open, unshaded site for the seedbed, otherwise the seedlings will become spindly as they develop. Cultivating the soil to create a friable tilth will minimize root damage when transplanting. A seedbed protected by a coldframe gives ideal conditions for early sowings.

Prepare the seedbed and make furrows as for sowing *in situ*. Rows can be closely spaced: about 4–5in (10–13cm) apart for leeks and about 8–9in (20–22cm) for brassicas. Sow seeds thinly and evenly along the rows and cover with soil.

*Aftercare:* Whether seed have been sown *in situ* or in a seedbed, it must not be allowed to dry out. If you have

## Transplanting leeks ...........................

Leeks need a long growing season, so they should be started early: either in plug trays, with three or four seeds to a cell, or in a seedbed in the garden. Transplant at the three-leaf stage.

▶ Leeks respond well when multi-sown, so those that were started in plug trays should be retained in their clumps when transplanted into the garden. Plant them 9in (22cm) apart and very slightly deeper than they were in the trays, aligning them against some string.

▶ Leeks that have been sown singly, in a seedbed, should be transplanted into dibbled holes, 6–8in (15–20cm) deep and 4–6in (10–15cm) apart, against a guide line. Water the hole gently, allowing the soil to fall in around the seedling.

to water before the seedlings emerge, keep the soil surface continuously moist to prevent it forming a hard crust. Covering with plastic can help to avoid this. Thin in stages, starting when the seedlings are large enough to handle, and ending with the final spacing.

*Transplanting:* Plants sown in a seedbed for later transplanting should be thinned to about 3in (7.5cm) apart so that their roots are not able to intertwine. Most seedlings are ready for transplanting when they have formed three or four true leaves, four to five weeks after sowing.

Water the seedlings before lifting and if the planting site is dry, water that too. Handling them only by their leaves, transfer them to a hole slightly bigger than the rootball and plant them with the lowest leaves just above the soil surface. Firm them in, then water in thoroughly. Shade them from hot sun until well established. (*For transplanting leeks, see below left.*)

### Pregermination

Sometimes it is advantageous to pregerminate seed before you sow it – for instance, when the temperature outside is too low for germination. It also acts as a check on the viability of the seed, so that valuable space in the garden is not wasted.

Sprinkle the seeds evenly on moist absorbent paper at the bottom of a plastic, lidded box. Put the lid on the box, and keep it warm (*for temperatures required for individual vegetables, see the plant directory, pp. 200–209*). When a tiny rootlet emerges, after few days in some cases, the seed can be sown in the normal way.

Large pregerminated seeds can be picked out when ready and sown individually. Handle them carefully, using tweezers if necessary. Smaller seeds can be fluid-sown, since this minimizes damage to their rootlets and helps to ensure even spacing.

### Sowing in containers

Sowing seeds in containers under glass or in a heated propagator rather than directly in the ground is useful in cold areas for vegetables that are not fully hardy and/or need a long growing season. This method can also be used for early crops of hardy vegetables. As you are creating a controlled environment for germination, it is also appropriate for seed that is scarce or expensive.

### Pregermination

This technique is sometimes used for small pregerminated seed, because it is an easy way to sow large numbers of seeds and it also protects the rootlets on the pregerminated seeds.

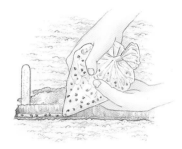

Smaller seeds can be mixed with a special gel to protect the rootlets. Wait until over 50 per cent of the seeds have germinated, then wash them off the paper into a sieve. Mix the seeds into the gel, and put the mixture in a plastic bag. Cut off a small corner, then pour the mixture into the furrow. Cover with soil in the usual way.

Seeds can be sown in pots or seed trays and then thinned out and potted on. Sow the seed thinly and evenly in a container of moist seed-germination

### Sowing in plug trays

In this method, each cell is filled individually with proprietary "blocking medium" to within ½in (1cm) of the top. Holes are then made in the center of each cell using the blocking tool.

▲ **1** Sow seed individually into each cell and cover it as for sowing and pricking out. (If you expect germination to be poor, sow several seeds per cell or block and nip out any excess seedlings that emerge.) For certain crops, however, you can sow several seeds per cell and allow all the seedlings to grow on.

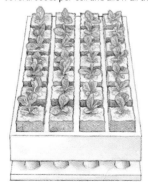

◄ **2** When the seedlings are ready to be transplanted, water them well. Then press them out from underneath the plug tray, using the blocking tool. Plant out the seedlings as one unit, but at a wider spacing than would be used for individual plants. Despite the uneven spacing, each plant will develop normally.

medium and, except for very fine seeds, cover with a thin layer of sieved medium. Put a sheet of glass or poleythylene over the container to prevent it drying out, and keep it in a warm place (*for temperatures, see the plant directory, pp. 202–211*).

Remove the cover as soon as the seedlings have germinated. When they are large enough to handle, prick them out into pots or trays of potting medium and gradually harden them off (*see below*).

*Sowing in plug trays or peat pellets:* Raising seedlings in plug trays or peat pellets saves time on pricking out. It also keeps root disturbance to the seedlings to a minimum and is useful for crops that normally resent transplanting, such as carrots and Chinese cabbage.

Plug trays are typically molded plastic or polystyrene trays consisting of a number of cells that vary in size according to the crop. They often come with blocking tool that are used to push out the seedling and rootball when they are ready for transplanting. Peat pellets are made with a proprietary peat material plus binding agent. They need to be soaked in water before use.

*Aftercare and hardening off:* For optimum development, seedlings must be kept in warm, light conditions. Transplant or pot them on as soon as the roots fill the containers, and do not allow overcrowding or they will be starved of nutrients and may not crop well.

Seedlings raised in a greenhouse need to be acclimatized to the harsher conditions outside ("hardened off") before they can be planted out. Move the containers outdoors into a coldframe or under plastic row covers, and increase ventilation gradually for an increasingly longer period each day. Alternatively, place the plants outside, first for a few hours in good weather, and gradually for longer. The hardening off period should ideally be about 10–14 days.

*Transplanting:* Ideally, choose a dull, still day for transplanting. Water plants in a seedbed thoroughly, then loosen the ground with a fork and pull the plants out gently, aiming to keep root damage to the minimum. If necessary, shade the newly transplanted vegetables from the sun with fabric row covers or newspaper when they are first planted out. Conditions are less critical for seedlings in pots and plug trays,

## Planting out asparagus crowns ................

Good drainage is vitally important for asparagus, so always choose a well-drained site and create a bed with a raised ridge of soil along its center.

With guide lines, mark out trenches 12in (30cm) wide for each row of the asparagus bed, allowing 12in (30cm) between rows. Dig each trench, 8in (20cm) deep, between the guides, and create a raised central mound. Set the crowns, 15in (38cm) apart, on the raised mound, with their roots spread widely and evenly around the crowns and into the bottom of each trench.

because these suffer less root damage. Keep all plants well watered until they are well established and growing away well. Thereafter, most need water only during prolonged dry periods.

## VEGETATIVE METHODS

There are a few perennial vegetables that may be increased by seed, but which are more commonly propagated by vegetative means.

Rhubarb is usually propagated by division, since seed-raised plants are so variable in quality. In autumn, after the leaves have died down, lift the crown and cut off sections of the rhizome (sets), each with at least one growing point or bud. Return the parent to the ground and plant the sets with the bud just below ground level (light soils) or just above (heavy soils).

Globe artichokes can be lifted in spring and divided like herbaceous perennials.

Garlic can be increased by detaching and growing on individual cloves from a parent bulb, but should only be grown from guaranteed disease-free stock. Lift the bulbs in fall and separate the cloves. Those with a diameter of at least ½in (1cm) produce the best plants. Plant them to twice their depth, 7in (18cm) apart. In cold areas, plant them in plug trays kept in a sheltered place through the winter. Plant out the next spring.

Asparagus can be raised cheaply from seed sown *in situ* in spring, but if you wish to do this, choose an all-male F1 hybrid seed strain, since male plants are higher yielding. For quicker cropping, buy one-year-old crowns for spring planting.

Asparagus beds can be productive for many years, but you must be patient in the early years. Only a few spears should be cut in the first year after planting crowns. In the second year, the bed can be cropped for up to six weeks. In the third and subsequent seasons, spears may be cut for a period of about two months.

## Special techniques

Some plants need special treatments after sowing and planting out to produce attractive and well-flavored crops. Rhubarb, for example, produces sweeter, more tender stems earlier in the season if forced: that is, grown under a light-excluding cover such as a forcing pot. The stems of leeks and bulb-like stems of Florence fennel are both blanched by excluding light, in this case by earthing up. Self-blanching cultivars of celery are available, but even these are sweeter if light is excluded from the stems by packing with straw. Trench celery can be blanched (*see right*).

Chicory and endive are blanched because otherwise the green leaves have a distinctive, rather bitter flavor that is too strong for most palates. The blanching technique differs slightly for the cut or curled leaved and the borad leaved types (*see instructions below right*).

## Blanching chicory . . . . . . . . . . . . . . . . . . . . . . . . . . . .

Chicory leaves tend to be somewhat bitter for most palates, but their blanched, leafy shoots, or "chicons," are a popular vegetable. These are obtained by digging up mature chicory plants in fall and forcing their roots by potting them up and blanching them.

▲ **I** Dig up one or more mature chicory plants in autumn and trim off all leaves to within 1in (2.5cm) of the crown. Also remove the root tips and any side roots. Fill a pot with moist potting medium and insert the prepared roots vertically, so that their crowns are just visible above the surface.

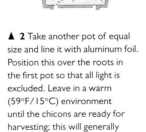

▲ **2** Take another pot of equal size and line it with aluminum foil. Position this over the roots in the first pot so that all light is excluded. Leave in a warm (59°F/15°C) environment until the chicons are ready for harvesting; this will generally be in 3–4 weeks' time.

## Blanching celery . . . . . . . . . . . . . . . . . . . . . . . . . . . . . .

Much celery is now self-blanching, but trench celery still needs to be blanched artifically. This relatively easy process is started when the celery plant is 12in (30cm) tall.

◀ **Blanching celery on the flat**
In well-prepared, flattened soil, plant celery plants *in situ*, after all risk of frost has passed. Once the stems are 12in (30cm) high, wrap a light-excluding paper collar around them, keeping it in place with raffia or soft twine, and allowing the leaves to emerge from the top. Wrap a second collar around the plant as more stem develops.

◀ **Blanching celery in a trench**
Prepare the soil very well and then make a trench 12in (30cm) deep and 15in (38cm) wide. Fill the trench with copious organic matter and then level it before planting. Plant the celery seedlings. As they develop, tie their stems with twine or raffia and keep adding soil to just below the plants' leaves. Ensure that no soil falls between the stems.

## Blanching endive . . . . . . . . . . . . . . . . . . . . . . . . . . . . . .

Endive leaves can be sweetened by blanching, either partially or totally, but cover only a few leaves at a time as they tend to deteriorate rapidly after such treatment.

▶ **Partial blanching**
Select an almost mature plant and, when dry, cover the central rosette with a large, domed plate. Harvest the leaves after 10 days.

▶ **Total blanching**
When the leaves are dry, loosely wrap a mature endive in soft twine or raffia and cover it beneath a bucket that excludes all light. Leave until the foliage has turned almost white and is ready for harvesting.

# DIRECTORY OF
# VEGETABLES

*Below are key tips on a selection of vegetables that are particularly suitable for propagation. Unless otherwise specified, follow the detailed instructions given under the propagation section on pages 196–201.*

## Key

Sp    Spacing
TH    Time to harvest
🌱    By seed

## *Allium cepa* (Onion)       Alliaceae

A staple crop grown for its bulb, used fresh and from store. Onions are grown from seed or sets (small individual bulbs) and need a long growing season. Most cultivars are started in spring for summer use and winter storage. Some are planted in fall, to give an early crop the next year.

### PROPAGATION

Sets – push sets into the ground, with tips just protruding. Plant maincrop late winter to mid-spring (or late spring if heat-treated); overwintering sets, early to late fall.
🌱 at min. 45°F (7°C), max. 75°F (24°C). Multi-sow in plug trays, six per cell, or in a seed tray for pricking out. Sow maincrop late winter-early spring under glass; plant out mid- to late spring. For sowing *in situ*, sow in furrows, ½–¾in (1–2cm) deep. Sow maincrop in early to mid-spring, under polyethylene film or plastic covers. Sow overwintering onions in late summer. The exact timing of sowing for overwintering crops is critical and depends on the locality; plants should reach a height of 6–8in (15–20cm) to overwinter successfully. Delay thinning until spring.
Sp multi-sown plugs, 10–12in (25–30cm) apart each way. Single plants or maincrop sets, 2–4in (5–10cm) apart, in rows 10in (25cm) apart, or 6–7in (15–18cm) apart each way. Overwintering sets, 4–6in (10–15cm) apart, in rows 12in (30cm) apart.
TH 5–6 months for maincrop onions.

## *Allium cepa* (Scallion)       Alliaceae

Scallions or salad onions can be used fresh from mid-spring to mid-fall with successional sowing. They are usually sown *in situ*, but sowing in plug trays can be useful for early-spring and midsummer sowings.

### PROPAGATION

🌱 in succession, early spring to early summer, at min. 45°F (7°C), max. 75°F (24°C). Multi-sow in plug trays with up to 10 seeds per cell. Sow thinly in furrows, ½–¾in (1–2cm) deep, 10cm (4in) wide. Sow hardy cultivars in mid- to late summer for an early spring crop.

Sp multi-sown plugs, 6in (15cm) apart each way. Seeds ¾–1¼in (2–3cm) apart.
TH 8–12 weeks for summer crops.

## *Allium esculentum* (Shallot)       Alliaceae

Shallots are grown for their bulbs and traditionally they are raised from sets, which multiply into a cluster of bulbs by the end of the season. A few cultivars, however, can be grown from seed.

### PROPAGATION

Sets – mid-spring. Choose small sets, about ¾in (2cm) in diameter, for planting. Put them in shallow furrows or push them into the soil, so that only their tips are showing. Firm the soil around them.
Sp 6in (15cm) apart, in rows 9in (22cm) apart, or 7in (18cm) apart each way.
🌱 mid-spring. Sow in a furrow, about ½in (1cm) deep and 4in (10cm) wide.
Sp about ½–¾in (1–2cm) apart to get single bulbs; larger spacings will encourage clusters of two or three bulbs to develop.
TH 5 months.

## *Allium porrum* (Leek)       Alliaceae

A very hardy vegetable grown for its white stems, cropping from early fall to spring of the following year, depending on cultivar. Leeks need a long growing season. Those used as mini-vegetables can be sown *in situ*.

### PROPAGATION

🌱 germinate at min. of 45°F (7°C). Sow in a seed tray and prick out into boxes, or plug trays with 3–4 per cell. Sow early cultivars in gentle heat, late winter to early spring; later cultivars early to late spring. Alternatively, sow in furrows, 1in (2.5cm) deep, early to late spring. Protect with a frame or cover. When 8in (20cm) tall, transplant into dibbled holes, 6–8in (15–20cm) deep. Water in and allow hole to fill naturally with soil.

*Allium cepa* 'Ailsa Craig'

*Allium cepa*

*Allium sativum*

*Apium graveolens*

*Beta vulgaris* var. *rapaceum*

Sp multi-sown plugs, 9in (22cm) apart each way. Single transplants, 3–6in (7.5–15cm) apart, in rows 12in (30cm) apart, or 4–6in (10–15cm) apart each way. Mini-leeks, ½in (1cm) apart, in rows 6in (15cm) apart. TH 7–9 months for mature crops, 13 weeks for mini-leeks.

### *Allium sativum* (Garlic)          Alliaceae

Garlic is grown for its strong-flavored bulb, but the leafy shoots can also be eaten. It is very hardy. The individual cloves, split from the parent bulb, are used for propagation. Best planted in fall, as it benefits from a cold period and long growing season; spring-planted, garlic may not mature satisfactorily.

#### PROPAGATION

Cloves – late fall or early spring, *in situ*, 2–4in (5–10cm) deep, pointed end up. For best results, cloves should be at least ½in (1cm) across. Alternatively, plant in plug trays, one clove per cell, in late fall and overwinter in a coldframe. Plant out when sprouted in mid- to late spring.
Sp 7in (18cm) apart each way.
TH 7–9 months.

### *Apium graveolens* var. *dulce*

(Celery)          Umbelliferae

Celery is grown for its crisp stems. Both trench celery and self-blanching celery are propagated in the same way. Trench celery is hardier but more difficult to grow.

#### PROPAGATION

🌱 late winter to early spring, at 55–60°F (13–16°C), and max. 66°F (19°C), in seed trays or plug trays. Do not cover seed, as it needs light to germinate. Grow on in deep boxes, or singly in pots, and plant out in

early summer. Temperatures below 50°F (10°C) cause bolting later: do not plant out too early, and use crop covers if necessary. Sp self-blanching celery 6–11in (15–27cm) each way; closer spacing will give more slender stems. Plant trench celery 12–18in (30–45cm) apart. TH 8 months.

### *Apium graveolens* var. *rapaceum*

(Celeriac)          Umbelliferae

A winter vegetable that is grown for the knobbly swollen root. It needs a long growing season.

#### PROPAGATION

🌱 late winter to early spring, at 55–60°F (13–16°C), or in a coldframe in mid-spring, in seed trays or plug trays; germination can be slow and erratic. Grow on in deep plugs or singly in pots and plant out in early summer. Exposure to temperatures below 50°F (10°C) can cause bolting later so do not plant out too early, and use crop covers if necessary.
Sp 12–15in (30–38cm) each way.
TH 8 months.

### *Asparagus officinalis* (Asparagus)

Liliaceae

A perennial grown for its young shoots in late spring and early summer. One-year-old crowns are widely available. Asparagus can also be grown from seed. Modern F1 hybrids are most productive. Traditionally sown in a seedbed outdoors, plug tray-sowing under glass is a good method of raising small numbers of F1 hybrids.

#### PROPAGATION

🌱 early spring, pre-soak overnight and sow at 55–60°F (13–16°C), singly in plug trays,

or in a seed tray and prick out into pots. Germination can be slow. Plant out in early to midsummer, in holes 4–5in (10–12.5cm) deep; fill in the holes in fall. In a seedbed, sow in furrows, 1in (2.5cm) deep, in mid-spring; thin to 3in (7.5cm) apart; transplant the largest crowns in mid-spring the next year.
Sp 12–18in (30–45cm) apart, in rows 30in (75cm) apart, or in beds of 2–3 rows 15in (38cm) apart.
TH cut a few spears the first spring after planting plugs, or one-year-old crowns of F1 hybrids. Leave other cultivars for another year. Next season, cut spears over about 4 weeks; in subsequent seasons, increase to 6 weeks.

### *Beta vulgaris* (Beet)          Chenopodiaceae

A summer root crop with round or cylindrical roots; the leaves are also good to eat. Usually has seed in clusters; monogerm cultivars have single seeds. Exposing young plants to cold can cause bolting later, so use bolt-resistant cultivars for early sowings.

#### PROPAGATION

🌱 germinate at min. 45°F (7°C). Multi-sow in plug trays, 3 seed clusters per cell, early spring to midsummer. Thin to 4 or 5 seedlings. Or, sow in furrows, ½–¾in (1–2cm) deep. For small beet to be used fresh, sow early spring to midsummer, using plastic covers. For maincrop for winter storage, sow late spring or early summer.
Sp multi-sown plugs, 9in (22cm) each way. Small young beet, 1in (2.5cm) apart, in rows 8in (20cm) apart, or 2in (5cm) each way; to encourage early cropping, use maincrop spacing. Maincrop, 3in (7.5cm) apart, in rows

*Beta vulgaris* subsp. *cicla*

*Brassica oleracea* Acephala Group

*Brassica oleracea* Botrytis Group

8in (20cm) apart, or 5–6in (13–15cm) apart each way.
TH 9–11 weeks for small young beet; 12 weeks for maincrop.

### Beta vulgaris subsp. cicla

(Spinach beet)　　　　　Chenopodiaceae

A biennial with more drought tolerance than true spinach (*Spinacia oleracea*). It is winter-hardy, and will not run to seed until the spring after sowing. It will grow in light shade.

**PROPAGATION**

germinate at min. 45°F (7°C). Sow singly in plug trays, or in furrows ¾in (2cm) deep, in mid- to late spring. Do not sow or plant out too early, as exposing young plants to cold can cause bolting. For an extra winter crop, sow in mid- to late summer.
Sp 10in (25) apart, in rows 15in (38cm) apart, or 12in (30cm) apart each way.
TH 10–12 weeks; can be harvested until late fall, during mild winters, and until late spring the following year.

### Beta vulgaris var. flavescens

(Swiss chard)　　　　　Chenopodiaceae

Swiss chard is a type of spinach beet with thick white midribs and stems. The stems and leaves can be used as separate vegetables. It is less hardy than leaf beet, and usually needs protection with plastic covers to overwinter. Propagation – as for Spinach beet (*above*).

### Brassica napus Napobrassica
### Group (Swede)　　　　Cruciferae

A brassica grown for its relatively hardy winter root. Normally sown *in situ*, but can be successfully sown in plug trays if outdoor sowing conditions are poor. Roots can be

used from the ground, but early cultivars are best lifted and stored in early winter.

**PROPAGATION**

germinates at min. 41°F (5°C). Sow thinly, or row-sow in furrows ¾in (2cm) deep, in late spring or early summer. It is essential to thin the seedlings early in order to allow good roots to develop.
Sp 9–12in (22–30cm) apart, in rows 15–18in (38–45cm) apart, or 12–14in (30–35cm) apart each way.
TH 18–20 weeks.

### Brassica napus Rapifera Group

(Turnip)　　　　　Cruciferae

Turnips are a quick-growing root crop; the tops are also useful for eating as spring greens. Successional sowings of suitable cultivars will give crops for summer harvest and winter storage. They are nearly always sown *in situ*.

**PROPAGATION**

germinates at min. 41°F (5°C). Sow in furrows ¾in (2cm) deep; early thinning is essential for good roots to develop. Make the first sowings of summer turnips in early spring with protection, followed by outdoor sowings every 3–4 weeks from mid-spring to late summer. Summer sowings may bolt quickly in hot and dry conditions. Sow hardy cultivars, for use in winter, from mid- to late summer.
Sp for summer crops, 4–6in (10–15cm) apart, in rows 9in (22cm) apart, or 6in (15cm) apart each way. For storage, 10–12in (25–30cm) apart, in rows 12in (30cm) apart, or 10in (25cm) apart each way.
TH 6–8 weeks for summer cultivars, 10–12 weeks for winter cultivars.

### Brassica oleracea Acephela Group

(Kale)　　　　　Cruciferae

A hardy winter brassica used for leaves in fall and winter, and young shoots in spring. There are curly and broad-leaved types. Traditionally grown in a seedbed for transplanting, but can be sown in plug trays; dwarf cultivars can be sown *in situ*. Some curly kales are also used for mini-vegetables and as a seedling crop.

**PROPAGATION**

germinates at min. 41°F (5°C). Sow singly in plug trays from mid- to late spring. Plant out early to midsummer. Alternatively, sow in furrows, ¾–1in (2–2.5cm) deep, for trans planting in midsummer; thin to 3in (7.5cm).
Sp 12–30in (30–75cm) apart each way, the closer spacing for dwarf cultivars, 6–8in (15–20cm) each way for mini-vegetables.
TH 7 months for leaves from mature plants; 18 weeks for mini-vegetables.

### Brassica oleracea Botrytis Group

(Cauliflower)　　　　　Cruciferae

Cauliflowers are grown for their large heads, which are usually white although purple and green types are available. They are grouped according to the season that the heads form. With successional sowing they can be harvested from early spring to early winter in most regions, and all year in mild areas. They are mostly large and slow-growing and are sown in a seedbed or containers for transplanting. Some cultivars can be closely spaced to produce a quick crop of mini-cauliflowers for sowing *in situ*.

**PROPAGATION**

germinates at min. 41°F (5°C). Sow in a seed tray for pricking out, or singly in plug

*Brassica oleracea* Capitata Group 'Firensa'

*Brassica oleracea* Gemmifera Group

*Brassica oleracea* Gongylodes Group

trays. Or, sow in furrows ¾–1in (2–2.5cm) deep, thinning to 3in (7.5cm) apart. Transplant when still young (after about six weeks) to minimize transplanting shock.

### Early-summer cauliflowers

Sow in early fall with protection and overwinter under plastic covers or frames. Alternatively, sow in containers under glass at 61°F (16°C) in late winter; plant out early to mid-spring.
Sp 18in (45cm) apart in rows 24in (60cm) apart, or 20in (50cm) apart each way.
TH 5–7 months.

### Summer and fall cauliflowers

Sow mid-spring to early summer, planting out in early to late summer.
Sp 18–24in (45–60cm) apart, in rows 24–28in (60–70cm) apart, depending on cultivar, or 20–25in (50–65cm) apart each way. For mini-cauliflowers, 5–6in (12–15cm) apart each way.
TH conventional crop, 5–6 months. For mini-cauliflowers, 12–16 weeks.

### Winter-/spring-heading cauliflowers

Sow early summer, plant out late summer.
Sp 24–30in (60–70cm) apart each way.
TH 10 months.

## *Brassica oleracea* Capitata Group

(Cabbage)                              Cruciferae

Leafy brassicas, usually forming dense heads, cabbages are grouped according to their season of maturity; cultivars can be selected to harvest fresh all the year round. They are usually transplanted from containers, or from a seedbed. The spacing that is used affects the size of the heads, the recommended spacing giving moderate-sized heads and a good yield for the particular cultivar.

For smaller heads, use closer spacings.

### PROPAGATION

🌱 germinates at min. 41°F (5°C). Sow singly in plug trays, or in a seed tray for pricking out. Alternatively, sow in furrows ¾–1in (2–2.5cm) deep, thinning to 3in (7.5cm) apart.

### Spring cabbage

Sow midsummer in cool areas, late summer in mild areas; plant out early to mid-fall.
Sp 12in (30cm) apart each way; or 4in (10cm) apart, in rows 12in (30cm) apart. Use two out of every three plants as 'spring greens' before they heart up.
TH 8–9 months.

### Summer and autumn cabbage

There are early- and later-maturing cultivars, and two or more sowings are needed for a continual supply. Red cabbage is included in this group. For a first crop, sow an early cultivar in gentle heat in early spring, for planting out mid- to late spring. Sow in mid- to late spring in the open, for planting out late spring to midsummer.
Sp 14–18in (35–45cm) apart each way.
TH 14–20 weeks.

### Winter cabbage

There are several types, varying in hardiness. Dutch white 'coleslaw' cabbages are included in this group. Sow in the open in late spring, plant out mid- to late summer.
Sp 20in (50cm) apart each way.
TH 4–5 months; most will stand *in situ* for several months, but Dutch white cabbages should be lifted before the first hard frosts.

## *Brassica oleracea* Gemmifera Group

(Brussels sprout)                      Cruciferae

A hardy winter vegetable grown for the buttons (swollen buds) on the stem. There

are early, mid-season, and late-maturing cultivars. The traditional method of sowing is in a seed bed for transplanting, but plants can also be raised in plug trays.

### PROPAGATION

🌱 germinates at min. 41°F (5°C). Sow singly in plug trays, early to late spring depending on cultivar. Plant out deeply, or plants may keel over later on. Or, sow in a seedbed, ¾–1in (2–2.5cm) deep, and thin to 3in (7.5cm) apart. Sow early cultivars in early spring, protected by a frame or plastic covers; plant out in late spring. Later cultivars can be sown in mid- to late spring outdoors, planting out in early summer.
Sp dwarf cultivars 24–30in (60–75cm) apart each way; tall cultivars 36in (90cm) each way.
TH 8–10 months; most cultivars crop for 2–3 months.

## *Brassica oleracea* Gongylodes Group (Kohlrabi)                      Cruciferae

A quick-maturing brassica grown for its swollen stem, although the leaves can also be eaten. Successional sowings are needed for crops throughout summer and fall. Use green cultivars (which are quicker to mature) for early sowings and hardier purple cultivars for fall use. Modern cultivars stand for longer in good condition.

### PROPAGATION

🌱 germinates at min. 41°F (5°C), but exposing young plants to cold, below 50°F (10°C), can cause bolting. Sow singly in plug trays in late winter or early spring, in gentle heat; plant out under plastic covers. Alternatively, sow in drills ½in (1cm) deep, mid-spring to late summer.
Sp 4–6in (10–15cm) apart, in rows 12in

*Brassica oleracea* Italica Group

*Brassica rapa* Pekinensis Group

*Capsicum annuum* 'Apache'

(30cm) apart, or 9in (22cm) apart each way. Use closer spacings for tender mini-roots. TH 9–12 weeks.

## *Brassica oleracea* Italica group

(Calabrese)                                    Cruciferae

A quick-maturing type of broccoli producing large green heads and smaller side shoots. It is not very hardy. Some cultivars mature more quickly than others, but successional sowing is necessary for continual cropping from mid-summer to mid-fall. It does not transplant well, so sow *in situ* or in plug trays.

**PROPAGATION**

🌱 germinates at min. 5°C (41°F). Sow singly in plug trays in early spring to midsummer, under glass, and plant out under plastic covers for earliest crops. Alternatively, sow in furrows ¾–1in (2–2.5cm) deep, mid-spring to midsummer.
Sp 6in (15cm) apart, in rows 12in (30cm) apart, or 9in (22cm) apart each way. Closer spacing will suppress side shoots.
TH 11–14 weeks, depending on cultivar.

## *Brassica oleracea* Italica Group

(Sprouting broccoli)                          Cruciferae

A hardy, overwintering brassica grown for its white or purple flowering shoots in spring. Purple types are the hardiest. The plants are large and widely spaced. They are usually transplanted from a seedbed, but can also be grown in plug trays. There are early- and late-maturing types. See also Calabrese (*above*).

**PROPAGATION**

🌱 germinates at min. 41°F (5°C). Sow in furrows ¾–1in (2–2.5cm) deep, thinning to

3in (7.5cm) apart, or sow singly in plug trays, in mid- to late spring. Plant out early to midsummer.
Sp 2ft (60cm) apart each way.
TH 11 months; individual plants usually crop for 6–8 weeks.

## *Brassica rapa* Pekinensis Group

(Chinese cabbage)                            Cruciferae

Quick-growing leafy brassica for late summer or early fall use. It is are only marginally frost-tolerant. Some modern cultivars are less prone to bolting. All dislike transplanting, so sow in plug trays or *in situ*.

**PROPAGATION**

🌱 germinates at min. 50°F (10°C), but see below. For early crops, sow bolt-resistant cultivars singly in plug trays, in late spring or early summer, maintaining a minimum temperature of 64°F (18°C) for the first 3 weeks after germination to reduce the risk of premature bolting. Plant out under plastic covers or crop covers. Make main sowings without heat in early to late summer, and at the end of summer for transplanting under cover for a cut-and-come-again crop. Row-sow in furrows ¾–1in (2–2.5cm) deep, early to late summer.
Sp for mature heads, 12–14in (30–35cm) apart each way.
TH 9–10 weeks.

## *Capsicum annuum*

(Pepper, sweet and chilli)                    Solanaceae

Frost-tender plants grown for their fruits, which come in a variety of shapes and colors. In cool temperate areas, peppers are raised under glass with some heat, and are

best cropped in a greenhouse or polytunnel, or under plastic covers. Dwarf cultivars grow well in pots.

**PROPAGATION**

🌱 sow singly in small pots or plug trays, or in a seed tray for pricking out, in early spring, at min. 70°F (21°C). Pot on into the final pot at 4in (10cm) high. Do not plant outside until danger of frost is past.
Sp 12in (30cm) apart for dwarf cultivars, 15–18in (38–45cm) for standard cultivars.
TH 5–6 months.

## *Cichorium endivia* (Endive)   Compositae

A slightly bitter salad plant sometimes blanched to make it sweeter. Cropping depends on the cultivar: the frilly types of endive are more suited to summer cropping, and the hardier broad-leaved types to fall and protected winter cropping. Endive can also be grown as a seedling crop. It is often sown *in situ*, but sowing in plug trays is useful for both early and midsummer sowings.

**PROPAGATION**

🌱 in furrows ½–¾in (1–2cm) deep, or singly in plug trays, early spring–late summer, giving protection to early and late crops.
Sp 10–15in (25–38cm) apart each way, depending on cultivar.
TH 12–16 weeks.

## *Cichorium intybus* (Chicory)

Compositae

Chicories are grown for their leaves, which are mainly used as salad crops from late summer to spring. There are three common types; all are sown singly in plug trays, or *in situ*, in furrows, ½–¾in (1–2cm) deep.

*Cichorium endivia*

*Cucumis sativus*

*Cucurbita pepo* 'Hundredweight'

### Red chicory

Most red chicories are for late summer or fall cropping, but some cultivars hardy and will overwinter under plastic covers or in a greenhouse. They can be treated as cut-and-come-again crops.

**PROPAGATION**

 sow early cultivars late spring to midsummer. Sow hardy cultivars in mid- to late summer for a protected winter crop.
Sp 8–14in (20–35cm) apart each way, depending on cultivar.
TH 12–14 weeks.

### Sugarloaf chicory

Forms dense conical heads when mature, usually in late summer and fall. Reasonably hardy. Can also be grown as a cut-and-come-again crop, or as a leafy seedling crop sown spring to early fall.

**PROPAGATION**

early to midsummer; cover late sowings to protect from frost.
Sp 10–12in (25–30cm) apart each way for mature heads.
TH 12–14 weeks.

### Whitloof chicory

The roots of Whitloof chicory, produced in summer, are forced in darkness in winter and early spring to produce tight white heads.

**PROPAGATION**

early to midsummer; cover late sowings to protect from frost.
Sp 9in (22cm) apart, in rows approximately 12in (30cm) apart.
TH lift roots for forcing from late winter to early spring, or, on light soils, force *in situ*. Cut off the leaves and cover the stumps with 6in (15cm) of soil.

### *Cucumis sativus* (Cucumber)

Cucurbitaceae

Tender, summer-fruiting plants, most of which trail or climb. Greenhouse cucumbers have long, slender, smooth fruits, and need warm conditions to flourish; modern all-female cultivars are easier to cultivate, needing no cross-pollination by hand. Ridge cucumbers are grown outside or in frames, as for winter squash (*see right*).

**PROPAGATION**

sow singly ½in (1cm) deep, in pots, at min. 75°F (24°C) in late spring for growing on in a frost-free greenhouse. Grow on at 66°F (19°C). Plant in permanent positions at 12in (30cm) high.
Sp 18in (45cm) apart, trained on string or netting.
TH 10–12 weeks.

### *Cucurbita pepo* (Marrow, zucchini, and summer squash)

Cucurbitaceae

Frost-tender crops grown for their fruits in summer. Marrows traditionally have green cylindrical fruits; other types are called summer squashes. There are both bush and trailing cultivars. Zucchini are the small immature fruits of some bush types, and many F1 hybrid cultivars have been bred specifically for zucchini production. All dislike root disturbance and are best sown in pots or *in situ*. Initial protection is usually needed.

**PROPAGATION**

sow singly in pots, at min. 55°F (13°C) in late spring, 4 weeks before last expected frost. Sow seeds on their edge with their flat sides vertical. Transplant when risk of frost has passed, first protecting with plastic

covers or crop covers. Or, hill-sow ½in (1cm) deep, in late spring or early summer, and protect with jars or plastic covers.
Sp bush types, 3ft (90cm) apart each way; trailing types, at least 4ft (1.2m) apart if allowed to sprawl, or 18in (45cm) if trained up trellis or netting.
TH 8–10 weeks for zucchini; a few weeks longer for large marrows.

### *Cucurbita pepo, C. maxima* (Pumpkin and winter squash)

Cucurbitaceae

Frost-tender plants grown for their late-summer fruits, stored for winter use. The large orange Hallowe'en pumpkins are the most familiar, but there are many other types. Most plants are trailing, but there are a few bush cultivars.

**PROPAGATION**

germination and sowing: As for Zucchini (*see left and above*).
Sp 4ft (1.2m), up to 6ft (1.8m) for very vigorous types.
TH about 16–20 weeks; harvest before first frosts. Fruit can be stored for up to 6 months.

### *Cynara scolymus* (Globe artichoke)

Compositae

A large perennial plant, grown for its immature flower buds in summer. Plants can be started off from seed, or from offsets (rooted suckers) of mature plants. Seed-raised plants will be variable in quality.

**PROPAGATION**

Offsets – spring or fall. In spring, slice suckers away from the parent plant with a sharp knife, and plant out immediately, about 2in (5cm) deep, in their final positions.

*Cynara scolymus*

*Daucus carota* 'Nantes'

*Foeniculum vulgare* var. *azoricum*

Alternatively, take offsets in fall and overwinter them in pots in a greenhouse. ❦ under glass, singly in small pots or in a seed tray for pricking out. Sow early spring for planting out early summer. Alternatively, sow in a seed bed, in furrows, about 1in (2.5cm) deep, in mid-spring for transplanting in early to midsummer.
Sp 3–4ft (90–120cm) apart each way.
TH plants grown from seed or offsets usually crop in late summer of their first year. Mature plants produce their main heads in early summer.

### *Daucus carota* (Carrot)        Umbelliferae
Moderately hardy root crop used from the ground throughout summer and fall, and usually stored for winter. Early cultivars have relatively short (sometimes round) roots and are mostly used fresh; maincrop cultivars are larger and longer, and are the best for storage. Carrots are generally sown *in situ*, but some early cultivars can be sown in plug trays. Pre-germination and protection with plastic covers or crop covers can help early sowings; row covers left in place until harvest also protect against carrotfly.

**PROPAGATION**
❦ at min. 44°F (7°C); slow to germinate. Multi-sow round, early cultivars in plug trays, four seeds per cell, under glass in late winter; plant out with protection. Alternatively, sow in furrows ½–¾in (1–2cm) deep. Sow early cultivars in early spring with protection, and in the open from mid-spring to midsummer for continual cropping. Sow maincrop cultivars from late spring to midsummer.
Sp multi-sown plugs, 3–4in (7.5–10cm) apart

each way. Early protected crops can be sown in 3–4in (7.5–10cm) bands, with seeds ¾in (2cm) apart; wider spacings encourage quicker growth. In furrows, 6in (15cm) apart with plants 1½–2in (4–5cm) apart.
TH early cultivars 10–12 weeks, maincrop cultivars 5–6 months.

### *Foeniculum vulgare* var. *azoricum*
(Florence fennel)            Umbelliferae
Annual summer vegetable grown for its bulbous stem base. The leaves can be used like those of herb fennel. The plants tend to bolt in lengthening days, or if checked by cold or dry conditions, so crops maturing in late summer are often more successful. Modern cultivars are less prone to bolting. Young plants dislike root disturbance, so sow *in situ* or in plug trays.

**PROPAGATION**
❦ in succession, from late spring to late summer, in furrows, ½–¾in (1–2cm) deep. Use bolt-resistant cultivars if sowing before midsummer. Protect late-sown plants with plastic covers.
Sp 12in (30cm) each way.
TH 3–4 months.

### *Helianthus tuberosus*
(Jerusalem artichoke)        Compositae
A hardy, easy-to-grow vegetable producing knobbly, edible tubers. These will keep growing if left from year to year, but must be lifted and replanted to get good crops.

**PROPAGATION**
Tubers – plant late winter to mid-spring, 4–6in (10–15cm) deep, in sun or partial shade. Small smooth tubers about the size of a hen's egg are best for planting. These can

generally be saved from the previous year's crop without risk of carrying disease.
Sp about 12in (30cm) apart, in single rows; 18in (45cm) each way for block planting.
TH 7–8 months; can be harvested from the ground throughout winter.

### *Lactuca sativa* (Lettuce)        Compositae
This popular salad crop can be grown nearly all year round, provided the right cultivars are used and winter crops are protected from the cold. Soft **Butterhead** lettuces mature quickly but also run to seed rapidly in hot weather. **Crisphead** lettuces are slower to mature and to bolt, and are good for summer use. **Cos** types are usually the hardiest; some cultivars can be sown for winter as well as spring and summer use. Loose-leaf, salad bowl lettuces, used as a cut-and-come-again crop, are very versatile all year round. All types of lettuce can be sown *in situ*, in a see bed, in seed trays, or in peat pellets or plug trays. Pellets or trays are useful for early and midsummer sowings. Transplanting from a seedbed is not advisable in midsummer. Lettuce can also be grown as a seedling crop.

**PROPAGATION**
❦ germinates well at low temperatures, but not above 77°F (25°C) for butterheads, or 85°F (29°C) for crispheads. Sow in a seed tray for pricking out, or singly in plug trays or peat pellets. Sow early spring to late summer; use gentle heat for first sowings and transplant under plastic covers to get an early crop. Sow late summer to mid-fall for winter cultivars in a greenhouse. Outdoors, sow in a seedbed for transplanting, or *in situ* in furrows ½–¾in

*Lactuca sativa* 'Lakeland'

*Lycopersicon esculentum* 'Gardener's Delight'

*Phaseolus coccineus* 'Pole Star'

(1–2cm) deep, from mid-spring to midsummer. In hot conditions, sowings are best made in containers or *in situ*, using soil-cooling film. Alternatively, sow in the late afternoon so that the heat-sensitive period of germination occurs in the evening.

Sp varies according to cultivar, from 6in (15cm) apart each way for small cos cultivars; 10in (25cm) each way for standard butterheads; 15in (38cm) each way for large crispheads.

TH 8–16 weeks for headed lettuces, depending on the season; loose-leaf cultivars can be harvested a few weeks earlier, before they are mature.

## *Lycopersicon esculentum*

(Tomato)                    Solanaceae

One of the most versatile of the summer fruiting vegetables. There are both bush and tall types. Suitable cultivars can be grown in a warm sheltered spot outdoors in most regions, but greenhouse, polyethylene tunnel, or plastic-cover protection gives earlier crops and higher yields.

### PROPAGATION

❦ Sow singly in plug trays or small pots, or in a seed tray for pricking out, at min. 68°F (20°C). Sow late winter to early spring for greenhouse crops, early to mid-spring for outdoor crops. Grow at 50°F (10°C), potting on as necessary. Plant out after about 8 weeks, but wait until there is no risk of frost. Alternatively, sow pre-germinated seed of bush tomatoes *in situ* in late spring, 1in (2.5cm) deep, cover with growing medium, and protect with plastic covers.

Sp tall cultivars 15–18in (38–45cm) apart; dwarf types 10–12in (25–30cm) apart; bush types 18–36in (45–90cm) apart.
TH 5–6 months.

## *Pastinaca sativa* (Parsnip)    Umbelliferae

A hardy winter root crop, which needs a long growing season in order to produce large roots. Parsnips are normally sown *in situ*, but can also be sown successfully in plug trays. The seed deteriorates rapidly in storage, so be sure to use fresh seed every year.

### PROPAGATION

❦ at min. 45°F (7°C); germination is very slow. Sow singly in plug trays, late winter to late spring; plant out seedlings before the tap root starts to develop. Alternatively, row-sow in furrows ½–¾in (1–2cm) deep, in early to late spring.

Sp depends on cultivar and size of root required. For small roots, 2in (5cm) apart, in rows 8in (20cm) apart, or 4in (10cm) apart each way. For long roots, 8in (20cm) apart, in rows 12in (30cm) apart, or 10in (25cm) each way.
TH 6–8 months.

## *Phaseolus coccineus*

(Runner bean)               Leguminosae

Frost-tender bean grown for its long, flat pods, from midsummer to early fall. Most are climbing, but there are some dwarf cultivars which give earlier crops but lower yields. Runner beans may be sown *in situ*, but sowing in containers is useful in cold areas.

### PROPAGATION

❦ germinate at min. 50°F (10°C). Sow in large plug trays under glass, in mid-spring,

and plant out after last frosts. Alternatively, hill-sow, 2–3in (5–7.5cm) deep, mid- to late spring. Protect with plastic covers initially (easier with dwarf cultivars).

Sp 6in (15cm) apart, in double rows 12–15in (30–38cm) apart for climbing types; 9in (22cm) apart each way for dwarf types.
TH 8–12 weeks.

## *Phaseolus vulgaris* (Green bean)

Leguminosae

Frost-tender bean usually grown for its pods, in early summer to early fall. The shelled beans can also be eaten. There are climbing and dwarf types, the pods varying in color and shape according to cultivar. Dwarf types crop earlier but give lower yields, and two successional sowings are usually needed for continual cropping. All types can be sown *in situ*, but sowing in containers is useful for early sowings and in cool areas.

### PROPAGATION

❦ germinate at min. 50°F (10°C). Sow singly in large plug trays; dwarf cultivars, early to mid-spring under glass for planting out under plastic covers; other cultivars in mid- to late spring. Alternatively, hill-sow 1½–2in (4–5cm) deep, in mid-spring under plastic covers or late spring to early summer in the open. Sow dwarf cultivars in midsummer for a late crop.

Sp 9in (22cm) apart each way for dwarf cultivars; 6in (15cm) apart, in double rows 12–15in (30–38cm) apart, for climbing types.
TH 8–10 weeks.

## *Pisum sativum* (Pea)    Leguminosae

Ordinary garden peas are grown for the peas inside the pod; petit pois produce very small,

*Pisum sativum* 'Oregon Sugar Pod'

*Raphanus sativa* 'Crystal Ball'

*Rheum rhaponticum*

sweet peas. In the flat-podded mangetouts and succulent, round-podded sugar peas, the whole pod is edible. There are tall and dwarf cultivars of all types. All peas are hardy; round-seeded peas will overwinter in mild areas to give an early crop. The summer harvest can be extended by sowing early, second-early, and maincrop cultivars at the same time. Sow successively for continual cropping of all dwarf peas throughout summer. Peas are usually sown *in situ*. Pre-germination and sowing under plastic covers or crop covers are useful for the production of earlier crops.

**PROPAGATION**

🌱 germinate at min. 45°F (7°C). Sow 1–1½in (2.5–4cm) deep, in 6in (15cm) wide furrows. Sow earlies in early spring with protection; in open ground, sow from mid-spring to midsummer. Midsummer sowings of dwarf cultivars may give a late crop in good seasons. In mild areas, sow hardy types in mid- to late fall for overwintering.
Sp 2–3in (5–7.5cm) apart, in wide furrows, 2–4ft (60–120cm) apart, depending on the height of the cultivar. Grow dwarf leafless cultivars in blocks, at spacings of 5in (12.5cm) apart.
TH earlies, dwarf mangetouts, and sugar peas: 11–12 weeks. Maincrops 13–14 weeks.

### *Raphanus sativa* (Radish)    Cruciferae

Radishes are cultivated mainly for their roots, although they are also grown for their seed pods or as a seedling crop. There are three main types: small, quick-growing radishes for salad use, large oriental types, and hardy winter radishes. They are usually sown *in situ*, but oriental types can be sown in plug trays.

**PROPAGATION**

🌱 germinates at min. 43°F (6°C). Sow salad radishes in drills, ½in (1cm) deep, from late winter to early fall. Protect early and late sowings from the cold, and make summer sowings in light shade. Thin seedlings when they are still small.
Sp ¾–1in (2–2.5cm) apart, in furrows 4–6in (10–15cm) apart.
TH 3–5 weeks; crops run to seed quickly.

#### Oriental radishes

🌱 sow singly in plug trays, and transplant before tap roots start to develop. Or, sow in furrows, ½–¾in (1–2cm) deep. Sow late spring to late summer depending on the cultivar. Use bolt-resistant cultivars for sowings before midsummer.
Sp 4in (10cm) apart, in rows 10in (25cm) apart.
TH 7–8 weeks.

#### Winter radish

🌱 row-sow in furrows ½–¾in (1–2cm) deep, in mid- to late summer. Protect or lift in late fall for winter storage.
Sp 8in (20cm) apart, in furrows 12in (30cm) apart.
TH 12 weeks.

### *Rheum rhaponticum* (Rhubarb)    Polygonaceae

A perennial plant grown for its leaf stalks, picked early spring to midsummer, depending on cultivar. A few cultivars can be raised from seed, but rhubarb is normally grown from sets taken from established plants.

**PROPAGATION**

Sets – late fall to early spring. Lift the parent plant and divide into pieces 4in (10cm) across, each with at least one bud.

🌱 early to mid-spring, in a seed tray. Prick out into small pots and plant out in early summer. Sow *in situ* in mid-spring, in furrows 1in (2.5cm) deep, 12in (30cm) apart, thinning to 6in (15cm) apart. Transplant in fall or the following spring.
Sp 30–36in (75–90cm) apart.
TH pull leaf stalks sparingly in the second growing season after planting. In subsequent years, harvest the leaf stalks from spring to midsummer.

### *Scorzonera hispanica* (Scorzonera)    Compositae

This is a perennial vegetable that is grown for its long, black-skinned roots, which are similar in flavor to those of salsify. It is propagated from seed, but any roots that are not big enough to use in the first winter can be left in the ground to allow another year's growth in order to thicken up. Sowing, spacing, and harvesting as for Salsify (see *opposite page*).

### *Solanum melongena* (Eggplant)    Solanaceae

Frost-tender plants, most with long, shiny, purple fruits. Closely related to tomatoes and peppers, but needing slightly higher temperatures to crop and grow well. In cool temperate climates, they are grown under glass or plastic covers.

**PROPAGATION**

🌱 sow singly in small pots or plug trays, or in a seed tray for pricking out, in early spring, at min. 70°F (21°C). Pot on successively, into final pots when 4in (10cm) high.
Sp 15–18in (38–45cm).
TH 5–6 months.

Solanum melongena

Solanum tuberosum

Vicia faba 'Bonny Lad'

## Solanum tuberosum (Potato) Solanaceae

Frost-tender crop grown for its tubers. Cultivars are grouped into earlies, second-earlies, and maincrop, according to the time they take to mature. All are grown from seed tubers, preferably certified as virus-free stock, which are allowed to sprout in a cool, light, frost-free place for 6–8 weeks before planting; the shoots should then be about ¾–1in (2–2.5cm) long.

### PROPAGATION

Seed tubers – earlies, from early to mid-spring. Protect young leaves from late frost by earthing up or covering with straw. Grow under plastic covers or crop covers for an extra-early crop. Plant later cultivars in mid- to late spring. Place seed tubers in a furrow 4–6in (10–15cm) deep, or in individual planting holes made with a trowel, 2in (5cm) below the soil surface. Sp early potatoes 12in (30cm) apart, in rows 15–20in (38–50cm) apart, others 15in (38cm) apart, in rows 30in (75cm) apart.
TH 15 weeks for earlies, 17 weeks for second-earlies, 18–20 weeks for maincrop.

## Spinacia oleracea (Spinach)
Chenopodiaceae

True spinach is a quick-growing, leafy annual. It is moderately hardy and grows best in cool spring and fall days, tending to bolt in hot weather. It can also be grown as a seedling crop. Traditionally, round-seeded types were used for summer crops and prickly-seeded types for fall crops. Modern cultivars are generally suitable for all sowings. Normally sown in situ.

### PROPAGATION

🌰 in furrows 2cm (¾in) deep, in early spring under protection in a warm site. Continue successional sowings in the open as required, making any early-summer sowings in light shade. Sow in late summer to early fall for a winter crop.
Sp 6in (15cm) apart, in rows 12in (30cm) apart, or 9in (22cm) each way.
TH 6–10 weeks, depending on the season.

## Tragopogon porrifolius (Salsify)
Compositae

A hardy biennial usually grown for its thin winter roots; the flower buds and blanched shoots can also be eaten the next spring.

### PROPAGATION

🌰 sow in situ. Sow thinly or row-sow in furrows, 1in (2.5cm) deep, from mid- to late spring. Germination can be erratic.
Sp 4–6in (10–15cm) apart, in rows 12in (30cm) apart, or 7–9in (18–22cm) apart each way.

## Vicia faba (Broad bean) Leguminosae

A hardy crop, usually grown for the green or white beans inside the pods, although very young pods can be eaten whole. The pods can be picked throughout the summer from successional sowings. Sowing in plugs for transplanting in warmer weather is useful for early crops in cold areas.

### PROPAGATION

🌰 germinates at min. 41°F (5°C). Sow in deep boxes in early winter for early-spring planting; late winter for mid-spring planting. Alternatively, hill-sow, in situ, 1½–2in (4–5cm) deep. For an early-summer crop, sow hardier cultivars in mid-fall or early winter; protect with plastic covers during the coldest months. In cold areas, sow quick-maturing cultivars in early spring. Sow mid- to late spring for a later crop.
Sp in double rows, 9in (22cm) apart with 18–24in (45–60cm) between rows, or in blocks, 9–12in (22–30cm) apart each way.
TH 7–8 months from fall sowings; 12–16 weeks for spring-sown plants.

## Zea mays (Corn) Gramineae

Tall frost-tender plant grown for its sweet ears formed in late summer. 'Supersweet' cultivars produce sweeter corn but the plants are less vigorous and not so hardy. There are also cultivars for producing "mini" corn. These are ears that are picked when they are only a few inches long and eaten whole, raw or cooked. Corn does not transplant well, so sow in situ. Use fast-maturing cultivars in colder areas and start the plants off under glass with heat.

### PROPAGATION

🌰 sow singly in plug trays in mid-spring, at min. 50°F (10°C). Plant out after risk of frost has passed. Protect with plastic covers or crop covers at first. Or, hill-sow, 1in (2.5cm) deep, in late spring, providing protection from the cold, or in early summer in the open.
Sp 12in (30cm) each way for short cultivars, 15–18in (30–45cm) each way for tall cultivars. Plant in blocks to assist pollination. For mini-corn, 6in (15cm) apart; in single rows; pollination is not necessary.
TH 16–20 weeks.

211

# SEED BOTANY AND HYBRIDIZING

## SEED BOTANY

The seeds of all plants share the same basic structure, consisting of an embryo and a food reserve, protected by a seed-coat or testa. The embryo is the growing point, from which develops the roots, shoots and leaves. The food reserve is often stored within the seed leaves, and these also serve to cushion the embryo from physical damage, including excessive heat and cold.

Flowering plants are classified as Monocotyledons or Dicotyledons, based on whether the seeds have one or two seed leaves. When a seed germinates, it is often the seed leaves which appear above ground first, the true leaves following later. This is called epigeal germination. However, in plants like sweet peas the seed leaves remain below the ground – hypogeal germination.

Commercial seed is stored in a partially dehydrated state and may remain viable for months or even years, until the right conditions are present to stimulate germination. Not surprisingly, the necessary conditions vary depending on the species, and especially on its native habitat. Seeds of tropical natives usually require higher temperatures than those from temperate zones.

A further complication is the dormancy shown by the seeds of some species. In these, even in the presence of suitable warmth and moisture, germination will not occur until dormancy is 'broken'.

This may require exposure to darkness or to light, a period of low temperatures, alternation of low and high temperatures or the gradual decay of the seed coat which prevents the uptake of water or contains germination-inhibiting chemicals. Gardeners have developed various techniques to overcome seed dormancy in a controlled and predictable way.

Seed dormancy is a survival aid for the plant. It prevents the seed from germinating when conditions would be unfavorable for the subsequent survival of the seedling.

## HYBRIDIZING

Gardeners often assume that cultivars and hybrids are the same thing, but this is not true. Cultivars are selected variants within a plant, and may be chance seedlings, mutations ('sports'), or simple or complex hybrids. They seldom breed true to type, because the development of a seed involves a recombination of the parent's genetic material. Hybrids are the offspring of two plants of the same or closely related species differing in one or more genes.

Plants are hybridized in the hope of combining desirable characteristics – for example, flower color, growth habit, or disease resistance – from each parent. The plant breeder chooses the parent plants with such qualities in mind.

To create a hybrid, the grower transfers pollen from the stamens of one plant to the stigma of the seed parent. To prevent the risk of the seed parent being fertilized by its own pollen, its stamens are removed before they mature. There is still a possibility of pollen from a plant other than the intended parent arriving, perhaps wind-borne, or via a visiting insect. Enclosing the flower in a paper bag before and after pollination helps to prevent this.

Hybrid plants are often more expensive, for various reasons. Firstly, the element of chance, even in planned hybridization, means that the desired flower color, or disease resistance, may or may not appear among the progeny and, in woody plants, it may be several years before this can be assessed.

The breeding of highly developed plants, such as roses or rhododendrons, involves the sowing of thousands of hybrid seeds, growing them on for some years, and perhaps grafting them to accelerate maturity, perhaps only to find that the particular crossing has not achieved the desired result.

Among hybrids, F1 hybrids are often encountered. The term indicates the 'first filial generation', that is, the direct progeny of the act of hybridization. Subsequent generations may show a decline in the desirable qualities of the F1 plants. This is avoided only by repeating the original crossing for each generation, hence the high cost of such seeds.

# PROPAGATING MEDIA

## POTTING AND PROPAGATION MEDIA

Garden soil does not make a good growing or sowing medium for plants in containers. It is prone to compaction, thus depriving the roots of oxygen and impeding good drainage. It is also likely to contain legion pests and diseases. Always use a good quality, purpose-made medium that is designed to have a durable physical structure, with free drainage and air movement as well as being able to retain moisture. The ingredients used to make up the mix will be free of pests, diseases, and weeds. The level of nutrients added to the mix will depend on its use.

## LOAM-BASED AND SOILLESS MEDIA

Loam-based media are the more traditional potting and propagation mixes. Clay loam from pasture land has an ideal structure and texture, combined with a pH between 6.0 and 6.5. Peat and grit are used to provide drainage and aeration.

Because of the difficulty in obtaining the correct type of loam, in recent years, loam-based mixes have been superseded by peat-based ones. Peat is lightweight with good aeration, drainage and water retention, but many peat bogs have been over-exploited to such an extent that its universal use can no longer be considered sustainable.

Alternatives to peat include coir (coconut fiber) and granulated, composted softwood bark. Both have slightly different drainage characteristics and such mixes need more careful attention to watering than more traditional types.

To aid re-wetting and aeration, coarse sand may be added to all types of mix. Perlite and vermiculite can be used instead of coarse sand; both are lightweight and sterile. Perlite absorbs and holds large quantities of air and moisture. Vermiculite has similar properties, but also retains plant nutrients.

Proprietary mixes of both types are easily obtained from your local garden center. Potting mixes contain a balanced level of fertilizers to sustain growth for 4–6 weeks. After this period, extra nutrients can be applied as a liquid fertilizer when watering. Propagating mixes contain low nutrient levels, because high levels damage the developing roots of seedlings and cuttings.

For gardeners who wish to make their own mixes there are a number of tried and tested formulae:

## LOAM-BASED MEDIA

### Seed Sowing Mix

2 parts by volume sterilized loam
1 part by volume sphagnum-moss peat
1 part by volume lime-free, coarse sand or grit
To each cubic foot add a base dressing (lb/ft³):
Ground chalk or limestone, 1.3lb (0.6kg)
Superphosphate, 2.6lb (1.2kg)

### Cutting Mix

1 part by volume sterilized loam
1 part by volume moss peat
1 part by volume lime-free coarse sand or grit
To each cubic foot add a base dressing thus: (lb/ft³)
Superphosphate 2.6lb (1.2kg)
Ground chalk or limestone 1.3lb (0.6kg)

## SOILLESS MIXES

### Seed Sowing Mix

5 parts by volume sphagnum moss peat
1 part by volume vermiculite
To each cubic foot add a base dressing (lb/ft³):
Potassium nitrate 0.8lb (0.4kg)
Superphosphate 1.6lb (0.75kg)
Ground chalk or limestone 6.6lb (3.0kg)

### Cutting Mix

4 parts by volume sphagnum-moss peat
1 part by volume coarse sand or Perlite
To each cubic foot add a base dressing (lb/ft³):
Superphosphate 2.6lb (1.2kg)
Ground chalk or limestone 1.3lb (0.6kg)

Once prepared, all mixes are ready for immediate use. Store them only for a short period, as otherwise they might break down and damage roots when used later.

# PROPAGATION BY SEASON

## MID- TO LATE WINTER

Now completely dormant, selected cultivars of *Betula* can be grafted, followed by the softer-wooded Rhododendrons after the first of the year; these in turn are followed by the harder-wooded varieties into early spring.

Conifers to side-veneer graft onto understock include *Abies*, *Cedrus*, and *Picea*.

Some plants that are raised from seed annually, including fibrous-rooted and tuberous begonias and bedding pelargoniums, need relatively high temperatures and a long growing season. Begin sowing 6–8 weeks after the first of the year.

Early sowing also applies to shrubby plants that need to be large enough to survive the following winter; rhododendrons and other ericaceous plants sown in late winter will be sturdy seedlings by the fall.

## EARLY SPRING

Layering of shrubs, if not carried out in the fall, can be done in early spring: the flowing sap makes branches as pliable as they would have been then. These include: *Abelia*, *Clematis*, *Cotinus*, *Hamamelis*, *Kolkwitzia*, *Magnolia*, *Paeonia*, *Rhododendron*, and *Wisteria*.

Some houseplants can be divided now: *Achimenes*, *Aspidistra*, *Chlorophytum*, and seed sown of others, *Agave*, *Aloe*, *Philodendron*, *Pittosporum*, and *Sinningia*.

Cuttings of *Achimenes*, small-leaved begonias, and *Euphorbia pulcherrima* (poinsettia) can be rooted now. The bulk of summer bedding should be sown at this time, leaving those that mature quickly until late spring.

## EARLY TO MID-SPRING

The first of the basal cuttings will be large enough now, including the Belladonna and Elatum groups of delphiniums (normally the earliest) followed by chrysanthemum (*Dendranthema*) then the tuberous-rooted dahlias.

When the soil has warmed and become workable, onion sets, shallots and potatoes can be planted with the first sowings of hardy vegetables and herbs.

On heavy soil, mid-spring is the best time to divide herbaceous perennials, including *Achillea*, *Aster* × *novi-belgii*, *Geum*, *Helenium*, and many others. On light, sandy soils, division during fall is usually more successful.

Conifer seed, of *Abies*, *Calocedrus*, *Cedrus*, *Larix*, *Pinus*, can be sown outside in prepared beds now. Some form of frost protection should be at hand, because this seed can germinate quickly and will be damaged by frost.

## MID- TO LATE SPRING

The first seedlings and plants propagated in early spring can be grown cooler to harden them off, making room to sow more tender plants for the vegetable garden, peppers and chillis, marrows, pumpkins, etc., and tomato.

Many of the herbaceous perennials will have made sufficient growth to be propagated from basal cuttings: *Achillea*, *Anaphalis*, *Anthemis*, *Aster*, *Echinacea*, *Erigeron*, *Nepeta*, and *Rudbeckia*.

Hardy annuals can be sown outside in shallow drill or broadcast, including *Calendula*, *Centaurea*, *Clarkia*, *Helianthus*, and *Linaria* – all plants that can be grown this way.

Most ferns divide readily at this time; there are some exceptions which are more difficult to propagate, but for the majority this is the easiest way.

Some of the water plants, in particular the marginals like *Acorus*, *Butomus*, *Calla*, *Menyanthes*, *Peltandra*, *Pontederia*, *Sagittaria*, *Saururus*, and *Sparganium,* can be divided as the water warms up.

## EARLY SUMMER

Softwood cuttings of deciduous shrubs and climbers now become available through to mid-summer, including *Abelia*, *Acer*, *Actinidia*, *Chimonanthus*, Knaphill azaleas, *Kolkwitzia*, and *Wisteria*.

Some of the deciduous fruiting plants that root from softwood cuttings include *Actinidia deliciosa*, *Morus*, *Vitis*, and *Passiflora*.

The spring-flowering *Cattleya* species are propagated by division, and stem bulbils are removed from *Ceropegia* to be potted up.

Stem tip cuttings offer another method of propagation. With the coming of good light at this time of the year, other houseplants can be propagated either by stem cuttings or leaf cuttings.

Now that pond water has warmed, the aquatic group of water plants are increased easily by division or cuttings, including *Ceratophyllum, Egeria, Elodea, Glyceria, Myriophyllum,* and *Nymphaea.*

## MID-SUMMER

Some of the bulbs are ready for division or separation of offsets now, including *Amaryllis belladonna, Anemone,* and *Colchicum.*

Strawberry runners offer a way of increasing stock for growing in pots or renewing old plants.

*Rubus,* including blackberries and hybrid berries, will readily produce a plant from tip-layering now.

Bud-grafting, either by chip- or T-budding, of fruit trees and ornamentals should begin now and carry on through to early autumn.

Selected cultivars of *Acer japonicum* and *A. palmatum* are side-veneer grafted onto seedling understocks.

Plants grown as biennials and used for spring bedding, including *Bellis, Erysimun,* and *Myosotis,* can be sown outside now and planted in the flowering sites later. Give attention to the alpines from now on. Many that do produce seed in cultivation will germinate more readily if the seed is sown as soon as ripe, including *Aethionema, Androsace, Campanula, Cyclamen, Dryas,* and *Pulsatilla.* Others can be grown from soft cuttings taken now: *Aubrieta, Aurinia, Campanula,*

*Dianthus,* and *Edraianthus.*

Herbaceous perennials that could be killed by a hard winter, such as penstemons, should also be propagated by stem cuttings to insure your stock.

Iris (rhizomatous) can be increased after flowering, by sectioning the rhizome with a fan of leaves attached.

Ericas, the heaths, root well through this season, including *E. arborea, E. carnea, E. × darleyensis* and *E. vagans.*

## MID- TO LATE SUMMER

Semi-ripe cuttings of many plants will be ready from now and on through to the fall, starting with: *Abelia, Cornus, Cotoneaster, Cytisus, Elaeagnus, Escallonia, Garrya, Hypericum, Ilex, Lavandula,* and *Mahonia* as well as the heathers, the many *Calluna* cultivars and *Daboecia.*

*Hamamelis* cultivars are grafted on to seedling understocks.

From now on lilies can be propagated from bulb scales.

## EARLY FALL

More semi-ripe cuttings to take from shrubs and conifers, including *Aucuba,* evergreen *Berberis, Buxus, Cotoneaster,* and *Mahonia.*

The conifers include *Chamaecyparis, Cryptomeria, Cupressus, Juniperus, Podocarpus, Prumnopitys, Taxus, Thuja* and *Tsuga.*

Among the bulbous plants, *Allium, Chionodoxa, Muscari* and *Scilla* can be lifted, divided and then replanted.

## MID-FALL

Stock of half-hardy tubers and herbaceous perennials for producing

basal cuttings in the spring, should be made as safe as possible in a frost free area.

On light, sandy soils, division of herbaceous perennials will be successful now; they have the chance to make good root growth while the soil still retains the warmth of summer.

## LATE FALL

Hardwood cuttings can be inserted just after leaf fall in a sheltered border outside, or in a frame.

Evergreen shrubs can be layered now, or left until spring.

## EARLY WINTER

The season for taking root cuttings begins. They are taken when weather conditions are right for the lifting of the stock plant (not too wet or cold, and not too dry).

Begin taking root cuttings of *Anemone, Papaver orientale* and the cultivars of *Phlox maculata* and *P. paniculata* to produce plants free of stem eelworm.

## MID-WINTER

Root cuttings of herbaceous perennials to propagate now include: *Acanthus, Echinops, Eryngium* and *Paeonia.*

*Aesculus parviflora, Embothrium, Romneya* and *Yucca,* can also be propagated in this way, provided that soil conditions are suitable for lifting roots. A similar technique can be used with *Bergenia* by sectioning rhizomes.

Fruiting and ornamental *Vitis* can be grown from single buds (vine-eyes) or hardwood cuttings.

215

# PESTS AND DISEASES

## PESTS

### Aphid

Most plants are susceptible. Aphids of various species are green, black, yellow or pink in color. They suck sap, cause distorted growth and transmit virus diseases. Infested plants become sticky with aphid excrement (honeydew) which allows sooty molds to develop. Select a pesticide specific for aphids that will not harm bees and other beneficial insects, including natural predators introduced to combat other greenhouse pests.

### Black vine weevil

Most plants are susceptible. The adult beetles are ¼–⅜in (7-9mm) long and have dull black, pear-shaped bodies. They are active between spring and autumn and eat irregular notches in leaf margins. Although unsightly, this is generally something the plants can tolerate. It is the grubs in the soil that kill plants by eating roots, corms and tubers. The grubs are up to ½in (10mm) long with plump, creamy white, legless bodies and brown heads. Plants grown in pots or other containers are more at risk than those in the open ground. Plant losses caused by the grubs occur between early fall and spring. Control is difficult as both adult weevils and grubs are tolerant of insecticides sold to amateur gardeners. Search for and destroy adult weevils at night during the summer. Control the grubs by

watering pathogenic nematodes, such as *Heterorhabditis megidis* or *Steinernema carpocapsae,* into the medium during late summer.

### Flea beetle

Brassicas, turnip, swede and some outdoor annuals and herbaceous perennials are susceptible. Beetles, ¹⁄₁₆–⅛in (2-3mm) long, eat rounded holes in the foliage and can kill seedlings. The beetles are black, sometimes with a yellow stripe along the wing cases, or they are metallic blue or green. Limit the numbers of overwintering adults by clearing away plant debris. Keep seedlings watered to speed their growth through this vulnerable stage. Chemical sprays or dusts are available if necessary.

### Fungus gnat or sciarid fly

Seedlings and soft cuttings under glass are susceptible. Grey-brown flies, ⅛in (3-4mm) long, run over the medium or fly up when disturbed. Small white maggots with black heads feed on seedling roots and bore into the base of cuttings. Suspend yellow sticky traps over seed trays to capture adult flies. Introduce the predatory mite, *Hypoaspis miles* to the medium. Chemical controls are available but may damage young seedlings.

### Red spider mite

Most plants are susceptible. Yellow-green, eight-legged mites, less than ¹⁄₁₆in (1mm) long, live on the

underside of leaves, causing a fine pale mottling of the upper surface. Chemical controls are available but pesticide-resistant strains are widespread. Treat with the predatory mite *Phytoseiulus persimilis*.

### Slug and snail

Most seedlings and cuttings are susceptible. These pests feed mainly at night, causing irregular holes in leaves and stems. A silvery slime trail is sometimes left on the plants. Use slug pellets with care; they can be harmful to children, pets, and slug-eating wildlife such as frogs, birds and hedgehogs. Alternatively use the pathogenic nematode, *Phasmarhabditis hermaphrodita*. Good hygiene and tidiness in greenhouses will remove slugs and their daytime hiding places.

### Whitefly

Most plants are susceptible, especially house- and greenhouse plants. White-winged insects, ¹⁄₁₆in (2mm) long, and flat, oval, pale green nymphs suck sap from the lower leaf surface. Infested plants become sticky with excreted honeydew on which sooty molds thrive. Chemical controls are available but pesticide resistance is common. The parasitic wasp, *Encarsia formosa*, kills the nymphs, causing them to turn black. With practice, adult numbers can be reduced by first disturbing them into flight, and then sucking them up with a vacuum cleaner.

# DISEASES

Many plant diseases can be controlled by chemical treatments, but prevention, as far as possible, is a far better option both in terms of expense and in reducing the chemical load on the environment. This is especially true in the case of edible crop plants.

The risk of disease attack and subsequent development is much reduced by maintaining a high standard of hygiene during propagation, as well as providing your plants with optimal growing conditions to keep them vigorous and healthy.

When selecting material for any form of vegetative propagation, whether it be bulbs, corms, tubers or cuttings, it is vital to ensure that the material is healthy and free from any disease. All forms of vegetative increase carry the risk of passing on any diseases to the resulting young plants.

## Botrytis (Gray mold)

This fungal disease frequently causes problems on protected plants under glass. In humid conditions, masses of spores develop which can be transported by air currents to infect other plant material. Gray mold can be prevented to a large degree by avoiding conditions favorable to its development; general good hygiene is important as well as avoiding overcrowding of plants. Good ventilation in frames and greenhouses, together with avoiding condensation or water splash on the leaf surfaces will reduce attacks.

Remove any affected material as soon as seen and spray the remaining plant material with an appropriate fungicide.

## Crown gall

Crown galls are knobbly swellings caused by a soil-borne bacterium and are found on roots, at the junction of roots and stems, and occasionally on the stems. On herbaceous material, they can provide a site of infection for rotting organisms that then penetrate the crown. Aerial galls are frequently found on roses and *Daphne* spp. and provided that they are not widespread can be cut out.

The disease is more prevalent on wet soils. No chemical control is available. Infection usually occurs when the bacteria penetrate small wounds. Try to avoid any damage to the roots or stems when transplanting or planting out and avoid planting in sites known to be affected.

## Damping-off

This is a common disease of seedlings both under glass and outdoors, especially on rapidly growing annual bedding plants. It is caused by a number of different fungi. Plants are attacked at or just above soil level before they fall over and die. This disease can largely be prevented by using sterilized media, avoiding over-watering, and ensuring good hygiene of all equipment and growing areas. Good ventilation and drenching with a copper-based fungicide will help to control disease attack.

Most seedlings are susceptible, especially if they are weak, over-crowded, or growing poorly.

## Downy mildew

This fungus can attack young plants and seedlings, causing yellowing of leaves in patches, followed by wilting and death. On the undersides of affected leaves it produces an off-white or purplish mold. Downy mildew is generally host specific in any genus. Remove affected plants or leaves and control with an appropriate fungicide.

## Leafy Gall

Infection by this bacterium results in a mass of stunted and often distorted, thickened shoots. Sweet peas, dahlias, chrysanthemums, and pelargoniums are commonly affected as well as on bedding plants causing blindness. Destroy diseased propagation stock plants.

## Powdery mildew

These fungi are among the most common air-borne plant diseases. A gray-white powdery coating develops on leaves, stems and other plant parts, distorting their growth. Plants are most susceptible when the soil is dry and development of the disease is promoted by humid, stagnant air around the foliage. It can be treated with fungicide, but repeat applications at two-week intervals throughout summer are likely to be necessary.

## Viruses

Viruses induce a wide range of symptoms, including stunting, spotting, mottling, or blotching. There are no chemical controls. Avoid the use of propagating material from affected plants and control the aphid vectors.

# GLOSSARY

**Acid** (of soil) Having a pH level of less than 7; lime-free or almost so.

**Alkaline** (of soil) Having a pH level above 7; limy.

**Alpine** 1. A plant native to mountains, usually above the tree line. 2. A small plant grown in a rock garden.

**Annual** A plant that completes its life cycle from germination to shedding seed in a single growing season.

**Basal cutting** A cutting taken from the base of a plant, usually a herbaceous plant, as growth begins in spring.

**Biennial** A plant that completes its life cycle in two seasons, producing leaf growth in its first year to flowers, set seed, and die in the second.

**Bleed** To lose sap from a wound, usually one made during pruning or when taking cuttings.

**Bolt** The rapid growth of a stem prior to flowering.

**Broadcasting** To scatter seed evenly over a seed bed.

**Bromeliad** A member of the family Bromeliaceae.

**Bud-grafting** A grafting technique that involves uniting a single bud from the *scion* with a *rootstock*; a general term that includes chip budding and T-budding.

**Bulb** A storage organ formed from tightly packed, fleshy scales. Also used generally for other underground storage organs including *corms* and *tubers.*

**Callus** Protective tissue that forms over a cut or wounded surface.

**Cambium** The layer of actively dividing cells just beneath the bark or *rind* of a plant.

**Chilling requirement** A period of time needed by a dormant plant below a given temperature to initiate the next season's flowers.

**Cloche** A temporary and moveable structure of glass or plastic used to protect young plants and/or to warm the soil.

**Cold frame** A low structure consisting of a timber, brick or concrete frame with the top and sometimes the sides glazed with glass or clear plastic.

**Corm** A solid underground storage organ developed from a swollen stem base. New corms are produced annually from buds on the parent corm.

**Crown** The part of a plant at or immediately below soil level from which leaves and shoots develop.

**Cultivar** A cultivated variety; a named selection of a species with distinctive characteristics that are retained when a plant is propagated by sexual or vegetative means.

**Cuttage** A generic term for bulb propagation techniques that involve cutting into the bulb.

**Cutting** A piece of a plant detached from a parent plant which forms roots and develops into a new plant that is usually identical to the parent plant.

**Division** A propagating technique that involves dividing a parent plant into smaller sections, each with roots and growth buds.

**Dormancy** The temporary cessation of growth and slowing down of metabolic processes,

usually as a mechanism to permit survival in adverse conditions.

**Drill** A straight, shallow furrow made for the sowing of seed.

**Earthing up** The process of drawing soil up around a plant's stem for blanching, or to encourage roots to grow from the stem.

**Epicormic shoot** A vigorous shoot that develops on or from the trunk. It arises from a latent bud beneath the bark, often in response to wounding or pruning. Also known as a water shoot.

**Eye 1.** A dormant growth bud, such as those found on dahlia tubers. **2.** A dormant bud in the leaf axils (see also *leaf-bud cutting*).

**F1 hybrid** Plants produced by crossing two selected, pure-breeding parents to produce vigorous, floriferous and uniform offspring.

**F2 hybrid** A plant which results from crossing selected F1 parents.

**Frond** The 'leaf' of a fern.

**Fungicide** A manufactured or naturally occurring chemical that controls or kills fungi.

**Grafting** A propagating technique whereby a *scion* from one plant is artificially united with the *rootstock* of another.

**Graft union** The point at which the *scion* and *rootstock* are united.

**Half-hardy** A plant that will not tolerate temperatures below freezing point, 32°F (0°C).

**Hardy** A frost-tolerant plant that survives outside year round without the need for special protection.

**Hardening off** The gradual acclimatization to the open garden of plants raised in a protected environment.

**Hardwood cutting** A cutting made from the fully ripened tissue of a woody plant.

**Heel cutting** A cutting taken with a heel or sliver of older wood at its base.

**Hybrid** The progeny of two plants of the same species differing in one or more genes, or closely related species.

**Internodal cutting** A cutting made with the basal cut between the *nodes*.

**Internode** The section of a stem between two *nodes*.

**Layering** A propagating technique whereby a stem is induced to root while attached to the parent plant.

**Leaf cutting** A cutting made from the whole or part of a leaf.

**Leaf-bud cutting** A cutting made of a short length of stem with a bud or pair of buds at the *node*.

**Mallet cutting** A cutting made from a side shoot with a plug of older wood at its base.

**Medium** A mixture or substrate in which plants are grown or propagated.

**Mutation** A genetic change which results in the production of tissues that are atypical; for example, when green-leaved plants produce col-ored or variegated shoots.

**Neutral** Of soils; with a pH of 7.

**Nodal cutting** A cutting made with the basal cut at or just below a node.

**Node** The point on a plant stem at which leaves, shoots, branches or flowers emerge.

**Offset** A young plant that arises at the base of its parent.

**Perlite** An expanded volcanic mineral that is granulated for use in propagating or other media.

**Petiole** A leaf stalk.

**pH** A scale of measurement of alkalinity and acidity, from 1–14, with 7 being the neutral point.

**Potting on** Transferring a plant from a smaller to a larger sized pot.

**Potting up** Transferring a seedling from its germinating medium or rooted cutting to an individual pot.

**Pricking out** Transferring a seedling from its germinating medium to a bed, tray, or pot to give them more space to grow.

**Reversion** The process whereby a *sport* or *mutation* returns to its original form.

**Rhizome** An underground stem, modified as a storage organ, which produces aerial growth at it tip and from buds along its length.

**Ripewood cutting** Alternative term for hardwood cutting (q.v) often applied to evergreens.

**Rooting hormone** A natural or manufactured chemical used to encourage root initiation.

**Rootstock** A plant that provides the root system for a grafted plant.

**Runner** A stem that creeps above ground and roots where its nodes touch the soil.

**Scarification** The process of nicking or abrading a hard seed coat so that water is imbibed and germination can proceed

**Scion** The bud or shoot that is grafted onto the rootstock of another.

**Semi-ripe cutting** A cutting made of the current season's growth just as it begins to harden at the base, usually in midsummer.

**Spore** The reproductive structure of non-flowering plants like fungi, ferns or mosses.

**Sport** An alternative term for *mutation*.

**Stem cutting** A general term for a cutting made from any part of a plant stem.

**Stem tip cutting** A cutting taken from a shoot tip, usually herbaceous, but also used for softwood cuttings of woody plants.

**Stock plant** A plant used as propagating material.

**Stolon** A spreading stem, usually above ground, that roots at its tip (see also *runner*).

**Softwood cutting** A cutting made from the current season's immature growth.

**Sport** See *mutation*.

**Stratification** Chilling or occasionally warming of seed to overcome dormancy.

**Sucker** 1. A shoot that arises from below the soil surface from the stem or roots. 2. A shoot that arises from below the *graft union* of a grafted plant.

**Top-dressing** A dressing applied to the soil surface around a plant's roots. Refers to fertilizers, new soil or other media, decorative or weed-suppressing materials.

**Tuber** A swollen, underground food storage organ derived from the root or stem base.

**Turion** A detached, overwintering bud produced by water plants.

**Variety** A naturally occurring variant of a wild species; abbr. var.

**Vegetative propagation** The propagation of plants by asexual methods, that is, by methods other than seed, resulting in genetically identical offspring.

**Widger** A spatula-like tool used for transplanting and pricking out seedlings.

# INDEX

# ACKNOWLEDGMENTS

The publishers would like to thank the following organizations and individuals for their kind permission to reproduce the photographs in this book:

**Key:** ALP: Andrew Lawson Photography; CNP: Clive Nichols Photography; GPL: Garden Picture Library; HSC: Harry Smith Collection; MaP: Mise au Point; PH: Photos Horticultural.

*Back jacket:* Jerry Harpur (Barnsley House); *front jacket:* ALP; 1 GPL (Mayer Le Scanff); 2 GPL (Gary Rogers); 4 GPL (J S Sira); 5 Jerry Harpur (Beth Chatto); 6–7 Jerry Harpur (Bank House, Borwick); 8–9 GPL (Lamontagne); 10 John Glover; 11 PH; 16 *bottom left* GPL (Ron Evans); 16 *bottom right* HSC; 17 *right* ALP; 17 *left* PH; 17 *center* HSC; 18 *left* ALP; 18 *center* PH; 18 *right* HSC; 19 *left* PH; 19 *center* HSC; 19 *right* HSC; 20 *center* GPL (Vaughan Fleming); 20 *left* GPL (Howard Rice); 20 *right* PH; 21 *center* GPL (Brian Carter); 21 *right* ALP; 21 *center* HSC; 22 *center* John Fielding; 22 *right* PH; 22 *left* PH; 23 *left* Jerry Harpur (Nick and Pam Coote); 23 *right* ALP; 23 *center* HSC; 24–5 GPL (Sunniva Harte); 24–5 CNP (Red Gables, Worcs); 26 CNP; 27 *top* GPL (Sunniva Harte); 34 *left* Hugh Palmer (RHS Wisley); 34 *right* PH; 35 *left* John Glover; 35 *center* ALP; 35 *right* PH; 36 *left* PH; 36 *right* PH; 36 *center* HSC; 37 *center* Eric Crichton; 37 *right* GPL (Cambridge Botanic Garden); 37 *left* ALP; 38 *right* Eric Crichton; 38 *center* John Fielding; 38 *left* John Glover; 39 *left* Eric Crichton; 39 *center* GPL (J S Sira); 39 *right* Hugh Palmer (RHS Wisley); 40 *center* GPL (J S Sira); 40 *right* PH; 40 *left* PH; 41 *right* John Glover; 41 *left* HSC; 41 *center* HSC; 42 *left* Eric Crichton; 42 *center* PH; 42 *right* PH; 43 *left* GPL (J S Sira); 43 *center* ALP; 43 *right* PH; 44 *right* Eric Crichton; 44 *left* ALP; 44 *center* HSC; 45 *right* Eric Crichton; 45 *center* John Fielding; 45 *left* GPL (Neil Holmes); 46–7 GPL (Jerry Pavia); 48 Eric Crichton; 49 GPL (J S Sira); 52 *left* Eric Crichton; 52 *right* John Glover; 53 *center* Eric Crichton; 53 *left* John Fielding; 53 *left* PH; 54–5 CNP (Mrs Frank's Garden, Steeple Aston, Oxon); 56 Eric Crichton; 57 PH; 64 *right* John Fielding; 64 *left* HSC; 65 *left* GPL; 65 *center* GPL (J S Sira); 65 *right* ALP; 66 *center* John Fielding; 66 *right* PH; 66 *left* HSC; 67 *left* John Glover; 67 *center* CNP (Beth Chatto Gardens, Essex); 67 *right* HSC; 68 *center* GPL (Philippe Bonduel); 68 *right* GPL (Philippe Bonduel); 68 *left* PH; 69 *right* GPL (John Glover); 69 *left* PH; 69 *center* PH; 70 *center* GPL (John Glover); 70 *right* ALP; 70 *left* HSC; 71 *center* ALP; 71 *left* CNP; 71 *right* PH; 72–3 ALP; 74 Jerry Harpur; 75 PH; 80 *left* GPL (Roger Hyam); 80 *right* HSC; 81 *right* CNP (Coates Manor Garden, Sussex); 81 *left* PH; 81 *center* PH; 82 *left* PH; 82 *center* HSC; 82 *right* HSC; 83 *center* John Glover; 83 *left* PH; 83 *right* PH; 84 *center* John Fielding; 84 *right* GPL (Juliette Wade); 84 *left* PH; 85 *center* John Fielding; 85 *right* PH; 85 *left* HSC; 86–7 GPL (Ron Evans); 88 Eric Crichton; 89 HSC; 94 *left* PH; 94 *right* HSC; 95 *left* Eric Crichton; 95 *center* GPL (Brian Carter); 95 *right* GPL (Jerry Pavia); 96 *center* John Glover (RHS Wisley); 96 *left* HSC; 96 *right* HSC; 97 *center* GPL (Brian Carter); 97 *left* CNP; 97 *right* PH; 98 *center* PH; 98 *right* PH; 98 *left* PH; 99 *left* John Glover; 99 *right* HSC; 99 *center* HSC; 100–101 GPL (Steven Wooster); 102 PH (Jane Sterndale-Bennett); 103 GPL (Howard Rice); 114 *right* John Glover; 114 *left* HSC; 115 *center* John Glover; 115 *left* HSC; 115 *right* HSC; 116 *left* John Fielding; 116 *right* CNP; 116 *center* HSC; 117 *left* John Fielding; 117 *right* John Glover; 117 *center* PH; 118 *center* GPL (Clive Nicholls); 118 *right* PH; 118 *left* PH; 119 *center* CNP; 119 *right* PH; 119 *left* PH; 120 *right* Eric Crichton; 120 *center* PH; 120 *left* PH; 121 *center* ALP; 121 *left* CNP; 121 *right* PH; 122 *left* John Fielding; 122 *right* John Fielding; 122 *center* PH; 123 *center* John Fielding; 123 *right* HSC; 123 *left* HSC; 124 *right* Eric Crichton; 124 *center* GPL (Jerry Pavia); 124 *left* GPL (Jerry Pavia); 125 *left* GPL (David Askham); 125 *right* PH; 125 *center* PH; 126 *right* ALP; 126 *left* PH; 126 *center* PH; 127 *center* John Glover; 127 *right* PH; 127 *left* HSC; 128–9 John Fielding; 130 GPL (John Ferro Sims); 131 GPL (Steven Wooster); 138 bottom *left* GPL (Bob Challinor); 138 bottom *center* PH; 139 top *center* PH; 139 top *right* PH; 139 top *left* PH; 140 *center* GPL (Brian Carter); 140 *right* John Glover; 140 *left* HSC; 141 *right* John Fielding; 141 *center* GPL (Densey Clyne); 141 *left* GPL (John Glover); 142 *left* John Fielding; 142 *center* GPL (John Glover); 142 *right* ALP; 143 *center* Eric Crichton; 143 *right* GPL (Brian Carter); 143 *left* PH; 144 *right* ALP; 144 *left* PH; 144 *center* HSC; 145 *center* John Fielding; 145 *right* John Glover; 145 *left* ALP; 146–7 Jerry Harpur (Mirabel Osler); 148 Neil Campbell-Sharp; 149 GPL (John Glover); 152 *center* Eric Crichton; 152 *left* GPL (Ron Sutherland); 153 *center* GPL (Densey Clyne); 153 *right* Hugh Palmer (RHS Wisley); 153 *right* PH; 154 *right* Eric Crichton; 154 *left* GPL (Howard Rice); 154 *center* GPL (J S Sira); 155 *right* GPL (J S Sira); 155 *center* ALP; 155 *left* HSC; 156 *center* Eric Crichton; 156 *right* PH; 156 *right* HSC; 157 *right* PH; 157 *center* PH; 157 *left* PH; 158–159 ALP; 160 John Glover; 161 GPL (Lynne Brotchie); 166 *center* GPL (Bob Challinor); 166 *left* CNP (Cerney House Garden, Glos); 167 *center* John Fielding; 167 *left* PH; 167 *right* PH; 168 *right* John Glover; 168 *left* Jerry Harpur; 168 *center* PH; 169 *left* ALP; 169 *center* CNP; 169 *right* PH; 170 *center* GPL (John Miller); 170 *left* GPL (John Neubauer); 170 *right* PH; 171 *right* John Glover; 171 *left* Jerry Harpur; 171 *center* PH; 174–5 MaP (Arnaud Descat); 186 *left* GPL (David Russell); 186 *center* GPL (Michel Viard); 187 *right* PH; 187 *left* PH; 187 *center* PH; 188 *right* GPL (Philippe Bonduel); 188 *center* GPL (Neil Holmes); 188 *left* PH; 189 *center* ALP; 189 *left* ALP; 189 *right* PH; 190 *center* GPL (Neil Holmes); 190 *left* PH; 191 *right* GPL (Michael Howes); 191 *left* ALP; 191 *center* PH; 194–5 GPL (Mayer/Le Scanff); 196 GPL (Ron Sutherland); 197 GPL (John Miller); 202 *center* John Fielding; 202 *left* HSC; 203 *left* John Fielding; 203 *center* John Fielding; 203 *right* John Fielding; 204 *center* HSC; 204 *right* HSC; 204 *left* HSC; 205 *left* John Fielding; 205 *right* John Glover; 205 *center* PH; 206 *right* John Fielding; 206 *left* HSC; 206 *center* HSC; 207 *left* CNP; 207 *right* PH; 207 *center* PH; 208 *left* John Fielding; 208 *right* PH; 208 *center* HSC; 209 *left* GPL (John Glover); 209 *center* John Glover; 209 *right* PH; 210 *center* John Fielding; 210 *right* PH; 210 *left* PH; 211 *left* Eric Crichton; 211 *right* PH; 211 *center* HSC.

The publishers would also like to thank the following individuals: David Ahsby; Peter Barnes; Caroline Bingham; Joanna Chisholm; Richard Dawes; Andrew Halstead; Andrew Jackson; Jodie Jones; Nicholas Morgan; Maggie O'Hanlon; Jane Royston; Sheila Seacroft; Jo Weeks; Sarah Widdicombe.